Fermentation as a Strategy to Improve the Functional and Nutritional Value of Foods

Fermentation as a Strategy to Improve the Functional and Nutritional Value of Foods

Guest Editors

Morena Gabriele
Laura Pucci

MDPI

Basel • Beijing • Wuhan • Barcelona • Belgrade • Novi Sad • Cluj • Manchester

Guest Editors

Morena Gabriele
Institute of Agricultural
Biology and Biotechnology
Italian National
Research Council
Pisa
Italy

Laura Pucci
Institute of Agricultural
Biology and Biotechnology
Italian National
Research Council
Pisa
Italy

Editorial Office
MDPI AG
Grosspeteranlage 5
4052 Basel, Switzerland

This is a reprint of the Special Issue, published open access by the journal *Foods* (ISSN 2304-8158), freely accessible at: https://www.mdpi.com/journal/foods/special_issues/3RKR08VEB7.

For citation purposes, cite each article independently as indicated on the article page online and as indicated below:

Lastname, A.A.; Lastname, B.B. Article Title. *Journal Name* **Year**, *Volume Number*, Page Range.

ISBN 978-3-7258-7504-7 (Hbk)
ISBN 978-3-7258-7505-4 (PDF)
https://doi.org/10.3390/books978-3-7258-7505-4

Contents

About the Editors

Morena Gabriele

Morena Gabriele is an Italian researcher at the Institute of Agricultural Biology and Biotechnology (IBBA) of the National Research Council (CNR) in Pisa. Her scientific activity focuses on food biochemistry and the study of functional and nutraceutical properties of plant-based foods, with particular attention to their impact on human health. Her main research areas include: Bioactive Compounds Characterization: identification and analysis of phytochemicals with potential health benefits; Antioxidant and Anti-inflammatory Activities: evaluation of the biological properties of food components and their role in preventing chronic diseases; Food and Health Relationship: investigation of molecular mechanisms linking diet to human health; Agri-food Valorization: development of innovative strategies to enhance the nutritional quality and functional value of plant-derived foods. The impact of her scientific contribution is documented in international peer-reviewed journals, where she has published extensively on topics related to food science, nutrition, and the role of diet in health promotion and disease prevention.

Laura Pucci

Laura Pucci is an Italian researcher at the Institute of Agricultural Biology and Biotechnology (IBBA) of the National Research Council (CNR) in Pisa. Her scientific activity primarily focuses on the study of the nutraceutical and functional properties of foods, with a specific emphasis on bioactive compounds and their impact on human health and metabolic diseases. Her main research areas include: Nutrition and Health: analyzing the effects of specific diets on inflammatory and metabolic pathologies; Food Matrix Characterization: enhancing agricultural products by identifying high-value-added molecules; Biochemistry and Oxidative Stress: investigating molecular mechanisms related to inflammation through dietary interventions. The impact of her scientific contribution is documented in major international databases, and her bibliometric indices, including her H-index and total citation count, establish her as a key figure in biochemical and nutritional research within the CNR.

Preface

The present reprint collects the contributions published in the Special Issue dedicated to fermentation as a strategy to enhance the functional and nutritional properties of foods. Fermentation is one of the oldest food processing techniques, yet it continues to represent a dynamic and innovative field of research with significant implications for human health and sustainable food systems.

The articles included in this Special Issue reflect the diversity of current research, ranging from the study of microbial communities to the identification of bioactive compounds and their physiological effects. The contributions also highlight the role of fermentation in improving food quality, safety, and shelf life.

This reprint is intended for researchers, food scientists, and professionals interested in the development of functional foods and the application of fermentation processes in modern food systems. We would like to thank all the authors and reviewers who contributed to the success of this Special Issue.

Morena Gabriele and Laura Pucci
Guest Editors

foods

MDPI

Article

Impact of Fermentation of Pumpkin Leaves and Melon Varieties with Lactobacillus Strains on Physicochemical Properties, Antioxidant Activity, and Carotenoid Compounds

Pretty Mhlanga [1], Sephora Mutombo Mianda [1] and Dharini Sivakumar [1,2,*]

[1] Department of Crop Sciences, Tshwane University of Technology, Pretoria 0001, South Africa; prettymhlanga99@gmail.com (P.M.); miandamutombos@tut.ac.za (S.M.M.)
[2] Centre for Nutrition & Food Sciences, Queensland Alliance for Agriculture and Food Innovation, The University of Queensland, Brisbane, QLD 4108, Australia
* Correspondence: sivakumard@tut.ac.za

Abstract: This study examined the impact of fermentation using *Lactiplantibacillus plantarum* (*L75*) and *Bifidobacterium longum* (*BF*) on the total soluble solids (TSS), pH, TA, LAB survival, color properties, ascorbic acid content, total phenolic content (TPC), carotenoid components, and antioxidant properties of smoothies made from melon varieties (Cantaloupe, Honeydew, and Watermelon) separately with pumpkin leaves (*Cucurbita moschata* and *Cucurbita pepo*). For all smoothies, pH (r = −0.74) and TSS (r = −0.79) were inversely and strongly correlated with LAB counts, while LAB counts were positively correlated with TA (r = 0.87). Fermentation time (24 to 72 h) significantly ($p < 0.05$) decreased the TSS (%), pH, and color properties of all smoothies fermented with *L75* or *BF*, while TA increased. Fermenting Cantaloupe melon and *C pepo* leaves with *L75* (CMCL75) for 24 h increased the ascorbic acid content to 3.8 mg/100 mL. The sensory panel scores were highest for Watermelon and *C. moschata* or C. pepo fermented with L75 or BF for 24 h. TPC concentration was highest in CMCL75 (70.76 mg of gallic acid per 100 mL) after 24 h. *C. pepo* leaves and Cantaloupe fermented with *L75* (CPCL75) showed the highest concentration of total carotenoids (70.38 mg/100 mL), lutein (2.53 µg/100 mL), cis β-carotene (25.43 µg/100 mL), and trans β-carotene (620.37 µg/100 mL). In contrast, CMCL75 showed the highest concentration of zeaxanthin (0.70 mg/100 mL). This study demonstrated the potential of fermenting Cantaloupe and pumpkin leaves together with the *L75* strain to produce non-dairy functional products.

Keywords: fermentation; underutilized leafy vegetables; Cucurbitaceae; lactic acid bacteria; functional compounds; sensory properties

1. Introduction

Melons belong to the Cucurbitaceae family, and the estimated global consumption is around 4.25 kg per person. The World Health Organization recommends consuming more than 400 g of fruits and vegetables daily to improve overall health and reduce the risk of various non-communicable diseases [1]. Melons provide several health benefits thanks to antioxidants like lycopene and vitamin C. Lycopene has been linked to the prevention of numerous chronic conditions, including cancer and diabetes [2]. Carotenoids, such as β-carotene found in Cantaloupe melons, are converted by the body into vitamin A (retinol). Lycopene primarily exists in its all-E-isomer configuration in foods. During processing, lycopene can undergo Z and E isomerization due to oxidation and heat. Reports indicate that only the E-isomers of 5Z-lycopene are biologically active. Given its polyene structure and electron-rich properties, lycopene is suitable for electrophilic reactions, able to scavenge free radicals, and is highly reactive [3]. Additionally, Honeydew melons are a good source of vitamin B (thiamine), potassium, and vitamin C. They are low in calories, containing only 36 calories per 100 g [4]. Watermelon is one of Africa's 101 most important orphan crops. Considering the nutrient-dense benefits of orphan or underutilized crops and their integration into diverse food systems, orphan crops have recently become a topic of interest. Since ancient times, pumpkin leaves have been eaten in Africa. In addition to being nutritious and functional, pumpkin leaves contain phenolics, flavonoids, and beta-carotene, and are active antioxidants [5]. In a 100 g serving, fresh pumpkin leaves contain 38 mg of magnesium, 39 mg of calcium, 15.90 mg of iron, 3.15 g of protein, and no cholesterol [6]. Starch-based foods in developing countries are low in micronutrients, so consuming leafy vegetables, such as pumpkin leaves, boosts daily micronutrient intake.

Approximately 40% of Watermelons are lost by producers and marketers, and some lose more than 60% [7]. Moreover, improper packaging materials can facilitate spoilage microorganism proliferation by trapping water released by pumpkin leaves during respiration or transpiration [8]. Post-harvest losses harm the economy, environment, and health. Moreover, the United Nations calls on people to reduce food loss by half by 2030 to improve global food security and the population's health. In addition, food waste contributes 6% to global carbon emissions [9]. A significant amount of pressure is also being put on small- and medium-scale retailers by the cost of energy and the transition from coal to renewable energy to maintain cold chain infrastructure [10].

The thermal processing of fruit and vegetable juices for microbial protection is a widely used preservation technique. The thermal processing of juices has a detrimental impact on important nutritional compounds, such as vitamins like ascorbic acid, thiamine, and folic acid [11]. Juice extraction, however, eliminates insoluble dietary fibers, despite their health benefits [12]. In this way, smoothies provide consumers with the nutritional benefits of fruits and vegetables. A smoothie is a semi-liquid nutrient-dense product consisting of fruit and vegetables [13].

Furthermore, lactic acid fermentation is an effective strategy for reducing post-harvest losses and preserving nutrients. Dairy probiotics may contain allergens, and a significant portion of carbon dioxide emissions from food production comes from animal products [14], prompting the food industry to explore alternative microbial delivery systems. This fermentation process helps preserve food, extend shelf life, enhance nutritional value, and improve flavor and texture. Research by Moloto et al. [15] shows it can also recover bioactive compounds from underutilized vegetables. Various melons can be used as dietary foods or natural flavorings. The global food and beverage market is projected to reach USD 7464.2 billion from 2022 to 2027 [16].

The use of lactic acid bacterium (LAB) cultures enhances the biological value of fermented foods. Additionally, incorporating indigenous fruits or vegetables into smoothies can improve their nutritional content. LAB fermentation boosts antioxidant activity by releasing bioactive compounds through the metabolism and biotransformation of phenolic compounds. *Lactiplantibacillus plantarum* (*L. plantarum*) is commonly added to fermented foods to enhance their quality and health benefits [17]. Meanwhile, *Bifidobacterium longum* (*B. longum*), a non-pathogenic, Gram-positive, catalase-negative, rod-shaped bacterium, is naturally found in the gastrointestinal tract of humans and is frequently added to various food products [18]. This study aims to reduce food waste from melons and pumpkin leaves while rescuing bioactive compounds, particularly carotenoids, through fermentation technology utilizing specific lactobacillus strains for the benefit of consumers. However, limited information is available regarding the changes in bioactive compounds—especially carotenoid components—and antioxidant activity in different types of melons after fermentation with leafy vegetables.

To develop fermented smoothies, it is essential to choose melon varieties based on their physical and sensory characteristics. Therefore, this study investigates the effects of fermentation by two lactobacillus strains—*Lactiplantibacillus plantarum* and *Bifidobacterium longum*—on the physicochemical properties, bioactive compounds (total phenols, total carotenoid content, and specific carotenoid components), antioxidant activities, and sensory properties of smoothies made from pumpkin leaves (*Cucurbita moschata* and *Cucurbita pepo*) and the melon varieties *Citrullus lanatus* and *Cucumis melo*, separately.

2. Materials and Methods

2.1. Chemicals

This study was conducted with chemicals and reagents supplied by Merck Life Science (Pty) Ltd. (Johannesburg, South Africa).

2.2. Creating Smoothies with Plant Material

This study utilized two accessions of pumpkin leaves, namely *Cucurbita moschata* and Cucurbita pepo, alongside three varieties of melons: Watermelon, Cantaloupe, and Honeydew. The pumpkin leaves were sourced from the Tshiombo Cooperative in Venda, located in the Limpopo province of South Africa, while the melons were purchased from a commercial farm within the same province. To prepare the pumpkin leaves, they were cleaned with tap water to remove excess soil. After that, the leaves were immersed in a 0.01% sodium hydroxide solution, rinsed again with sterile water, sliced, and dried. The melons were peeled and diced using stainless steel knives in a sterile environment. Smoothies were prepared by blending and homogenizing 800 g of each type of melon fruit with 200 g of each pumpkin leaf accession (*C. moschata* or *C. pepo*) separately, along with 500 mL of distilled water. This process was carried out in a Russell Hobbs blender (700 W). Following this, the smoothies underwent pasteurization for 30 min in a water bath (WBE28A12E, PolyScience, Niles, IL, USA) at 63 °C. After pasteurization, the smoothies were cooled on ice before being inoculated with lactic acid bacterium (LAB) strains, specifically *L. plantarum* (L75) and *B. longum* (BF).

2.3. Smoothie Starter Cultures and Fermentation

Lactobacillus (LAB) strains (*Lactiplantibacillus plantarum* (*L. plantarum*) (*L75*) and *Bifidobacteria longum* (*B. longum*) (*BF*)) were obtained from the Fruit and Vegetable Technology culture collection at Tshwane University of Technology, Pretoria, South Africa. The strains were revived and multiplied by sequential suspension in 9 mL of de Man, Rogosa, and Sharpe (MRS) broth and incubated anaerobically for 48 h at 30 °C, as reported by Nirina et al. [19]. In sterile distilled water, the resulting cells were re-suspended after centrifugation at 12,000× *g* for 5 min at 4 °C. Spectrophotometer (UV-1750 UV-Vis spectrometer, Shimadzu, Suzhou, China) was used to determine culture concentrations (turbidity method). The pre-culture concentration was calibrated to 0.05 McFarland standard (1–5 × 10^8 CFU/mL) at a 660 nm wavelength. Subsequently, a 1% inoculum was introduced into smoothies, as outlined by Managa et al. [20]. The smoothies were inoculated with each strain separately. The smoothies were incubated (fermented) at 37 °C for 2 h, 24 h, 48 h, and 72 h. Fermented smoothies were stored at −20 °C until analysis after fermentation. The non-inoculated smoothies served as the control. The fermentation was performed in triplicate.

2.4. Survival of LABs

At different fermentation time intervals, the viability of the LAB strains and the safety of the smoothies were assessed. Pour plate counts were conducted, as described by Managa et al. [20], to enumerate the colonies on medium plates. Concisely, 100 µL of smoothie was fused with 900 µL of sterile distilled water and then further diluted to a factor of 10^{-6}. Concisely, 100 µL of the diluted sample was placed on suitable plate media, i.e., MRS agar, nutrient agar (NA), xylose lysine deoxycholate agar (XLD), and Eosin-methylene blue agar (EMB) to measure the growth of surviving LABs, aerobic bacteria, *Salmonella*, and *E. coli*, respectively. Plated Petri dishes were incubated at 37 °C for 48 h for MRS, and 24 h for NA, XLD, and EMB. Cell viability was expressed in CFU/mL of smoothie, as described by Managa et al. [20].

2.5. Assessment of Physicochemical Parameters

The pH of both inoculated and non-inoculated treatments was evaluated using a digitally calibrated pH meter (Melter-Toledo Instruments Co., Shanghai, China), as outlined by Managa et al. [20]. Total acidity (TA) measurements were taken using 0.1 N NaOH and the phenolphthalein titration method [20]. Total soluble solids (TSSs) were assessed with a pocket refractometer (Agato pocket PAL-2, Tokyo, Japan). Distilled water was used to calibrate the refractometer before each test to enhance the accuracy of the results. Color values for all treatments were recorded using a colorimeter (CR-400 Chroma Meter, Konica Minolta Sensing, Shanghai, China), which was calibrated with a white standard tile. The color values were expressed in the CIEL*a*b* color space, where L* indicates lightness or luminosity (0 = dark and 100 = bright), a* represents the amount of green (−) and red (+) color, and b* measures the amount of blue (−) and yellow (+) color. The color change or difference was calculated using the *L*a*b** readings and expressed as ΔE [20]. In this context, *L*0*, a*0, and b*0 represent the color parameters for fresh melon (Watermelon, Cantaloupe, or Honeydew melon) and pumpkin leaf (*C. moschata* and *C. pepo*) smoothies, while *L**, *a**, and *b** denote the corresponding values after fermentation.

$$\Delta E = \sqrt[2]{(L^* - L_0^*)^2 + (a^* - a_0^*)^2 + (b^* - b_0^*\ )^2}$$

2.6. Ascorbic Acid Content

The samples were analyzed for ascorbic acid content using the metaphosphoric acid and 2,6-dichloroindophenol dye titration method, as outlined by Mashitoa et al. [21]. Titration was performed using a 10 mL aliquot of the solution against the dye solution, with a pink color indicating the titration endpoint. The ascorbic acid concentration was expressed in milligrams per 100 g on a fresh weight basis. To calculate the ascorbic acid content, Equation (1) was utilized.

$$\text{Ascorbic acid (mg/100 g)} = \text{Titre} \times \text{Dye factor} \times \text{volume made} \times 100 \quad (1)$$

$$(\text{Aliquot of extract taken} \times \text{Weight of sample})$$

2.7. Total Phenolic Content (TPC)

The Folin–Ciocalteu (F-C) assay [22] was employed to determine the total phenolic content (TPC) in smoothies. For each treatment, 2 mL of smoothie was collected and homogenized with 2 mL of 80% methanol. The mixture was vortexed and sonicated for 10 min at 30 °C. In conical-bottom Eppendorf tubes (LABOCARE™, Strathavon 2031, Johannesburgh, South Africa), 100 µL of the extracted samples were combined with 750 µL of F-C reagent and allowed to react for 5 min. Following this, sodium carbonate (Na_2CO_3) was added to the mixture, which was then incubated for 2 h. The samples were plated onto 96-well microplates (BMG LABTECH GmbH, Spectro Star Nano, Ortenberg, Germany), and the absorbance was measured at 750 nm using a spectrophotometer (UV-1750 UV-Vis spectrometer, Shimadzu, Suzhou, China). The concentration of TPC was calculated using the calibration curve equation $Y = 0.0019x - 0.1733$, with an R^2 value of 0.8511, employing 1 µM gallic acid as the standard.

2.8. Total Carotenoid Content and Carotenoid Profile

As outlined by Djuikwo et al. [23], the smoothies were analyzed for total carotenoid content (TCC) and individual carotenoids. A 2 mL sample of each smoothie was mixed with 5 mL of acetone and 5 mL of 95% ethanol containing 0.1% butylated hydroxytoluene (BHT). The mixture was vortexed and sonicated for 10 min at 30 °C using a vortex mixer (Benchmark Scientific Inc., Taiwan) and a DC-150H MRC ultrasonic cleaner. After the initial sonication, 10 mL of 20% potassium hydroxide (KOH) was added to the solution, which was then sonicated for an additional 10 min. To separate the carotenoids, 10 mL of a hexane/dichloromethane solution (70:30 *v*/*v*) containing 0.1% BHT was added and sonicated for 5 min at 30 °C. Following this, 5 mL of sodium chloride (NaCl) solution was added, and the mixture was centrifuged at 3900× *g* for 5 min. Under a nitrogen stream, the extracted samples were collected three times. To reconstitute the fresh extracts for profiling and total carotenoid content analysis, 2 mL of a methanol/MTBE solution (50:50 *v*/*v*) was added. An aliquot of 200 μL was then placed in 96-well microplates (BMG LABTECH GmbH, Spectro Star Nano, Ortenberg, Germany) for total carotenoid content analysis, with the data expressed in mg per 100 mL for TCC.

The HPLC (Shimadzu, LC-2030C 3D, Tokyo, Japan) analysis for the quantification of individual carotenoids was conducted as described by Bhengu et al. [24] without any modification. An aliquot of 1000 μL was withdrawn from the reconstituted samples and transferred into HPLC vials. The separation of compounds was performed on an InertSustain C30 column (3.0 × 250 mm, 3 μm, GL Sciences, Tokyo, Japan) using a gradient elution of 0.1% of formic acid in methanol (solvent A) and methyl tert-butyl ether (solvent B) at a flow rate of 0.3 mL/min. Chromato-grams were read at 460 nm. The quantification of carotenoids was achieved using calibration curves constructed with lutein, β-carotene, lycopene, and zeaxanthin, and is expressed in mg/100 mL.

2.9. *Antioxidant Activity*

2.9.1. Ferric-Reducing Antioxidant Power (FRAP)

The FRAP assay was performed as described by Benzie and Strain [25] with some modifications outlined by Llorachet al. [26]. The working solution of FRAP was prepared by mixing 25 mL of acetate buffer with 2.5 mL of TPZ (2,4,6-tripyridyl-S-triazine) solution and 2.5 mL of ferric achloride. The mixture was shaken at 37 °C using a water bath (WBE28A12E, PolyScience, Niles, IL 60714-4516, USA). For the final FRAP analysis, 75 μL of extracted samples were mixed with 1425 μL of FRAP working solution and incubated for 30 min in the dark at room temperature. After the incubation time lapsed, 200 μL of each sample treatment was plated on the 96-well microplates (BMG LABTECH GmbH, Spectro Star Nano, Ortenberg, Germany) and the absorbance was measured at 593 nm on a spectrophotometer (UV-1750 UV-Vis spectrometer, Shimadzu, Suzhou, China). Trolox was used as the standard for FRAP and the results were expressed as mM TEAC/mL on a fresh weight basis.

2.9.2. 1,1-Diphenyl-2-Picrylhydrazyl (DPPH)

The capability of the fermented extracts to scavenge the radicals of DPPH was determined utilizing the spectrophotometric methods [15]. To prepare the DPPH solution, 6 mg of DPPH was dissolved in 25 mL of solvent (80% methanol), while a working solution was prepared by homogenizing 10 mL of DPPH solution with 50 mL of solvent. DPPH's scavenging ability was further determined by plating different concentrations of the extracted samples (10–100 μL), solvent (90–0 μL), and constant concentration (200 μL) of DPPH on the 96-well microplates (BMG LABTECH GmbH, Spectro Star Nano, Ortenberg, Germany). The plated solutions were incubated for 10 min in the dark; thereafter, the absorbance of the mixture was read at 593 nm utilizing a spectrophotometer (UV-1750 UV-Vis spectrometer, Shimadzu, Suzhou, China). The solvent was used as a blank. The scavenging activity % RSA of the samples was calculated using the formula below, and the results were expressed as IC_{50} μL/mL of pumpkin leaf–melon smoothies on a fresh weight basis:

$$\text{Scavenging } (\%) = [A_a - (A_b - A_c)]/A_a \times 100$$

where A_a is the absorbance of the DPPH solution without adding the solvent; A_b is the absorbance of the mixture solution containing both the sample and DPPH; and A_c is the absorbance of the blank solution with DPPH.

2.9.3. 2,2-Azinobis-(3-Ethyl-Benzothiazoline-6-Sulfonic Acid) (ABTS)

The $ABTS^+$ cation radical was created by combining 5 mL of ABTS solution with 145 mL of phosphate buffer solution, which was left to incubate for 12 to 16 h in the dark at 25 °C [15]. Before ABTS analysis, a precise 100 μL of the sample extract was diluted into 900 μL of solvent. Following samples dilution, different concentrations of the samples (10–100 μL), solvent (90–0 μL), and uniform concentration (200 μL) of $ABTS^+$ cation radical were plated on 96-well microplates and read on a spectrophotometer at an absorbance of

593 nm. Like DPPH, the solvent was used as a blank and the scavenging activity % RSA of the samples was calculated using the formula described in Section 2.9.2. The results of ABTS were expressed as the IC_{50} µL/mL of pumpkin leaf–melon smoothies on a fresh weight basis.

2.10. Sensory Analysis

The panel consisted of 40 untrained individuals, both males and females between the ages of 18 and 53. The panelists were all familiar with the taste of pumpkin leaves and melons. Before the session, the panel was briefed on how smoothies are made. Each panelist received a coded smoothie from the different fermentation hours in shot glasses, still water, sensory forms, and pens. Additionally, the samples were presented in a completely random order throughout the session. The smoothies were tested for color, taste, aroma, and overall acceptability. Following each test, water was used to clean the mouth. Using a scale of 0–9 with 0 being extremely disliked, 5 being neither like nor dislike, and 9 being extremely like, intensity scores were assigned based on individual panelists' degree of liking a product.

2.11. Statistical Analysis

This experiment utilized a completely randomized design, featuring three replicates for each treatment, and was conducted twice. The statistical analysis was performed using GenStat 64-bit release 22.1 (Hempstead, England). A two-way analysis of variance (ANOVA) was conducted for the survival of lactic acid bacteria (LABs), total soluble sugars, pH and titratable acidity (TA), color properties, and ascorbic acid levels. In contrast, total phenolics, carotenoid content, carotenoid components, and antioxidant scavenging activities were analyzed using one-way ANOVA. Fisher's protected least significant difference (LSD) test was employed for all analyses to determine significant differences at a threshold of $p < 0.05$. Additionally, mean scores from sensory analysis were calculated, and significant variations were assessed using ANOVA, followed by Fisher's LSD test, to identify differences at $p < 0.05$.

3. Results and Discussion

3.1. LAB Survival

According to Figure 1, LAB activity was not present in all the controls, as they were not inoculated with the fermentation cultures. The different melon and pumpkin leaf smoothies were inoculated with *L75* or *B. Longum* at 6 log cfu /mL. However, the population of the inoculated smoothies ranged between log 4.3 and log 9.8 cfu/mL of LABs during the fermentation process and further peaked at 24 h of fermentation. According to our findings, LAB survival was also affected by increased TA (lactic acid) when fermentation time was increased from 48 to 72 h. Overall, smoothies inoculated with *L75* had the highest colony counts. *L75* populations in different melon and pumpkin leaf smoothies after 24 h were shown as fold increase compared to the inoculated LAB counts as follows: in *C. moschata* + Watermelon + *L75*(CMW75)-1.27-fold increase, *C. moschata* + Cantaloupe + *L75* (CMCL75)-1.26-fold increase, *C. moschata* + Honeydew + *L75* (CMHD75)-1.26-fold increase, *C. pepo* + Honeydew + *L75* (CMHD75)-1.23-fold increase, and *C. pepo*+Honeydew + *B. longum* (CPHDBF)-1.23-fold increase. However, the increase was not significant ($p < 0.5$). On the other hand, although *C. pepo* + Cantaloupe + *L75* (CPCL75), *C. pepo* + Watermelon + *B. longum* (CPWBF), and *C. pepo* + Watermelon combined with *L75* (CPW75) increased lactic acid bacterium (LAB) counts with increases of approximately 1.19-fold, 1.15-fold, and 1.14-fold. However, CPCL75, CPWBF, and CPW75 exhibited significantly lower increases in LAB counts compared to CMW75, CMCL75, CMHD75, and CPHDBF smoothies ($p < 0.05$).

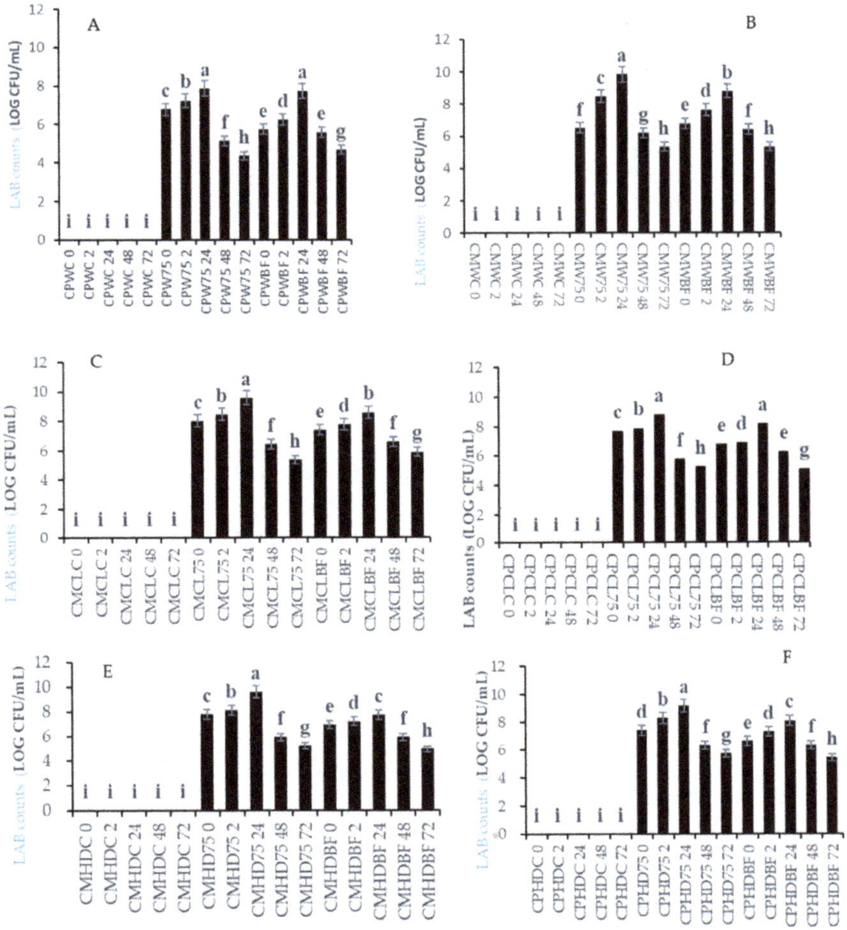

Figure 1. The population of *Lactiplantibacillus plantarum* (*L75*) and *Bifidobacterium longum* (*BF*) in smoothies of different melon types and pumpkin leaves (*Cucurbita moschata* and *Cucurbita pepo*) after 2, 24, 48, and 72 h of fermentation. Mean values were calculated based on three replicate samples. Different letters in the same bar refer to statistical difference at $p < 0.05$ according to Fisher's protected LSD test. (**A**) *Cucurbita pepo* + Watermelon smoothies fermented with LABs; (**B**) *Cucurbita moschata* + Watermelon smoothies fermented with LABs; (**C**) *Cucurbita moschata* + Cantaloupe smoothies fermented with LABs; (**D**) *Cucurbita pepo* + Cantaloupe smoothies fermented with LABs; (**E**) *Cucurbita moschata* + Honeydew melon smoothies fermented with LABs; (**F**) *Cucurbita pepo* + Honeydew melon smoothies fermented with LABs. *Keys: Cucurbita moschata* + Watermelon (CMWC), *Cucurbita moschata* + Watermelon + *L75* (CMW75), *Cucurbita moschata* + Watermelon + *B. longum* (CMWBF), *Cucurbita pepo* + Watermelon Control (CPWC), *Cucurbita pepo* + Watermelon + *L75* (CPW75), *Cucurbita pepo* + Watermelon + *B. longum* (CPWBF), *Cucurbita moschata* + Cantaloupe Control (CMCLC), *Cucurbita moschata* + Cantaloupe + *L75* (CMCL75), *Cucurbita moschata* + Cantaloupe + *L75* (CMCL75), *Cucurbita pepo* + Cantaloupe Control (CPCLC), *Cucurbita pepo* + Cantaloupe + *L75* (CPCL75), *Cucurbita pepo* + Cantaloupe + *B. longum* (CPCLBF), *Cucurbita moschata* + Honeydew Control (CMHDC), *Cucurbita moschata* + Honeydew + *L75* (CMHD75), Cucurbita moschata + Honeydew + *B. longum* (CMHDBF), *Cucurbita pepo* + Honeydew Control (CPHDC), *Cucurbita pepo* + Honeydew + *L75* (CMHD75), and *Cucurbita pepo* + Honeydew + *B. longum* (CPHDBF).

Thus, combinations such as *C. moschata* + Watermelon, *C. moschata*+ Cantaloupe, and *C. pepo* + Honeydew, as well as *C. pepo* + Cantaloupe, demonstrated higher cell counts, exceeding 9.1 log cfu/mL (for *L75* and *B. longum*), than the *C. pepo* + Watermelon smoothies. According to Mandha et al. [27], Lactobacillus strains fermented for 24 h showed an increase from 8 to 9 log cfu/mL during the lactic acid fermentation of Watermelon juice. They reported a significant growth ($p < 0.05$) with *L75*, which reached 9.23 log cfu/mL. Furthermore, Fonteles et al. [28] noted an increase in the viability of *L. casei* to 8.3 log cfu/mL in Cantaloupe juice. This indicates that melons serve as a suitable matrix for the survival of LABs.

According to Mandha et al. [27], LABs fermented for 24 h increased to 8–9 log cfu/mL during the lactic acid fermentation of Watermelon juice. Furthermore, Fonteles et al. [28] reported an increase in *L. casei* viability in Cantaloupe juice cells of 8.3 log cfu/ mL. Therefore, melons are a suitable LAB matrix for the survival of LABs. Moreover, Mashitoa et al. [21] showed that the fermentation with *L75* and leaves of the sweet potatoes of orange-fleshed 'Bophelo' and 'Beauregard' increased the *L75* survival after 24 h, which ranged from 8 to 10 log cfu/ mL. On the contrary, the leaves of purple-fleshed '08.21P' and cream-fleshed 'Ndou' displayed an increase in *L75* to log 8 cfu/ mL and 10 log cfu/mL after 48 h. The differences in survival of *L75* and fermentation hours again depend on the substrate used for fermentation and the build-up of metabolites. Therefore, this study shows that LAB survival is dependent on the type of melons and the variety of pumpkin leaves. For a food to be labelled as a probiotic, it must possess log 6 cfu/gm or mL of probiotic microorganisms [29]. To be effective, fermented probiotic products should contain a high concentration of viable bacterium cells [29]. Also, 100 mL or grams of a product must ideally contain around 7 $\log^{-10}$ cfu/ mL viable cells [30]. According to our results, we obtained viable cells at a probiotic dosage of above log 7 cfu/ mL for the three types of melons and pumpkin leaves fermented by *L75* or *B. longum*. The pH dropped as the LAB population increased, indicating a negative correlation (r = −0.74). Contrary to this, the LAB population increased with increasing TA and showed a positive correlation (r = 0.87). However, LAB survival may decrease during fermentation due to the continuous decrease in pH and the increase in acidic environments, since acids hinder bacterial viability and growth by inhibiting enzymatic reactions, leading to acidity in the cytoplasm and requiring more energy to maintain intracellular pH [31]. Furthermore, the decrease in sugars during fermentation may have slowed down the fermentation process. The data collected indicate that no pathogens, fungi, or bacteria grew in any of the plated media, indicating that the smoothies are safe for consumption.

3.2. Total Soluble Sugars (TSSs)

The sugar index is often used to measure total soluble solids (TSSs), a quality indicator related to sweetness [32]. At 0 h of fermentation, all three non-inoculated melon type (Cantaloupe, Honeydew melon, and Watermelon) and pumpkin leaf (*C. moschata*/*C. pepo*) smoothies had higher TSSs than their fermented samples (Table 1). There was a significant drop in TSSs (%) with increasing fermentation time (24 to 72 h) for the three types of melon (Cantaloupe, Honeydew, and Watermelon) and pumpkin leaves (*C. moschata*/*C. pepo*) fermented with *L75* or *B. longum*, but non-inoculated smoothies (0 h) of these respective samples showed higher TSS levels. A reduction in TSSs (%) compared to the 0 h control samples was shown for all types of melon and pumpkin leaves with respect to 24, 48, and 72 h in Table 1. The bioconversion of sugar into lactic acid and the consumption of sugar for the growth and metabolism of bacteria can lead to a reduction in TSSs. For example, Managa et al. [20] reported a decrease in TSSs in chayote-pineapple smoothie fermented with LAB strains *L75* and *Weissella cibaria* (*W64*). Additionally, Kaprasob et al. [33] found that *L. plantarum* consumes sugar more rapidly than other bacteria during the fermentation of cashew apple juice. This observation supports the claim of decreased TSSs in *L. plantarum*-fermented mango juice after 72 h [34]. A decrease in TSSs suggests that the culture can exploit sugars in smoothies of melon types for the growth of both LAB strains. In contrast, the fermentation of the three types of melons and pumpkin leaves (*C. moschata/C. pepo*) with *B. longum* for 72 h significantly ($p < 0.05$) decreased the TSSs. This indicates that some probiotic strains were found to be capable of growth in different fruit matrices [34]. Kumar et al. [34] showed that the sugar utilization pattern can be used to determine whether bacteria rapidly consumed the substrate (sugars) in the fruit matrix.

Table 1. Changes in physicochemical parameters during fermentation with different lactobacillus strains on pumpkin leaf (*Cucurbita moschata or C. pepo)* and melon smoothies.

	***Cucurbita moschata* Leaf and Watermelon Smoothies Fermented with LABs**			
	Fermentation Time (h)	**TSS (°Brix)**	**pH**	**TA (Equivalent Lactic Acid g/100 g)**
CMWC	0	9.63 ± 0.058 [a,*]	5.69 ± 0.007 [a]	1.333 ± 0.058 [k]
CMWC	2	9.20 ± 0.100 [b] (−4.5%) **	5.54 ± 0.010 [d]	1.66 ± 0.006 [j] (+24.5%) ***
CMWC	24	8.50 ± 0.100 [d] (−11.7%)	5.47 ± 0.000 [e]	1.67 ± 0.116 [j] (+25.3%)
CMWC	48	8.10 ± 0.173 [f] (−15.9%)	4.94 ± 0.006 [h]	1.77 ± 0.058 [i] (+32.8%)
CMWC	72	7.86 ± 0.058 [g] (−18.4%)	4.84 ± 0.006 [i]	2.30 ± 0.000 [g] (+72.5%)
CMW75	0	9.20 ± 0.000 [b]	5.61 ± 0.000 [c]	1.83 ± 0.058 [h]
CMW75	2	8.36 ± 0.058 [de] (−9.1%)	5.28 ± 0.000 [g]	3.33 ± 0.058 [f] (+81.9%)
CMW75	24	7.90 ± 0.100 [g] (−14.1%)	4.34 ± 0.000 [h]	4.26 ± 0.153 [e] (+132.8%)
CMW75	48	7.56 ± 0.116 [h] (−17.8%)	3.89 ± 0.000 [k]	5.53 ± 0.058 [c] (+202.2%)
CMW75	72	7.33 ± 0.058 [i] (−20.3%)	3.76 ± 0.006 [m]	5.80 ± 0.000 [b] (+216.9%)
CMWBF	0	9.00 ± 0.100 [c]	5.63 ± 0.06 [b]	1.86 ± 0.058 [h]
CMWBF	2	8.33 ± 0.058 [e] (−7.4%)	5.31 ± 0.06 [f]	3.36 ± 0.018 [f] (+80.6%)
CMWBF	24	7.567 ± 0.058 [h] (−15.9%)	4.30 ± 0.010 [j]	4.66 ± 0.056 [d] (+150.5%)
CMWBF	48	7.16 ± 0.153 [j] (−20.4%)	3.88 ± 0.056 [k]	5.76 ± 0.078 [b] (+209.7%)
CMWBF	72	6.80 ± 0.010 [k] (−24.4%)	3.76 ± 0.016 [m]	5.92 ± 0.070 [a] (+218.3%)
	Cucurbita pepo leaf and Watermelon smoothies fermented with LABs			
	Fermentation Time (h)	TSS (°Brix)	pH	TA (Equivalent Lactic Acid g/100 g)
CPWC	0	8.03 ± 0.058 [a]	5.34 ± 0.05 [a]	1.06 ± 0.116 [m]
CPWC	2	7.66 ± 0.018 [b] (−4.6%)	4.98 ± 0.006 [c]	1.53 ± 0.058 [l] (+44.3%)
CPWC	24	7.30 ± 0.100 [d] (−9.1%)	4.81 ± 0.012 [d]	1.70 ± 0.100 [k] (+60.4%)
CPWC	48	6.90 ± 0.0200 [f] (−14.1%)	4.75 ± 0.006 [e]	1.98 ± 0.058 [j] (+86.8%)
CPWC	72	6.50 ± 0.040 [h] (−19.1%)	4.70 ± 0.022 [f]	2.30 ± 0.010 [i] (+117.0%)
(CPW75)	0	7.53 ± 0.058 [c]	5.29 ± 0.050 [b]	1.80 ± 0.010 [k]
CPW75	2	7.36 ± 0.068 [d] (−2.3%)	4.66 ± 0.012 [g]	2.86 ± 0.058 [f] (+58.9%)
CPW75	24	7.03 ± 0.059 [e] (−6.6%)	4.08 ± 0.010 [h]	4.83 ± 0.058 [d] (+168.3%)
CPW75	48	6.43 ± 0.048 [h] (−14.6%)	3.71 ± 0.010 [i]	5.46 ± 0.058 [c] (+203.3%)
CPW75	72	6.00 ± 0.010 [j] (−20.3%)	3.62 ± 0.006 [j]	6.48 ± 0.116 [a] (+260.0%)
CPWBF	0	7.63 ± 0.018 [b]	5.30 ± 0.074 [b]	1.53 ± 0.028 [l]
CPWBF	2	7.36 ± 0.055 [d] (−3.5%)	4.71 ± 0.066 [e]	2.26 ± 0.018 [h] (+47.7%)
CPWBF	24	6.90 ± 0.060 [f] (−9.6%)	4.08 ± 0.0076 [h]	4.53 ± 0.098 [e] (+196.1%)
CPWBF	48	6.60 ± 0.080 [g] (−13.5%)	3.71 ± 0.036 [i]	5.40 ± 0.0240 [c] (+252.9%)
CPWBF	72	6.20 ± 0.008 [i] (−18.7%)	3.65 ± 0.010 [j]	5.98 ± 0.012 [b] (+290.8%)
	Cucurbita moschata leaf and Cantaloupe smoothies fermented with LABs			
	Fermentation Time (h)	TSS (%)	pH	TA (Equivalent Lactic acid g/100 g
CMCLC	0	6.30 ± 0.01 [a]	6.66 ± 0.010 [a]	1.20 ± 0.050 [m]
CMCLC	2	5.70 ± 0.02 [b] (−9.5%)	6.44 ± 0.015 [c]	1.35 ± 0.150 [l] (+12.5%)
CMCLC	24	5.13 ± 0.06 [d] (−18.6%)	6.38 ± 0.000 [d]	1.67 ± 0.120 [k] (+39.2%)
CMCLC	48	4.73 ± 0.06 [ef] (−24.9%)	6.09 ± 0.006 [e]	1.69.100 [k] (+40.8%)
CMCLC	72	4.20 ± 0.10 [g] (−33.3%)	5.85 ± 0.000 [f]	1.82 ± 0.01 [j] (+51.7%)

Table 1. *Cont.*

	Cucurbita moschata leaf and Cantaloupe smoothies fermented with LABs			
	Fermentation Time (h)	TSS (%)	pH	TA (Equivalent Lactic acid g/100 g
CMCL75	0	5.73 ± 0.12 ^{b}	6.52 ± 0.002 ^{b}	2.33 ± 0.050 ^{h}
CMCL75	2	4.77 ± 0.12 ^{e} (−16.7%)	5.27 ± 0.006 ^{h}	3.87 ± 0.050 ^{f} (+66.1%)
CMCL75	24	4.23 ± 0.06 ^{g} (−26.2%)	4.32 ± 0.000 ^{i}	4.21 ± 0.100 ^{d} (+80.7%)
CMCL75	48	3.80 ± 0.10 ^{h} (−33.7%)	4.04 ± 0.000 ^{k}	4.50 ± 0.010 ^{b} (+93.1%)
CMCL75	72	3.53 ± 0.06 ^{I} (−38.4%)	3.77 ± 0.001 ^{n}	4.97 ± 0.120 ^{a} (+113.3%)
CMCLBF	0	5.53 ± 0.06 ^{c}	6.56 ± 0.006 ^{b}	2.23 ± 0.116 ^{i}
CMCLB	2	4.63 ± 0.06 ^{f} (−19.4%)	5.32 ± 0.006 ^{g}	3.50 ± 0.030 ^{g} (+56.9%)
CMCLBF	24	4.27 ± 0.05 ^{g} (−22.8%)	4.27 ± 0.000 ^{j}	4.14 ± 0.100 ^{e} (+85.6%)
CMCLBF	48	3.83 ± 0.06 ^{h} (−30.7%)	4.19 ± 0.000 ^{l}	4.30 ± 0.013 ^{c} (+92.8%)
CMCLBF	72	3.60 ± 0.01 ^{I} (−34.9%)	3.81 ± 0.006 ^{m}	4.57 ± 0.058 ^{b} (+104.9%)
	Cucurbita pepo leaf and Cantaloupe melon smoothies fermented with LABs			
	Fermentation Time (h)	TSS (%)	pH	TA (Equivalent Lactic Acid g/100 g
CPCLC				
CPCLC	0	5.3 ± 0.01 ^{a}	6.43 ± 0.001 ^{a}	1.53 ± 0.060 ^{l}
CPCLC	2	5.1 ± 0.02 ^{b} (−3.8%)	6.24 ± 0.002 ^{d}	1.59 ± 0.060 ^{l} (+3.9%)
CPCLC	24	4.9 ± 0.01 ^{e} (−7.5%)	5.87 ± 0.001 ^{e}	1.67 ± 0.050 ^{k} (+9.1%)
CPCLC	48	4.9 ± 0.03 ^{e} (−7.5%)	5.76 ± 0.004 ^{f}	1.74 ± 0.060 ^{j} (+13.7%)
CPCLC	72	4.4 ± 0.10 ^{I} (−17.0%)	5.49 ± 0.003 ^{g}	1.78 ± 0.040 ^{j} (+16.3%)
CPCL75	0	5.0 ± 0.06 ^{d}	6.42 ± 0.001 ^{c}	2.03 ± 0.030 ^{h}
CPCL75	2	4.8 ± 0.03 ^{f} (−4.0%)	5.43 ± 0.009 ^{h}	3.97 ± 0.120 ^{d} (+95.6%)
CPCL75	24	4.6 ± 0.04 ^{g} (−8.0%)	4.32 ± 0.002 ^{i}	4.10 ± 0.020 ^{c} (+102.0%)
CPCL75	48	4.500 ± 0.02 ^{h} (−10.0%)	4.24 ± 0.003 ^{i}	4.27 ± 0.060 ^{b} (+110.0%)
CPCL75	72	4.000 ± 0.03 ^{j} (−20.0%)	3.78 ± 0.006 ^{k}	4.43 ± 0.060 ^{a} (+118.2%)
CPCLBF	0	5.03 ± 0.06 ^{c}	6.43 ± 0.002 ^{b}	1.90 ± 0.030 ^{i}
CPCLBF	2	4.8 ± 0.01 ^{f} (−4.6%)	5.40 ± 0.001 ^{h}	2.53 ± 0.070 ^{g} (+33.2%)
CPCLBF	24	4.5 ± 0.01 ^{h} (−10.5%)	4.27 ± 0.001 ^{i}	3.00 ± 0.050 ^{f} (+57.9%)
CPCLBF	48	4.5 ± 0.02 ^{h} (−10.5%)	4.10 ± 0.002 ^{j}	3.78 ± 0.060 ^{e} (+98.9%)
CPCLBF	72	3.97 ± 0.03 ^{j} (−21.1%)	3.81 ± 0.001 ^{k}	3.90 ± 0.070 ^{d} (+105.2%)
	Cucurbita moschata leaf and Honeydew melon smoothies fermented with LABs			
	Fermentation time (h)	TSS (%)	pH	TA (Equivalent Lactic Acid g/100 g
CMHDC	0	5.80 ± 0.01 ^{a}	6.44 ± 0.006 ^{a}	1.06 ± 0.05 ^{k}
CMHDC	2	4.90 ± 0.02 ^{c} (−15.5%)	6.37 ± 0.005 ^{c}	1.70 ± 0.08 ^{j} (+60.4%)
CMHDC	24	4.30 ± 0.01 ^{f} (−25.9%)	5.93 ± 0.001 ^{d}	1.78 ± 0.01 ^{j} (+67.9%)
CMHDC	48	4.00 ± 0.03 ^{h} (−31.0%)	5.80 ± 0.004 ^{e}	2.01 ± 0.06 ^{I} (+89.6%)
CMHDC	72	3.87 ± 0.05 ^{I} (−33.3%)	5.82 ± 0.006 ^{e}	2.06 ± 0.06 ^{I} (94.3%)
CMHD75	0	5.40 ± 0.01 ^{b}	6.40 ± 0.002 ^{b}	2.46 ± 0.05 ^{h}
CMHD75	2	4.50 ± 0.01 ^{e} (−16.7%)	5.56 ± 0.001 ^{f}	3.30 ± 0.06 ^{g} (+34.1%)
CMHD75	24	3.86 ± 0.05 ^{i} (−28.5%)	4.20 ± 0.002 ^{h}	3.90 ± 0.02 ^{f} (+58.5%)
CMHD75	48	3.40 ± 0.01 ^{l} (−37.0%)	3.94 ± 0.006 ^{i}	4.36 ± 0.06 ^{e} (+77.2%)
CMHD75	72	3.10 ± 0.02 ^{n} (−52.6%)	3.78 ± 0.001 ^{k}	5.70 ± 0.10 ^{a} (+131.7%)
CMHDBF	0	4.80 ± 0.01 ^{d}	6.36 ± 0.002 ^{c}	2.46 ± 0.08 ^{h}
CMHDBF	2	4.20 ± 0.02 ^{g} (−12.5%)	5.23 ± 0.005 ^{g}	3.95 ± 0.06 ^{f} (+60.6%)
CMHDBF	24	3.80 ± 0.05 ^{j} (−20.8%)	4.19 ± 0.006 ^{h}	4.90 ± 0.02 ^{d} (+99.2%)
CMHDBF	48	3.53 ± 0.05 ^{k} (−26.4%)	3.85 ± 0.001 ^{j}	5.26 ± 0.05 ^{c} (+113.8%)
CMHDBF	72	3.20 ± 0.02 ^{m} (−33.3%)	3.79 ± 0.061 ^{k}	5.33 ± 0.06 ^{b} (+116.7%)

Table 1. *Cont.*

	Cucurbita pepo leaf and Honeydew melon smoothies fermented with LABs			
	Fermentation time (h)	TSS(%)	pH	TA (Equivalent Lactic Acid g/100 g
CPHDC	0	4.53 ± 0.06 [c]	6.27 ± 0.001 [a]	1.08 ± 0.058 [n]
CPHDC	2	4.40 ± 0.01 [d] (−2.9%)	6.18 ± 0.002 [d]	1.52 ± 0.01 [m] (+40.7%)
CPHDC	24	4.16 ± 0.05 [f] (−8.2%)	5.87 ± 0.001 [f]	1.65 ± 0.01 [l] (+52.8%)
CPHDC	48	3.86 ± 0.06 [h] (−14.8%)	5.57 ± 0.004 [h]	2.00 ± 0.02 [k] (+85.2%)
CPHDC	72	3.60 ± 0.07 [j] (−20.5%)	4.91 ± 0.006 [i]	2.26 ± 0.05 [j] (109.3%)
CMHD75	0	5.13 ± 0.05 [a]	6.23 ± 0.002 [b]	3.46 ± 0.05 [i]
CMHD75	2	4.80 ± 0.01 [b] (−6.4%)	5.96 ± 0.005 [e]	5.30 ± 0.06 [f] (+53.2%)
CMHD75	24	4.50 ± 0.02 [c] (−12.3%)	4.16 ± 0.005 [j]	6.80 ± 0.08 [e] (+96.5%)
CMHD75	48	4.20 ± 0.05 [f] (−18.1%)	4.05 ± 0.009 [k]	7.16 ± 0.08 [c] (+106.9%)
CMHD75	72	4.03 ± 0.11 [g] (−21.4%)	4.02 ± 0.008 [l]	7.36 ± 0.06 [a] (+112.7%)
CPHDBF	0	4.30 ± 0.07 [e]	6.21 ± 0.002 [c]	3.44 ± 0.05 [i]
CPHDBF	2	4.20 ± 0.05 [f] (−2.3%)	5.76 ± 0.001 [g]	4.70 ± 0.09 [h] (+36.6%)
CPHDBF	24	3.90 ± 0.06 [h] (−9.3%)	4.17 ± 0.006 [j]	5.20 ± 0.03 [fg] (+51.1%)
CPHDBF	48	3.70 ± 0.07 [i] (−13.9%)	3.97 ± 0.006 [m]	7.03 ± 0.07 [d] (+104.3%)
CPHDBF	72	3.10 ± 0.08 [k] (−27.9%)	3.81 ± 0.006 [n]	7.20 ± 0.01 [b] (+109.3%)

Mean values were calculated based on three replicate samples. Different letters in the same column refer to statistical difference at $p < 0.05$ according to Fisher's protected LSD test. * Standard deviation; −% decrease **; +% increase ***. *Keys: Cucurbita moschata* + Watermelon (CMWC), *Cucurbita moschata* + Watermelon + *L75* (CMW75), *Cucurbita moschata* + Watermelon + *B. longum* (CMWBF), *Cucurbita pepo* + Watermelon Control (CPWC), *Cucurbita pepo* + Watermelon + *L75* (CPW75), *Cucurbita pepo* + Watermelon + *B. longum* (CPWBF), *Cucurbita moschata* + Cantaloupe Control (CMCLC), *Cucurbita moschata* + Cantaloupe + *L75* (CMCL75), *Cucurbita moschata* + Cantaloupe + *L75* (CMCL75), *Cucurbita pepo* + Cantaloupe Control (CPCLC), *Cucurbita pepo* + Cantaloupe + *L75* (CPCL75), *Cucurbita pepo* + Cantaloupe + *B. longum* (CPCLBF), *Cucurbita moschata* + Honeydew Control (CMHDC), *Cucurbita moschata* + Honeydew + *L75* (CMHD75), Cucurbita moschata + Honeydew + *B. longum* (CMHDBF), *Cucurbita pepo* + Honeydew Control (CPHDC), *Cucurbita pepo* + Honeydew + *L75* (CMHD75), and *Cucurbita pepo* + Honeydew + *B. longum* (CPHDBF).

Different strains of *Bifidobacterium longum* have been found to ferment carbohydrates to varying degrees, with the majority being able to ferment fructose, glucose, sucrose, and raffinose [35]. Different types of melon may contain other soluble carbohydrates. However, an increase in TSSs at 2 h of fermentation might be due to the capability of complex carbohydrates present in the introduced fruit and vegetable commodities to biotransform into simple sugar molecules, thus increasing the values of TSSs [33]. Compared to Watermelon and pumpkin leaf smoothies, the following combinations showed TSS levels < 4.0 after fermentation: *C. moschata* + Honeydew + *B. longum* (CMHDBF) and *C. pepo* + Honeydew + *B. longum* (CPHDBF) after 24 h, 48 h, and 72 h of fermentation; *C. moschata* + Honeydew + *L75* (CMHD75) after 24 h, 48 h, and 72 h of fermentation; *C. moschata* + Cantaloupe + *L75* (CMCL75) after 48 h and 72 h of fermentation; *C. moschata* + Cantaloupe + *B. longum* (CMCLBF) after 72 h of fermentation; and *C. pepo* + Cantaloupe + *B. longum* (CPCLBF) after 48 h and 72 h of fermentation. Possibly due to substrate effects (fruit matrix), the nutrient composition influencing fermentation and TSSs decreases [35]. Hou et al. [36] reported that carbohydrate fermentation varied depending on substrate type and even fermentation time. The reduction in sugar content is unappealing to taste. However, the consumption of low-sugar smoothies can help regulate blood sugar levels for individuals with diabetes.

3.3. pH and TA

The pH value of lactic acid fermentation plays a crucial role in its physicochemical properties. Fermented smoothies made from different melon types and *C. moschata* or *C. pepo* showed a significant decrease in pH with increasing fermentation time (Table 1). The pH of non-inoculated smoothies was higher than that of inoculated smoothies. The following fermentation treatments showed a pH lower than 4 at 72 h: *C. moschata* + Watermelon + *L75* (CMW75), *C. pepo* + Watermelon + *L75* (CPW75), *C. moschata* + Cantaloupe

+ *L75* (CMCL75), *C. pepo* + Cantaloupe + *L75* (CPCL75), *C. moschata* + Honeydew + *L75* (CMHD75), *C. moschata* + Watermelon + *B. longum* (CMWBF), *C. moschata* + Cantaloupe + *B. longum* (CMCLBF), *C. pepo* + Cantaloupe + *B. longum* (CPCLBF), *C. moschata* + Honeydew + *B. longum* (CMHDBF), and *C. pepo* + Honeydew + *B. longum* (CPHDBF). The growth of LABs could have contributed to pH decrease, while pH was inversely and strongly correlated to LAB counts in melon types and *C. moschata* or *C. pepo* (r = −0.74, $p < 0.05$) smoothies fermented with *L75* and *B. longum*. The pH values dropped remarkably during the fermentation of chayote leaf (*Sechium edule*) and pineapple with lactic acid bacterium (LAB) strains such as *Lactobacillus plantarum (L75)*, *Weissella cibaria (W64)*, and their combination (*LW64 + 75*) from 3.46 to 2.50 [20]. Also, Mandha et al. [37] showed a similar decrease in pH in Watermelon and mango juices fermented by *Levilactobacillus (L.) brevis, Lacticaseibacillus (La.) casei,* and *Pediococcus (P.) pentosaceus*. The continuous decline in pH during fermentation could be attributed to the actions of LAB strains generating organic acids and CO_2 through the metabolic pathways of carbohydrates with resulting carbonic acids lowering pH during the metabolic process [38]. It is advantageous to reduce the pH level of food during fermentation to stimulate the growth of LABs and lactic acid producers, and to inhibit the growth of pathogenic bacteria *Listeria monocytogenes* and *Escherichia coli* O157:H7, while maintaining 4.6 pH for food safety [39]. In all treatments, TA levels increased with fermentation time, and the highest concentration of TA (Equivalent Lactic Acid g/100 mL) was observed at 72 h for all melon types and pumpkin leaves (*C. moschata*/*C. pepo*). Fresh Watermelon juice contains malic acid (dicarboxylic acid) and, during fermentation, it transforms into lactic acid and CO_2 [39]. A decline in pH value and TSSs was correlated with an increase in the TA of smoothies of melon types and pumpkin leaves (*C. moschata*/*C. pepo*).

3.4. Color Properties

Color is a crucial parameter that influences how consumers perceive and accept fruits and vegetables. Supplementary (S) Table S1 shows the color properties and color change (ΔE) values of smoothies. After 48 and 72 h of fermentation, the ΔE increased significantly ($p < 0.05$) in all three melon types and *C. moschata* or *C. pepo* leaves fermented with *L75* or *B. longum*. The color degradation may be due to the breakdown of pigments due to observed pH reductions and acidic conditions (Supplementary Table S1). As part of the CIELab three-dimensional space, there are three components: luminosity ($L^* = 0$ indicates black, and $L^* = 100$ implies colorless), red/green ($a^* > 0$ is in correspondence with red, and $a^* < 0$ is correlated with green), and the yellow/blue color component ($b^* > 0$ is in correspondence with yellow, and $b^* < 0$ is linked with blue) [34]. However, when Watermelon, *C. moschata*, or *C. pepo* leaves were fermented with *L75* or *B longum*, the brightness (L^*), yellowness (b^*), and redness (a^*) decreased with increasing fermentation time compared to the unfermented samples (0 h). Conversely, in Cantaloupe, *C. moschata*, or *C. pepo* leaves fermented with *L75* or *B longum*, the brightness (L*) and yellowness (b^*) decreased while the redness (a^*) decreased with increasing fermentation time compared to the unfermented samples (0 h). Contrary to this, the brightness (L^*) and yellowness (b^*) decreased when Cantaloupe or Honeydew melons and *C. moschata* or *C. pepo* leaves were fermented with *L75* or *B longum*. Additionally, the redness (a^*) decreased with increasing fermentation time compared to the unfermented samples (0 h).

Mannaga et al. [20] demonstrated that fermenting chayote leaves (Sechium edule) and pineapple with lactic acid bacterium (LAB) strains, specifically L75 and *Weissella cibaria* (W64), as well as their combination (LW64 + L75), impacted the color properties of the leaves. This effect was likely due to enzymatic oxidation during the fermentation process. Rocha and Morais [40] showed that the onset of browning reactions can result in a decrease in lightness and an increased a^* (redness) value. Moreover, Watermelon pigments (lycopene and β-carotene) and Cantaloupe (carotenoids) could have been broken down by heat during pasteurization (Section 2.2) [39]. In addition, fermentation may further degrade these carotenoids into aroma volatiles such as monoterpenes, geranyl acetone, and geraniol [39]. The color change in all smoothies was reduced at 24 h of fermentation compared to 48 h and 72 h.

3.5. Ascorbic Acid

Its electron-donating potential makes l-ascorbic acid (vitamin C) an effective antioxidant due to its nutritional and clinical properties [41]. There was a significant increase in ascorbic acid content in the following mixtures: *C. moschata* + Watermelon + *L75* (CMW75), *C. pepo* + Watermelon + *L75* (CPW75), *C. moschata* + Cantaloupe + *L75* (CMCL75), *C. pepo* + Cantaloupe + *L75* (CPCL75), *C. moschata* + Honeydew + *L75* (CMHD75), and *C. pepo* + Honeydew + *L75* (CPHD75) (Figure 2). These mixtures showed the highest concentration of ascorbic acid after 24 h of fermentation compared to the control groups and those inoculated and fermented with *B. longum* at 24, 48, and 72 h. The ascorbic acid content of all melon types and *C. moschata* or *C. pepo* fermented with *L75* or *B. longum* was significantly ($p < 0.05$) reduced at 72 h. Inoculated smoothies containing *B. longum* showed higher concentrations of AA after 24 h compared to their respective control groups. Specifically, the smoothies *C. pepo* + Watermelon + *B. longum* (CPWBF), *C. pepo* + Cantaloupe + *B. longum* (CPCLBF), and *C. moschata* + Honeydew + *B. longum* (CMHDBF) exhibited higher AA concentrations. However, the decrease in vitamin C content during the fermentation of the three types of melon and *C. pepo* and *C. moschata* leaf smoothies by *L75* and *B. longum* ($p > 0.05$) was consistent. This supports the hypothesis that *L75* and *B. longum* can consume ascorbic acid independently. Based on our data, we disagree with Hashemi et al. [42], who found that the concentration of ascorbic acid in *Citrus limetta* juice was not altered after fermentation with *L. plantarum LS5*. The ascorbic acid concentrations can be affected by storage and pasteurization treatment temperatures [42]. In their [39] investigation, the pasteurization temperature was at 80 °C for 5 min and stored frozen (−20 °C for 2 weeks) before analysis, and this could have affected the concentration of AA. On the other hand, the LABs could have metabolized the AA from the food matrix [42].

Furthermore, after 24 h of fermentation, *C. moschata* + Watermelon + *B. longum* (CMWBF) and *C. moschata* + Watermelon Control (CMWC) showed similar levels of AA. Similarly, after 24 h of fermentation, *C. moschata* + Cantaloupe + *B. longum* (CMCLBF) and *C. moschata* + Cantaloupe Control (CMCLC) exhibited similar levels of AA. Likewise, *C. moschata* + Honeydew + *B. longum* (CMHDBF) and *C. moschata* + Honeydew Control (CMHDC), after 24 h of fermentation, showed similar levels of AA. Even though ascorbic acid is abundant in melon and pumpkin leaf smoothies, the fermentation can slow down the loss of ascorbic acid, and a similar observation was shown by Xu et al. [42]. Ascorbic acid content rose in citrus 'Tarocco' juice, rising to 295 ± 25.3 g L^{-1} when fermented by *L. brevis* [43]. Furthermore, according to Xu et al. [43], citrus juice fermentation accumulated a high concentration of ascorbic acid (above 300 mg/L) with *L. brevis* 182 and *L. paracasei* 502.

Fermentation decreases vitamin C content due to its instability and easy oxidative decomposition, as well as the metabolic utilization of lactic acid bacteria [44]. In smoothies and juices, ascorbic acid degradation occurs due to the presence of ascorbic acid oxidase, which is not completely destroyed by pasteurization [44]. Several factors influence ascorbic acid stability, including redox state, metal ion concentration, and pH [44]. Furthermore, Urbienė and Mitkute [45] found that the growth of LABs and the increase in their population throughout fermentation had a minimal impact on changes in ascorbic acid, since oxidation was the main cause of degradation.

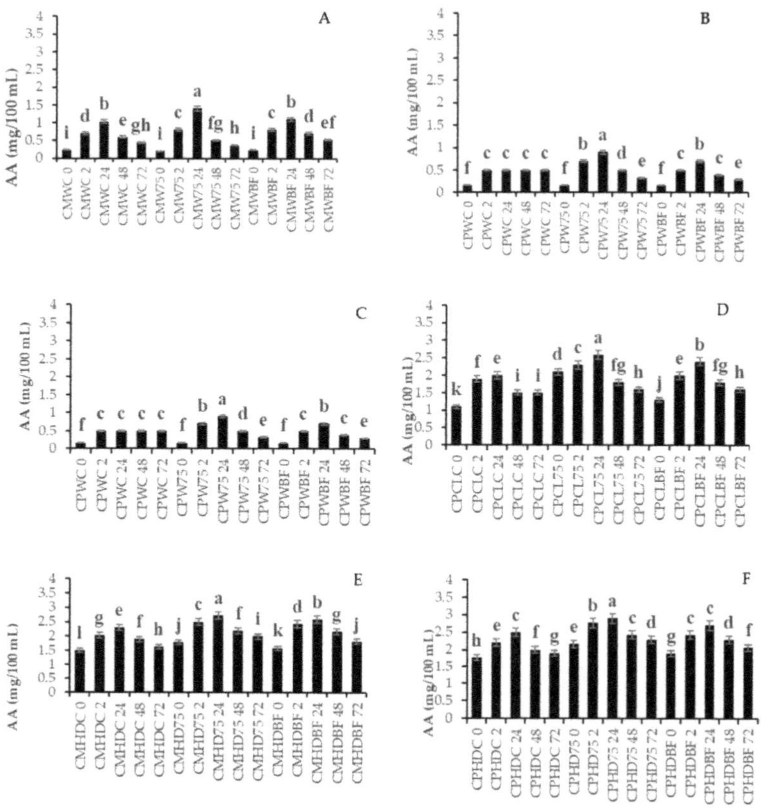

Figure 2. Changes in ascorbic acid concentration during fermentation with different lactobacillus strains on pumpkin leaf (*Cucurbita moschata or C. pepo)* and melon smoothies. (**A**) *Cucurbita moschata* and Watermelon smoothies fermented with LABs; (**B**) *Cucurbita pepo* + Watermelon LABS smoothies fermented with LABs; (**C**) *Cucurbita moschata* + Cantaloupe smoothies fermented with LABs; (**D**) *Cucurbita pepo* + Cantaloupe smoothies fermented with LABs; (**E**) *Cucurbita moschata* + Honeydew smoothies fermented with LABs; (**F**) *Cucurbita pepo* + Honeydew smoothies fermented with LABs. Mean values were calculated based on three replicate samples. Different letters in the same bar refer to statistical difference at $p < 0.05$ according to Fisher's protected LSD test. *Keys: Cucurbita moschata* + Watermelon (CMWC), *Cucurbita moschata* + Watermelon + *L75* (CMW75), *Cucurbita moschata* + Watermelon + *B. longum* (CMWBF), *Cucurbita pepo* + Watermelon Control (CPWC), *Cucurbita pepo* + Watermelon + *L75* (CPW75), *Cucurbita pepo* + Watermelon + *B. longum* (CPWBF), *Cucurbita moschata* + Cantaloupe Control (CMCLC), *Cucurbita moschata* + Cantaloupe + *L75* (CMCL75), *Cucurbita moschata* + Cantaloupe + *L75* (CMCL75), *Cucurbita pepo* + Cantaloupe Control (CPCLC), *Cucurbita pepo* + Cantaloupe + *L75* (CPCL75), *Cucurbita pepo* + Cantaloupe + *B. longum* (CPCLBF), *Cucurbita moschata* + Honeydew Control (CMHDC), *Cucurbita moschata* + Honeydew + *L75* (CMHD75), Cucurbita moschata + Honeydew + *B. longum* (CMHDBF), *Cucurbita pepo* + Honeydew Control (CPHDC), *Cucurbita pepo* + Honeydew + *L75* (CMHD75), and *Cucurbita pepo* + Honeydew + *B. longum* (CPHDBF).

3.6. Sensory Properties

In Figure 3, the sensory qualities of smoothies are presented. The sensory qualities of smoothies made with three types of melon and pumpkin leaves were affected by fermentation. The overall quality was determined based on the color, aroma, and taste of all the samples. It was observed that smoothies fermented with *L75* and *B. longum* retained color, aroma, taste, and acceptability 24 h after fermentation, similar to non-inoculated smoothies at 0 h. After 24 h of fermentation, CPCL75, CPCLBF, CMCL75, and CMCLBF achieved an overall quality score of 8.8. The CMW75, CMWBF, CPW75, and CPWBF fermented smoothies scored 9.0 and 8.9 for overall quality after 24 h. Moreover, CMHD75 and CMHDBF showed overall quality scores of 5.7 after 2 h of fermentation. Similarly, CPHD75 and CPHDBF showed scores of 2.6 and 5.3 after 24 h of fermentation, respectively. Some fermented smoothies from Honeydew melon scored around 5.3 and 5.7 (neither liked nor disliked) for overall acceptability, while the non-inoculated Honeydew smoothies scored around 2.3 (extremely disliked). Increased fermentation time adversely affected the sensory properties of the color, taste, and aroma, while negatively impacting the overall acceptability of these smoothies. Due to a change in the metabolic process of the cultures, increased fermentation time resulted in alterations in the smoothies' contents that negatively affected the final product's aroma, color, and taste. Lactic acid bacteria (LABs) primarily utilize glucose and fructose to produce organic acids and specific aromatic compounds that contribute to the flavor of the food [46]. According to the results, panelists may have been influenced by sweetness, pleasant aroma (due to volatile compounds), and color in making their decisions. In a sensory evaluation, reduced taste could be due to sourness associated with organic acids, while sweetness is linked to sugars such as glucose and fructose [46]. Sensory evaluation further confirmed the higher sugar (TSS) consumption and organic acid (lactic acid) production during 72 h fermentation.

3.7. Total Phenols

Figure 4 shows the effect of fermentation on TPC, and it increased between 31.45 mg GAE/100 mL and 70.17 mg GAE/100 mL when Cantaloupe and Watermelon were fermented with *L75* or *B. Longum*. When fermented with *B. Longum*, both *C. pepo* + Cantaloupe + *B. longum* (CPCLBF) and *C. moschata* + Cantaloupe + *B. longum* (CMCLBF) showed the highest level of phenolics (62.91 mg GAE/100 mL; 70.17 mg GAE/100 mL) compared to *C. moschata* + Watermelon and *C. pepo* + Watermelon fermented with *B. Longum* and *L75* (CMW75, CMWBF, CPW75, and CPWBF), as well as *C. moschata* + Cantaloupe (CMCL75) and *C. pepo* + Cantaloupe fermented with *L75* (CPCL75). The TPC was increased 1.33-fold by fermenting *C. moschata* + Cantaloupe with *B. longum* compared to their control samples. In general, TPC increased from 1.01- to 1.33-fold when fermented with LAB strains compared with their control samples. It is thought that the increase in TPC during fermentation is due to the ability of *L75* and *B longum* to deglycosylate glycosylated phenolic compounds in plant cells [47]. As substrates in food fermentation, polyphenols are hydrolyzed into smaller phenolic compounds with higher bioactivity and bioavailability by polyphenol-associated enzymes (tannases, esterases, phenolic acid decarboxylases, and glycosidases) [48]. Lactic acid bacteria secrete tannases, esterases, phenolic acid decarboxylases, or glycosidases during lactic acid fermentation and release the bound phenols to free phenols. In fermented foods, this biotransformation increases the amount and type of free phenols [48]. Different polyphenols undergo different biotransformation pathways by polyphenol-associated enzymes secreted by different microorganisms [48]. Furthermore, *Bifidobacterium longum* can deglycosylate a wide range of substrates due to their genome-encoded β-glycosidases [47].

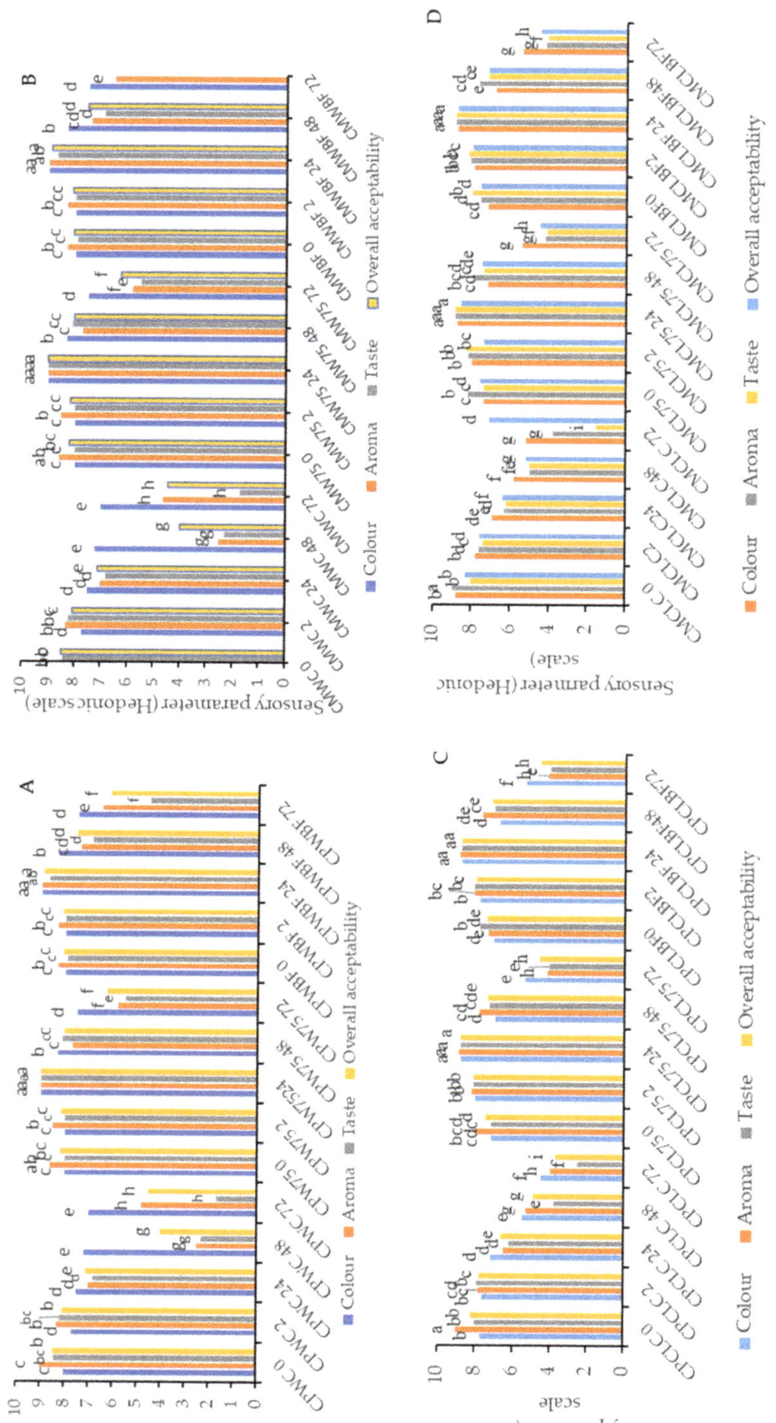

Figure 3. *Cont.*

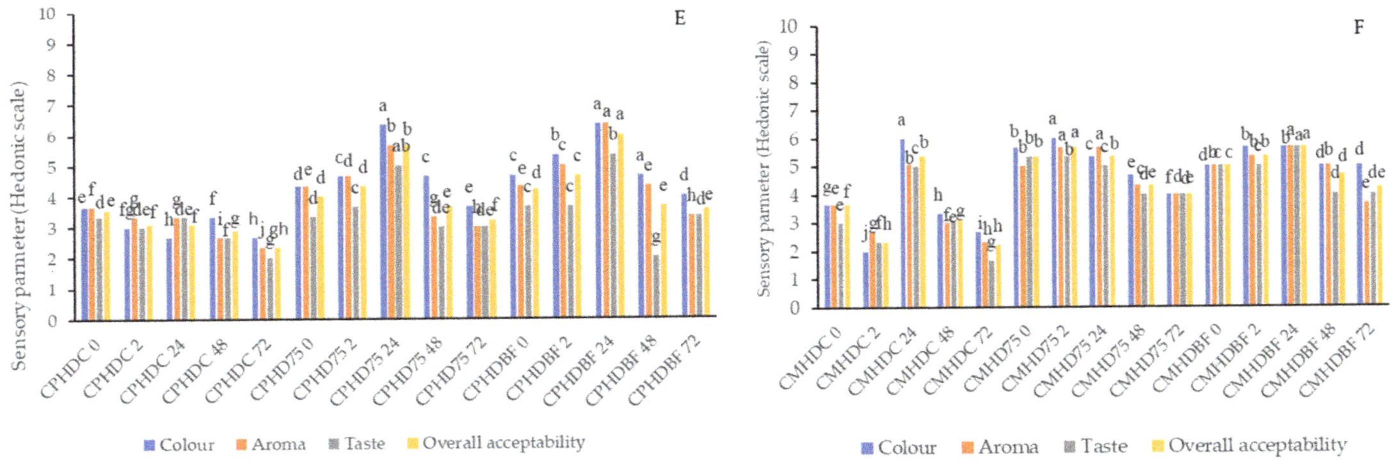

Figure 3. Sensory properties of pumpkin leaf (*Cucurbita moschata* or *C. pepo*) and melon smoothies fermented with different lactobacillus strains. Different letters in the same bar refer to statistical difference at $p < 0.05$ according to Fisher's protected LSD test. (**A**,**C**) *Cucurbita moschata* + Cantaloupe smoothies fermented with LABs; (**B**,**E**) *Cucurbita pepo* + Cantaloupe smoothies fermented with LABs; (**D**,**F**) *Cucurbita pepo* + Watermelon smoothies fermented with LABs; *Keys: Cucurbita moschata* + Watermelon (CMWC), *Cucurbita moschata* + Watermelon + *L75* (CMW75), *Cucurbita moschata* + Watermelon + *B. longum* (CMWBF), *Cucurbita pepo* + Watermelon Control (CPWC), *Cucurbita pepo* + Watermelon + *L75* (CPW75), *Cucurbita pepo* + Watermelon + *B. longum* (CPWBF), *Cucurbita moschata* + Cantaloupe Control (CMCLC), *Cucurbita moschata* + Cantaloupe + *L75* (CMCL75), *Cucurbita moschata* + Cantaloupe + *L75* (CMCL75), *Cucurbita pepo* + Cantaloupe Control (CPCLC), *Cucurbita pepo* + Cantaloupe + *L75* (CPCL75), *Cucurbita pepo* + Cantaloupe + *B. longum* (CPCLBF), *Cucurbita moschata* + Honeydew Control (CMHDC), *Cucurbita moschata* + Honeydew + *L75* (CMHD75),Cucurbita moschata + Honeydew + *B. longum* (CMHDBF), *Cucurbita pepo* + Honeydew Control (CPHDC),*Cucurbita pepo* + Honeydew + *L75* (CMHD75) and *Cucurbita pepo* + Honeydew + *B. longum* (CPHDBF).

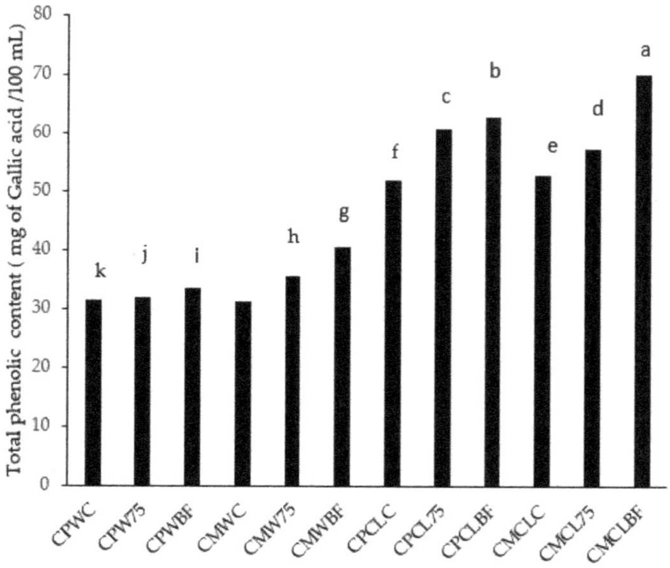

Figure 4. Changes in total phenolic content during fermentation with different lactobacillus strains on pumpkin leaf (*Cucurbita moschata or C. pepo)* and melon smoothies. Mean values were calculated based on three replicate samples. Different letters in the same bar refer to statistical difference at $p < 0.05$ according to Fisher's protected LSD test. *Cucurbita pepo* + Watermelon Control (CPWC), *Cucurbita pepo* + Watermelon + *L75* (CPW75), *Cucurbita pepo* + Watermelon + *B. longum* (CPWBF), *Cucurbita moschata* + Watermelon Control (CMWC), *Cucurbita moschata* + Watermelon + *L75* (CMW75), *Cucurbita moschata* + Watermelon + *B. longum* (CMWBF), *Cucurbita moschata* + Cantaloupe Control (CMCLC), *Cucurbita moschata* + Cantaloupe + *L75* (CMCL75), *Cucurbita moschata* + Cantaloupe + *B. longum* (CMCLBF), *Cucurbita Moschata* + Cantaloupe Control (CMCLC), *Cucurbita Moschata* + Cantaloupe+ *L75* (CMCL75), and *Cucurbita Moschata* + Cantaloupe + *B. longum* (CMCLBF).

3.8. Total Carotenoids (TCCs), Lutein, and β Carotene

In *C. pepo* + Cantaloupe fermented with *L75* (CPCL75), TCCs showed the highest concentration (26.14 mg/100 mL), followed by *C. pepo* + Cantaloupe fermented with *B. longum* (CPCLBF) (23.18 mg/100 mL), compared to all other fermented and unfermented smoothies (Table 2). Compared with the control counterparts, *C. pepo* and Cantaloupe fermented with *L75* (CPCL75) or *B. longum* (CPCLBF) increased their TCC content 1.85- and 1.64-fold, respectively. Contrary to this, fermentation with *L75* increased the TCCs for *C. moschata* and Watermelon (CMW75) 1.22-fold, but only 1.16-fold with *B. longum* (CMWBF). Moreover, when *C. pepo* and Cantaloupe are combined, fermentation with *L75* and *B. longum* results in a 1.85-fold and 1.64-fold increase, respectively, in TCCs as compared to their control counterparts. Antognoni et al. [49] showed an increase in the carotenoid content of wheat dough after fermentation with *L. plantarum* (29DAN, 83DAN). Among all fermented smoothies, *C. pepo* + Cantaloupe smoothie fermented with *L75* (CPCL75) showed the highest concentration of lutein (2.53 mg/100 mL), trans-β-carotene (620.37 mg/100 mL), and cis-β-carotene (28.20 mg/100 mL). However, *C. moschata* + Watermelon fermented with *B. longum* (CMWBF) demonstrated a 1.46-fold increase in lutein compared to their unfermented control samples. Watermelon and *C. pepo* fermented with *L75* (CPW75) or *B. longum* (CPWBF) showed a 3.26- and 3.09-fold increase, respectively, in trans-β-carotene compared to their control samples. Similar trends in the increase in cis β-carotene were observed as well. These results show that the addition of *lactobacillus* or *bifidobacterium* to Watermelon and *C. pepo* can increase the concentration of beneficial carotenoids. Meanwhile, *C. moschata* combined with Watermelon and fermented with *L75* (CMW75) showed a 7.57-fold increase in cis-β-carotene and a 4.97-fold increase in trans-β-carotene compared to the control samples. Also, *C. pepo* + Cantaloupe fermented with *L75* (CPCL75) showed a 1.57- and 2.41-fold increase in trans and cis-β-carotenes. Contrary to our findings, Szutowska et al. [50] showed that curly kale juice fermented with LAB strains reduced the amount of lutein and beta-carotene, respectively, by 40% and 34.8%, while the amount of carotenoid decreased by 17–31%. In addition, Odongo et al. [51] reported that Ethiopian kale fermented with LAB strains lost 98% of its lutein, which was also 75% of its carotenoids. Demir et al. [52] reported an increase in carrot juice carotene content following fermentation by *L. plantarum*, similar to our findings. Carotenoid content decreases during lactic acid fermentation due to the formation of volatile derivatives, which contribute to the flavor and aroma of fermented foods [53]. Meanwhile, Oloo et al. [54] investigated the influence of lactic fermentation (*L. plantarum* MTCC 1407 at 25 ± 2 °C for 48 h) on the accumulation of β-carotene in three varieties of orange-fleshed sweet potatoes (Zapallo, Nyathiodiewo, and SPK004/06) from Kenya and demonstrated the elevation of carotene content by 93.97%.

Table 2. Changes in total phenolic compounds, total carotenoids, and different components of carotenoids in pumpkin leaf and melon smoothies fermented by different lactobacillus strains.

	***Cucurbita pepo* and Watermelon Fermented Smoothie**				
	Total Carotenoids (mg/100 mL)	**Lutein (mg/100 mL)**	**Trans-β-Carotene (mg/100 mL)**	**Cis-β-Carotene (mg/100 mL)**	**Zeaxanthin (μg/mL) (mg/100 mL)**
CPWC	3.27 ± 0.11 ^{k,*}	1.55 ± 0.01 ^{l}	31.51 ± 0.02 ^{l}	1.02 ± 0.04 ^{k}	0.06 ± 0.05 ^{g}
CPW75	7.38 ± 0.02 ^{g} (2.26-fold) **	1.63 ± 0.05 ^{j} (1.05-fold)	102.64 ± 0.03 ^{i} (3.26-fold)	7.35 ± 0.1 ^{g} (7.21-fold)	0.19 ± 0.14 ^{f} (3.16-fold)
CPWBF	7.31 ± 0.12 ^{g} (2.24-fold)	1.83 ± 0.00 ^{h} (1.19-fold)	98.29 ± 0.05 ^{j} (3.11-fold)	2.10 ± 0.23 ^{i} (2.05-fold)	0.17 ± 0.00 ^{f} (2.83-fold)
	Cucurbita moschata and Watermelon fermented smoothie				
CMWC	4.12 ± 0.0 ^{j}	1.57 ± 0.00 ^{k}	47.39 ± 0.08 ^{k}	1.48 ± 0.32 ^{j}	0.08 ± 0.01 ^{g}
CMW75	5.01 ± 0.16 ^{h} (1.22-fold)	1.69 ± 0.00 ^{i} (1.07-fold)	358.60 ± 0.07 ^{g} (7.57-fold)	7.35 ± 0.42 ^{g} (4.97-fold)	0.36 ± 0.05 ^{e} (4.50-fold)
CMWBF	4.77 ± 0.02 ^{i} (1.16-fold)	2.30 ± 0.00 ^{c} (1.46-fold)	125.45 ± 0.04 ^{h} (2.65-fold)	3.40 ± 0.31 ^{h} (2.30-fold)	0.32 ± 0.01 ^{e} (4.0-fold)
	Cucurbita pepo and Cantaloupe fermented smoothie				
CPCLC	14.16 ± 0.07 ^{e}	2.07 ± 0.00 ^{g}	396.09 ± 0.00 ^{f}	11.69 ± 0.06 ^{f}	0.41 ± 0.03 ^{d}
CPCL75	26.14 ± 0.17 ^{a} (1.85-fold)	2.53 ± 0.00 ^{a} (1.25-fold)	620.37 ± 0.00 ^{a} (1.57-fold)	28.20 ± 0.12 ^{a} (2.41-fold)	0.65 ± 0.02 ^{b} (1.58-fold)
CPCLBF	23.18 ± 0.00 ^{b} (1.64-fold)	2.25 ± 0.00 ^{d} (1.09-fold)	401.57 ± 0.00 ^{d} (1.01-fold)	15.40 ± 0.10 ^{c} (1.32-fold)	0.55 ± 0.01 ^{c} (1.34-fold)
	Cucurbita Moschata and Cantaloupe fermented smoothie				
CMCLC	12.59 ± 0.02 ^{f}	2.09 ± 0.00 ^{f}	400.23 ± 0.00 ^{e}	12.70 ± 0.04 ^{e}	0.43 ± 0.02 ^{d}
CMCL75	17.70 ± 0.03 ^{c} (1.41-fold)	2.51 ± 0.10 ^{b} (1.19-fold)	454.45 ± 0.00 ^{b} (1.14-fold)	25.43 ± 0.17 ^{b} (2.00-fold)	0.70 ± 0.03 ^{a} (1.62-fold)
CMCLBF	16.98 ± 0.02 ^{c} (1.35-fold)	2.20 ± 0.20 ^{e} (1.04-fold)	443.11 ± 0.00 ^{c} (1.11-fold)	13.42 ± 0.09 ^{d} (1.06-fold)	0.55 ± 0.03 ^{c} (1.27-fold)

Mean values were calculated based on three replicate samples. Different letters in the same column refer to statistical difference at $p < 0.05$ according to Fisher's protected LSD test. * Standard deviation, ** fold increase or decrease.

Bartkiene et al. [55] showed tomato varieties Ronaldo and Cunero fermented with *L sakei* increased total carotenoids levels by 41.1 and 33.6%, respectively, compared to the untreated samples. However, the observed significant increase in this carotene could be due to a variety of reasons. In the fermentation of melon and pumpkin leaves, enzymatic activities and microbiological conversions (macromolecular changes) are involved, and several strains of *L. plantarum* produce carotenoids to protect against oxidative stress, or fermentation-induced structural changes could have enhanced the extraction of carotenoids [56]. In our research and that of other researchers, we found that factors such as carotenoid type, matrix, and fermentation conditions can greatly affect the bioavailability of carotenoid components. However, a comprehensive metabolomics study of carotenoids and their derivatives before and after fermentation with lactic acid has not been conducted. Carotenoid compounds contain health benefits due to their antioxidant properties, which are highly effective radical quenchers [56]. The prevalence of all-trans β-carotene in tissues and serum indicates that it may be the preferred form of β-carotene [57]. The smoothies made from various types of melon and pumpkin leaves were found to contain between 10 and 60 mg of lutein per 10 mL, while 100 mL of these smoothies contained about 2500 μg RAE of vitamin A. This suggests that consuming a cup of these smoothies will help meet the general daily recommendation of 6–10 mg of lutein and 900–700 μg RAE of vitamin A for adult men and women, respectively.

Cucurbita pepo + Watermelon Control (CPWC), *Cucurbita pepo* + Watermelon + *L75* (CPW75), *Cucurbita pepo* + Watermelon + *B. longum* (CPWBF), *Cucurbita moschata* + Watermelon Control (CMWC), *Cucurbita moschata* + Watermelon + *L75* (CMW75), *Cucurbita moschata* + Watermelon + *B. longum* (CMWBF), *Cucurbita moschata* + Cantaloupe Control (CMCLC), *Cucurbita moschata* + Cantaloupe + *L75* (CMCL75), *Cucurbita moschata* + Cantaloupe + *B. longum* (CMCLBF), *Cucurbita Moschata* + Cantaloupe Control (CMCLC), *Cucurbita Moschata* + Cantaloupe+ *L75* (CMCL75), and *Cucurbita Moschata* + Cantaloupe + *B. longum* (CMCLBF).

3.9. Antioxidant Activities

Cucurbita moschata + Watermelon + *L75* (CMW75) showed the highest antioxidant power (FRAP) (9.89 mM TEAC/mL) followed by *Cucurbita pepo* + Cantaloupe + *L75* (CPCL75) (7.76 mM TEAC/mL) and *Cucurbita moschata* + Cantaloupe + *L75* (7.40 mM TEAC/mL). CMW75 also showed a 1.23-fold increase in TPC compared to their unfermented control samples (Table 3). The increased FRAP activity could be due to the ability of antioxidant properties and phenols to typically scavenge free radicals [58], chelate metal ions, and act as reduction agents [58] in the fermented smoothies. Moreover, fermentation can produce new phenolic compounds while improving the concentration of the present compounds and enhancing the contribution of free soluble antioxidants.

Cucurbita pepo + Cantaloupe + *L75* (CPCL75) showed the highest DPPH radical scavenging activity (1.18 IC50 μL/mL) followed by *Cucurbita moschata* + Cantaloupe + *L75* (CMCL75) (2.66 IC50 μL/mL) (Table 3). In general, *L75*-fermented CMW75, CPCL75, and CMCL75 smoothies showed higher DPPH radical scavenging activity than *B. longum* fermented smoothies. There is, however, a similar level of DPPH scavenging activity in smoothies made with CMW75 and CMWBF. In addition, the melon and pumpkin leaf smoothies fermented with *L75* showed the highest ascorbic acid content. DPPH strongly correlates with AA; the increased DPPH scavenging activity could be attributed to the high concentration of AA available in the smoothies as it scavenges the free radicals, thus improving DPPH potency [12]. Additionally, a reduction in pH, as observed in Section 3.3, enhances the durability of ascorbic acid and decelerates its conversion to dehydroascorbic acid, thereby aiding in an increased DPPH scavenging ability [59].

A higher level of ABTS scavenging activity was observed in CMCL75 smoothies followed by CMCLBF smoothies (Table 3). Overall, the melon and pumpkin leaf smoothie combinations fermented with *L75* had significantly ($p < 0.05$) greater antioxidant power (FRAP) and DPPH scavenging activity than smoothies fermented with *B. longum*. Above all, the increased levels of antioxidant properties in these smoothies may be attributed to the higher levels of phytochemicals such as phenolics and AA for FRAP and DPPH and carotenoids for ABTS that are produced through microbial hydrolysis or biotransformation, leading to the breakdown of antioxidant polysaccharides and peptides in the smoothies [60]. Different chemical structures of the radicals and the mechanism of action of the molecules, as well as the type of phenolic compounds present in the smoothies, contribute to the variation in antioxidant scavenging potential. Carotenoids' structures influence their antioxidative activity. The carbonyl and hydroxyl groups play a significant role in the carotenoids' quenching capacity, which is due to their affinity to reactive oxygen species (ROS). Thus, carotenoids can be enhanced by adding more hydroxyl groups or by adding conjugated double bonds (C=C and C=O) to increase their antioxidant activity [61]. Carotenoids donate electrons to ROS and create carotenoid–radical cations by oxygenation, thus lowering the oxidation state of the ROS [62]. Also, carotenoids scavenge reactive oxygen species through the hydrogen atom transfer mechanism in which a proton and electron are moved between two substrates during a chemical reaction [48].

Table 3. Changes in antioxidant activities of pumpkin leaf and melon smoothies fermented by different Lactobacillus strains.

	Cucurbita pepo **Leaf and Watermelon Fermented Smoothie**		
	FRAP (mM TEAC/mL)	**DPPH (IC_{50} µL/mL)**	**ABTS (IC_{50} µL/mL)**
CPWC (Control)	6.4 ± 0.058 [i,*]	4.00 ± 0.03 [j]	1.72 ± 0.01 [h]
CPW75	7.00 ± 0.00 [f]	3.47 ± 0.02 [ef]	0.52 ± 0.03 [f]
CPWBF	6.92 ± 0.10 [g]	3.37 ± 0.01 [e]	0.72 ± 0.06 [f]
	Cucurbita moschata leaf and Watermelon fermented smoothie		
	FRAP (mM TEAC/mL)	DPPH (IC50 µL/mL)	ABTS (IC50 µL/mL)
CMW (Control)	6.57 ± 0.20 [i]	3.94 ± 0.06 [i]	1.69 ± 0.00 [h]
CMW75	9.89 ± 0.48 [a]	3.15 ± 0.03 [d]	0.45 ± 0.12 [d]
CMWBF	7.12 ± 0.19 [e]	3.17 ± 0.02 [d]	0.83 ± 0.29 [g]
	Cucurbita pepo leaf and Cantaloupe melon fermented smoothie		
	FRAP (mM TEAC/mL)	DPPH (IC50 µL/mL)	ABTS (IC50 µL/mL)
CPCLC (Control)	6.16 ± 0.10 [j]	3.86 ± 0.22 [gh]	1.84 ± 0.12 [i]
CPCL75	7.76 ± 0.04 [b]	1.18 ± 0.07 [a]	0.36 ± 0.04 [c]
CPCLBF	7.27 ± 0.185 [d]	2.80 ± 0.04 [c]	0.39 ± 0.04 [c]
	Cucurbita moschata leaf and Cantaloupe melon fermented smoothie		
	FRAP (mM TEAC/mL)	DPPH (IC50 µL/mL)	ABTS (IC50 µL/mL)
CMCLC (Control)	6.14 ± 0.11 [j]	3.77 ± 0.01 [g]	1.84 ± 0.06 [i]
CMCL75	7.40 ± 0.34 [c]	2.66 ± 0.02 [b]	0.17 ± 0.04 [a]
CMCLBF	6.67 ± 0.05 [h]	2.84 ± 0.04 [c]	0.23 ± 0.44 [b]

Mean values were calculated based on three replicate samples. Different letters in the same column refer to statistical difference at $p < 0.05$ according to Fisher's protected LSD test. * Standard deviation.

Cucurbita pepo + Watermelon Control (CPWC), *Cucurbita pepo* + Watermelon + *L75* (CPW75), *Cucurbita pepo* + Watermelon + *B. longum* (CPWBF), *Cucurbita moschata* + Watermelon Control (CMWC), *Cucurbita moschata* + Watermelon + *L75* (CMW75), *Cucurbita moschata* + Watermelon + *B. longum* (CMWBF), *Cucurbita moschata* + Cantaloupe Control (CMCLC), *Cucurbita moschata* + Cantaloupe + *L75* (CMCL75), *Cucurbita moschata* + Cantaloupe + *B. longum* (CMCLBF), *Cucurbita Moschata* + Cantaloupe Control (CMCLC), *Cucurbita Moschata* + Cantaloupe+ *L75* (CMCL75), and *Cucurbita Moschata* + Cantaloupe + *B. longum* (CMCLBF).

4. Conclusions

Smoothies made from fruits and vegetables are rich in vitamins, minerals, and bioactive compounds, such as carotenoids, along with fiber. Fermented products can also reduce post-harvest losses and food waste while improving food security. Probiotic non-dairy products are commonly produced by fermenting fruit and vegetable commodities by lactic acid bacteria to enhance their health-related functional characteristics and bioactive compounds. Developing non-dairy probiotic smoothies enriched with bioactive compounds opens new perspectives, as functional compounds contained in fermented fruit and indigenous vegetable products emphasize the phytonutrient content. This study comprehensively examined the impact of two different lactic acid bacterium (LAB) strains inoculation on the changes in physicochemical parameters, total phenols, total carotenoids and the components of carotenoids, and antioxidant activity in different types of melon and pumpkin leaf smoothies. This study confirmed the potential of fermenting Cantaloupe and pumpkin leaves together with the *L75* strain to produce non-dairy functional products.

Furthermore, these smoothies provide a significant amount of ascorbic acid, lutein, and β-carotene. Adding them to our daily diets will contribute to diversifying our diet, while also supporting overall health and meeting daily nutrient requirements. Additionally, fermentation with probiotics that produce high-antioxidant carotenoids can deliver substantial health benefits by enhancing digestion and reducing oxidative stress in the body, preventing cellular damage. However, further research is needed to enhance the application of fermentation to produce microbial carotenoids by obtaining a deeper understanding of less common carotenoid types that could be beneficial to consumers.

Supplementary Materials: The following supporting information can be downloaded at: https://www.mdpi.com/article/10.3390/foods13223562/s1, Table S1: Changes in color properties during fermentation with different lactobacillus strains on pumpkin leaf (*Cucurbita moschata* or *C. pepo)* and melon smoothies.

Author Contributions: P.M.: formal analysis, investigation, writing—original draft preparation, S.M.M.: investigation, methodology, data curation, validation, visualization, D.S.: conceptualization, supervision, project administration, and funding acquisition. All authors have read and agreed to the published version of the manuscript.

Funding: The authors gratefully acknowledge the SARChI Research Chair grant for Phytochemical Food Network to Improve Nutritional Quality for Consumers, supported by the NRF (National Research Foundation), grant number 98352.

Institutional Review Board Statement: Not applicable.

Informed Consent Statement: Informed consent was obtained from all subjects involved in the study.

Data Availability Statement: The original contributions presented in the study are included in the article, further inquiries can be directed to the corresponding author.

Conflicts of Interest: The authors declare no conflicts of interest.

References

1. World Health Organization. Healthy Diet. Available online: https://www.who.int/news-room/fact-sheets/detail/healthy-diet#:~:text=Eating%20at%20least%20400%20g,daily%20intake%20of%20dietary%20fibre (accessed on 1 November 2024).
2. Manivannan, A.; Lee, E.S.; Han, K.; Lee, H.E.; Kim, D.S. Versatile nutraceutical potentials of watermelon—A modest fruit loaded with pharmaceutically valuable phytochemicals. *Molecules* **2020**, *11*, 5258. [CrossRef] [PubMed]
3. Honda, M. *Z*-Isomers of lycopene and β-carotene exhibit greater skin-quality improving action than their all-*E*-isomers. *Food Chem.* **2023**, *421*, 135954. [CrossRef] [PubMed]
4. USDA. Database Entry for Honeydew Melon (Raw). Available online: https://fdc.nal.usda.gov/fdc-app.html#/food-details/169911/nutrients (accessed on 30 September 2024).
5. Mashitoa, M.F.; Shoko, T.; Slabbert, R.M.; Shai, J.L.; Sultanbawa, Y.; Sivakumar, D. A practical chemometric approach using UPLC–QTOF/MS tool to investigate three varieties of pumpkin species and in vitro bioactivities. *Food Biosci.* **2021**, *43*, 101229. [CrossRef]
6. Odhav, B.; Beekrum, S.; Akula, U.; Baijnath, H. Preliminary assessment of nutritional value of traditional leafy vegetables in KwaZulu-Natal, South Africa. *J. Food Compos. Anal.* **2007**, *20*, 430–435. [CrossRef]
7. Global Trade Magezine. 2020. Available online: https://www.globaltrademag.com/ (accessed on 30 September 2024).
8. Adepoju, A.O.; Ologan, O. Post-harvest loss along the watermelon value chain in the tropics. *Int. J. Veg. Sci.* **2021**, *27*, 461–471. [CrossRef]
9. Ritchie, H. Food Waste Is Responsible for 6% of Global Greenhouse Gas Emissions Published online at OurWorldinData.org. 2020. Available online: https://ourworldindata.org/food-waste-emissions (accessed on 6 July 2024).
10. Elisha, G.O.; Arnold, O.M.; Christian, U.; Huyskens-Keil, S. Postharvest treatments of African leafy vegetables for food security in Kenya: A review. *Afr. J. Hort Sci.* **2016**, *9*, 32–40.

11. Reddy, M.B.; Love, M. The impact of food processing on the nutritional quality of vitamins and minerals. *Adv. Exp. Med. Biol.* **1999**, *459*, 99–106. [PubMed]
12. Dreher, M.L. Whole fruits and fruit fiber emerging health effects. *Nutrients* **2018**, *12*, 1833. [CrossRef] [PubMed]
13. Teleszko, M.; Wojdyło, A. Bioactive compounds vs. organoleptic assessment of 'smoothies'-type products prepared from selected fruit species. *Int. J. Food Sci.* **2014**, *49*, 98–106. [CrossRef]
14. Cichońska, P.; Ziarno, M. Legumes and legume-based beverages fermented with lactic acid bacteria as a potential carrier of probiotics and prebiotics. *Microorganisms* **2022**, *10*, 91. [CrossRef]
15. Moloto, R.; Akinola, S.A.; Seke, F.; Shoko, T.; Sultanbawa, Y.; Shai, J.L.; Remize, F.; Sivakumar, D. Influence of fermentation on functional properties and bioactivities of different cowpea leaf smoothies during in vitro digestion. *Foods* **2023**, *12*, 1701. [CrossRef] [PubMed]
16. Food & Beverages Market Size Report, 2022–2027. Available online: https://www.industryarc.com/Research/Food-And-Beverages--Global-Market-Research-513301 (accessed on 30 September 2024).
17. Garcia-Gonzalez, N.; Battista, N.; Prete, R.; Corsetti, A. Health-promoting role of *Lactiplantibacillus plantarum* isolated from fermented foods. *Microorganisms* **2021**, *9*, 349. [CrossRef] [PubMed]
18. Son, J.K.; Jo, J.Y.; Jung, Y.J.; Lee, Y.R.; Lee, J.; Jeong, H.S. Fermentation and quality characteristics of yogurt treated with *Bifidobacterium longum*. *Nutrients* **2023**, *15*, 3490. [CrossRef] [PubMed]
19. Nirina, A.H.; Miora, R.; Vincent, P.; Hiol, A.; Remize, F.; Louisette, R. Phytochemical composition and antioxidant activity of fermented *Moringa oleifera* leaf powder. *Eur. J. Nutr. Food Saf.* **2017**, *7*, 8–84. [CrossRef]
20. Managa, M.G.; Akinola, S.A.; Remize, F.; Garcia, C.; Sivakumar, D. Physicochemical parameters and bioaccessibility of lactic acid bacteria fermented chayote Leaf (*Sechium edule*) and pineapple (*Ananas comosus*) smoothies. *Front. Nutr.* **2021**, *8*, 649189. [CrossRef]
21. Mashitoa, F.M.; Akinola, S.A.; Shoko, T.; Seke, F.; Sunette, L.M.; Sultanbawa, Y.; Sivakumar, D. Effect of lactic acid fermentation on the quality and phytochemical constituent in smoothies made from the leaves of different sweet potato (*Ipomoea batatas* L.) cultivars. *Int. J. Food Sci. Technol.* **2023**, *58*, 4697–4714. [CrossRef]
22. Singleton, V.L.; Orthofer, R.; Lamuela-Raventós, R.M. Analysis of total phenols and other oxidation substrates and antioxidants by means of folin-ciocalteu reagent. *Methods Enzymol.* **1999**, *299*, 152–178. [CrossRef]
23. Djuikwo, D.V.N.; Ejoh, R.A.; Gouado, I.; Mbofung, C.M.; Tanumihardjo, S.A. Determination of major carotenoids in processed tropical leafy vegetables indigenous to Africa. *Food Nutr. Sci.* **2011**, *2*, 8–9. [CrossRef]
24. Bhengu, N.B.; Mianda, S.; Maboko, M.M.; Sivakumar, D. The effects of nitrogen application and varietal variation on the product quality and in vitro bioaccessibility of bioactive compounds of baby spinach varieties grown in a soilless growth medium. *Foods* **2024**, *13*, 2667. [CrossRef]
25. Benzie, I.F.; Strain, J.J. The ferric reducing ability of plasma (FRAP) as a measure of "antioxidant power": The FRAP assay. *Anal. Biochem.* **1996**, *239*, 70–76. [CrossRef]
26. Llorach, R.; Tomás-Barberán, F.A.; Ferreres, F. Lettuce and chicory byproducts as a source of antioxidant phenolic extracts. *J. Agric. Food Chem.* **2004**, *52*, 5109–5116. [CrossRef] [PubMed]
27. Mandha, J.; Shumoy, H.; Devaere, J.; Schouteten, J.J.; Gellynck, X.; de Winne, A.; Matemu, A.O.; Raes, K. Effect of lactic acid fermentation of watermelon juice on its sensory acceptability and volatile compounds. *Food Chem.* **2021**, *358*, 129809. [CrossRef] [PubMed]
28. Fonteles, T.; Costa, M.G.M.; De Jesus, A.L.T.; Rodrigues, S. Optimization of the fermentation of cantaloupe juice by *Lactobacillus casei* NRRL B-442. *Food Bioprocess Technol.* **2011**, *5*, 2819–2826. [CrossRef]
29. Gul, S.; Durante-Mangoni, E. Unraveling the Puzzle: Health Benefits of Probiotics—A Comprehensive Review. *J. Clin. Med.* **2024**, *13*, 1436. [CrossRef]
30. Bustos, A.Y.; Taranto, M.P.; Gerez, C.L.; Agriopoulou, S.; Smaoui, S.; Varzakas, T.; El Enshasy, H.A.E. Recent advances in the understanding of stress resistance mechanisms in probiotics: Relevance for the design of functional food systems. *Probiotics Antimicrob. Prot.* **2024**. [CrossRef]
31. Shabala, L.; McMeekin, T.; Budde, B.B.; Siegumfeldt, H. *Listeria innocua* and *Lactobacillus delbrueckii* subsp. bulgaricus employ different strategies to cope with acid stress. *Int. J. Food Microbiol.* **2006**, *110*, 1–7. [CrossRef]
32. Jaywant, S.A.; Singh, H.; Arif, K.M. Sensors and Instruments for Brix Measurement: A Review. *Sensors* **2022**, *22*, 2290. [CrossRef]
33. Kaprasob, R.; Kerdchoechuen, O.; Laohakunjit, N.; Sarkar, D.; Shetty, K. Fermentation-based biotransformation of bioactive phenolics and volatile compounds from cashew apple juice by select lactic acid bacteria. *Process Biochem.* **2017**, *59*, 141–149. [CrossRef]
34. Kumar, B.V.; Sreedharamurthy, M.; Vijaya Sarathi Reddy, O. Probiotication of mango and sapota juices using *Lactobacillus plantarum* NCDC LP 20. *Nutrafoods* **2015**, *14*, 97–106. [CrossRef]
35. Pokusaeva, K.; Fitzgerald, G.F.; van Sinderen, D. Carbohydrate metabolism in Bifidobacteria. *Genes Nutr.* **2011**, *6*, 285–306. [CrossRef]
36. Hou, J.W.; Yu, R.C.; Chou, C.C. Changes in some components of soymilk during fermentation with bifidobacteria. *Food Res. Int.* **2005**, *33*, 393–397. [CrossRef]

37. Mandha, J.; Shumoy, H.; Matemu, A.O.; Raes, K. Evaluation of the composition and quality of watermelon and mango juices fermented by *Levilactobacillus brevis*, *Lacticaseibacillus casei* and *Pediococcus pentosaceus* and subsequent simulated digestion and storage. *Int. J. Food Sci.* **2022**, *57*, 5461–5471. [CrossRef]
38. Wang, Y.; Wu, J.; Lv, M.; Shao, Z.; Hungwe, M.; Wang, J.; Bai, X.; Xie, J.; Wang, Y.; Geng, W. Metabolism characteristics of lactic acid bacteria and the expanding applications in the food industry. *Front. Bioeng. Biotechnol.* **2021**, *9*, 612285. [CrossRef] [PubMed]
39. Laranjo, M.; Potes, M.E.; Elias, M. Role of starter cultures on the safety of fermented meat products. *Front. Microbiol.* **2019**, *10*, 853. [CrossRef] [PubMed]
40. Rocha, A.M.C.N.; Morais, A.M.M.B. Shelf life of minimally processed apple (cv. Jonagored) determined by colour changes. *Food Control* **2003**, *14*, 13–20. [CrossRef]
41. Farnworth, E.R.; Lagace, M.; Couture, R.; Yaylayan, V.; Stewart, B. Thermal processing, storage conditions, and the composition and physical properties of orange juice. *Food Res. Int.* **2001**, *34*, 25–30. [CrossRef]
42. Hashemi, S.M.B.; Khaneghah, A.M.; Barba, F.J.; Nemati, Z.; Shokofti, S.S.; Alizadeh, F. Fermented sweet lemon juice (*Citrus limetta*) using *Lactobacillus plantarum* LS5: Chemical composition, antioxidant and antibacterial activities. *J. Funct. Foods* **2017**, *38*, 409–414. [CrossRef]
43. Xu, H.; Feng, L.; Deng, Y.; Chen, L.; Li, Y.; Lin, L.; Liang, M.; Jia, X.; Wang, F.; Zhang, X.; et al. Change of phytochemicals and bioactive substances in lactobacillus fermented citrus juice during the fermentation process. *LWT-Food Sci. Technol.* **2023**, *180*, 114715. [CrossRef]
44. Saud, S.; Tang, X.; Fahad, S. The consequences of fermentation metabolism on the qualitative qualities and biological activity of fermented fruit and vegetable juices. *Food Chem. X.* **2024**, *21*, 101209. [CrossRef]
45. Urbiene, S.; Mitkute, D. Changes of vitamin C during milk fermentation. *Milchwissenschaft* **2007**, *62*, 130–132.
46. Hakim, B.N.A.; Xuan, N.J.; Oslan, S.N.H. A Comprehensive review of bioactive compounds from lactic acid bacteria: Potential functions as functional food in dietetics and the food industry. *Foods* **2023**, *12*, 2850. [CrossRef] [PubMed]
47. Nguyen, B.T.; Bujna, E.; Fekete, N.; Tran, A.T.; Rezessy-Szabo, J.M.; Prasad, R.; Nguyen, Q.D. Probiotic beverage from pineapple juice fermented with Lactobacillus and Bifidobacterium strains. *Front. Nutr.* **2019**, *6*, 54. [CrossRef] [PubMed]
48. Yang, F.; Chen, C.; Ni, D.; Yang, Y.; Tian, J.; Li, Y.; Chen, S.; Ye, X.; Wang, L. Effects of Fermentation on Bioactivity and the Composition of Polyphenols Contained in Polyphenol-Rich Foods: A Review. *Foods* **2023**, *12*, 3315. [CrossRef] [PubMed]
49. Antognoni, F.; Mandrioli, R.; Potente, G.; Saa, D.L.T.; Gianotti, A. Changes in carotenoids, phenolic acids and antioxidant capacity in bread wheat doughs fermented with different lactic acid bacteria strains. *Food Chem.* **2019**, *292*, 211–216. [CrossRef]
50. Szutowska, J.; Gwiazdowska, D.; Rybicka, I.; Pawlak-Lemańska, K.; Biegańska-Marecik, R.; Gliszczyńska-Świgło, A. Controlled fermentation of curly kale juice with the use of autochthonous starter cultures. *Food Res. Int.* **2021**, *149*, 110674. [CrossRef]
51. Odongo, G.A.; Schlotz, N.; Herz, C.; Hanschen, F.; Baldermann, S.S.; Neugart, S.; Trierweiler, B.; Frommherz, L.; Franz, C.M.; Ngwene, B.; et al. The role of plant processing for the cancer preventive potential of Ethiopian kale (*Brassica carinata*). *Food Nutr. Res.* **2017**, *61*, 1271527. [CrossRef]
52. Demir, N.; Acar, J.; Bahçeci, K.S. Effects of storage on quality of carrot juices produced with lactofermentation and acidification. *Eur. Food Res. Technol.* **2004**, *218*, 465–468. [CrossRef]
53. Mapelli-Brahm, P.; Barba, F.J.; Remize, F.; Garcia, C.; Fessard, A.; Khaneghah, A.M.; Sant'Ana, A.S.; Lorenzo, J.M.; Montesano, D.; Meléndez-Martínez, A. The impact of fermentation processes on the production, retention and bioavailability of carotenoids: An overview. *Trends Food Sci.* **2020**, *99*, 389–401. [CrossRef]
54. Oloo, B.O.; Shitandi, A.; Mahungu, S.; Malinga, J.B.; Rose, O.B. Effects of lactic acid fermentation on the retention of beta-carotene content in orange fleshed sweet potatoes. *J. Food Stud.* **2014**, *3*, 13–33. [CrossRef]
55. Bartkiene, E.; Vidmantiene, D.; Juodeikiene, G.; Viskelis, P.; Urbonaviciene, D. Lactic acid fermentation of tomato: Effects on cis/trans lycopene isomer ratio, b-carotene mass fraction and formation of l(+)- and d(−)-lactic acid. *Food Technol. Biotechnol.* **2013**, *51*, 471–478.
56. Garrido-Fernández, J.; Maldonado-Barragán, A.; Caballero-Guerrero, B.; Hornero-Méndez, D.; Ruiz-Barba, J.L. Carotenoid production in *Lactobacillus plantarum*. *Int. J. Food Microbiol.* **2010**, *140*, 34–39. [CrossRef] [PubMed]
57. Stahl, W.; Sundquist, A.; Hanusch, R.M.; Schwarz, W.; Sies, H. Separation of beta-carotene and lycopene geometrical isomers in biological samples. *Clin. Chem.* **1993**, *39*, 810–814. [CrossRef] [PubMed]
58. Hur, S.J.; Lee, S.Y.; Kim, Y.C.; Choi, I.; Kim, G.B. Effect of fermentation on the antioxidant activity in plant-based foods. *Food Chem.* **2014**, *160*, 346–356. [CrossRef] [PubMed]
59. Scalzo, R.L. Organic acids influence on DPPH scavenging by ascorbic acid. *Food Chem.* **2008**, *107*, 40–43. [CrossRef]
60. Zhou, Y.; Wang, R.; Zhang, Y.; Yang, Y.; Sun, X.; Zhang, Q.; Yang, N. Biotransformation of phenolics and metabolites and the change in antioxidant activity in kiwifruit induced by *Lactobacillus plantarum* fermentation. *J. Sci. Food Agric.* **2020**, *100*, 3283–3290. [CrossRef]

61. Shimidzu, N.; Goto, M.; Miki, W. Carotenoids as singlet oxygen quenchers in marine organisms. *Fish. Sci.* **1996**, *62*, 134–137. [CrossRef]
62. Gao, Y.; Focsan, A.L.; Kispert, L.D. Antioxidant activity in supramolecular carotenoid complexes favored by nonpolar environment and disfavored by hydrogen bonding. *Antioxidants* **2020**, *9*, 625. [CrossRef]

foods

MDPI

Article

Okra (*Abelmoschus esculentus* L.) Flour Integration in Wheat-Based Sourdough: Effect on Nutritional and Technological Quality of Bread

Francesca Valerio [1,*], Mariaelena Di Biase [1], Valentina Cifarelli [1], Stella Lisa Lonigro [1], Amina Maalej [2], Stella Plazzotta [3], Lara Manzocco [3], Sonia Calligaris [1,3] and Hana Maalej [4]

1 Institute of Sciences of Food Production, National Research Council, Via G. Amendola 122/O, 70126 Bari, Italy; mariaelena.dibiase@ispa.cnr.it (M.D.B.); valentina.cifarelli@ispa.cnr.it (V.C.); lisa.lonigro@ispa.cnr.it (S.L.L.); sonia.calligaris@uniud.it (S.C.)

2 Laboratory of Environmental Bioprocesses, Centre of Biotechnology of Sfax, Sfax 3018, Tunisia; maalejamina@yahoo.fr

3 Department of Agricultural, Food, Environmental and Animal Sciences, University of Udine, via Sondrio 2/A, 33100 Udine, Italy; stella.plazzotta@uniud.it (S.P.); lara.manzocco@uniud.it (L.M.)

4 Laboratory of Biodiversity and Valorization of Arid Areas Bioresources (BVBAA), LR16ES36, Faculty of Sciences of Gabes, University of Gabes, Gabes 6072, Tunisia; hana.maalej@univgb.tn

* Correspondence: francesca.valerio@ispa.cnr.it; Tel.: +39-0805929369

Abstract: The aim of this study was to develop an innovative sourdough using dehydrated okra (*Abelmoschus esculentus* L.) pod flour and to use it in the production of bread. Three different flours (sun-dried S, freeze-dried F, oven-dried O) were individually mixed at 9% with wheat flour (Dough Yield 300) and fermented (N_0: 8.0 $\log_{10}$ CFU/g) for 14 h, using *Lactiplantibacillus plantarum* ITM21B, *Weissella cibaria* C43-11 or *Leuconostoc mesenteroides* C43-2M. The results showed that after fermentation, the content of organic acids (lactic, acetic and propionic), exopolysaccharides (EPS), l-glutamic acid and total free amino acids (TFAA) increased and the high molecular weight proteins were converted into smaller proteins. Sourdough based on *Leuc. mesenteroides* and O flour (O_LeuMes) was selected to evaluate its applicability in bread making. It was included in the yeast-leavened bread formulation at 20 or 40% (0.6% and 1.21% *w/w* O flour replacement). The results showed that fermentation limited the negative effects of unfermented O flour on bread quality attributes, mainly the specific volume and firmness. Bread with O_LeuMes at 40% was improved in TFAA, EPS and l-glutamic acid content and showed a higher specific volume and lower moisture and firmness compared to bread with the unfermented O flour.

Keywords: lactic acid bacteria; fermentation; protein content; amino acids; organic acids; l-glutamic acid; polysaccharides; wheat bread

1. Introduction

Okra (*Abelmoschus esculentus* L.), commonly known as gumbo or lady's finger, belongs to the *Malvaceae* family and is an easily available, renewable and inexpensive natural source. It is consumed in the Mediterranean region and can be considered as a functional food since it is rich in minerals, protein, fat, phenolic compounds and polysaccharides [1,2]. In Tunisia, okra is cultivated in Djérid, a semi-desert region in the south-west of the country, in the areas of Sidi Daoud near La Marsa, and in the north-west of the country in Béja, Jendouba and Bizerte [3]. Given the growing interest in the Mediterranean diet, this plant could be valorized as a healthy ingredient applicable in many food products, especially in cereal-based foods, which represent up to 77% of the total caloric consumption in Africa [2,4,5]. Okra is rich in soluble dietary fiber in the form of gums and pectins, which help to lower serum cholesterol levels, thereby reducing the risk of heart disease [6]. The health-promoting capacity of okra is mainly related to its polysaccharides [7,8], which are typical pectic polysaccharides containing abundant rhamnogalacturonan I (RG I) with repeating disaccharide units $\rightarrow$ 4)-d-GalpAMe-(1 $\rightarrow$ 2)-l-Rhap-(1 $\rightarrow$ as backbone and neutral sugars of galactose and arabinose as side chains [9,10]. Okra polysaccharides possess significant bioactivities such as anti-fatigue [11], antioxidant [12], anti-hyperlipidemic [2] and anti-hyperglycemic effects [7] as well as interesting potential as prebiotics [13]. In addition, the polysaccharides in the seedless pod of okra form a slimy viscous system in water [14]. Due to its high moisture content, fresh okra is known to be highly perishable at room temperature, resulting in deterioration of its chemical, physical and biological properties. A dehydration process would therefore be required to allow its safe use as a food ingredient and to preserve its bioactive compounds. However, it should be emphasized that the technology used to remove moisture (e.g., sun drying, oven drying, freeze drying) could deeply affect the health and technological functionalities of dried plant flour [15].

Innovation in the bakery sector is currently exploring biotechnological processes, including the combination of alternative flours as sources of carbohydrates, proteins, vitamins and minerals to wheat flour, in order to improve the nutritional, functional and technological qualities of products [16]. White wheat bread has a high glycemic index (GI), i.e., it rapidly raises the postprandial blood glucose levels. For this reason, the bio-functional properties of okra fruits may be particularly relevant to enriching the nutritional properties of wheat bread by reducing its starch digestion rate and extension. Soluble dietary fibers, such as okra pod mucilage, could form viscous solutions or gels, affecting passage rate, viscosity and interactions with digestive enzymes and bile salts in the stomach and small intestine [17]. Recently, some studies have reported the potential of okra flour or derived hydrocolloids in the production of bread, with the ultimate aim of enriching bread with prebiotic polysaccharides useful for intestinal microflora, phenolic compounds, and micro- and macro-nutrients [18–21]. Other health benefits of okra consumption are also related to its phenolic content, since such compounds have beneficial biological activities, including the ability to modulate oxidative and inflammatory stress [17]. However, fortification of bread with ingredients containing phenolic compounds could slow down starch hydrolysis through the potential inhibition of digestive enzymes (i.e., α-amylase and glucosidases) [22] and glucose transporters in the intestinal brush border [23–25]. The application of okra derivatives in bread making could reveal other technological drawbacks, mainly related to the soluble fiber content of the polysaccharidic fraction, which retains water, thus negatively affecting wheat-gluten development during bread manufacturing [26]. As has been widely demonstrated, the textural and sensorial qualities of fiber-enriched breads can be significantly improved by sourdough fermentation, even in the presence of hydrocolloids such as pectin [27]. The sourdough process is a well-known ancient and sustainable technology characterized by the simultaneous presence of lactic acid bacteria (LAB) and yeasts in a dough which determines the transformation of food raw materials. During the process, several compounds are produced such as hydrocolloids, organic acids and enzymes, flavor and taste-active compounds and antimicrobial substances, which have technological, functional and health properties. The process can occur spontaneously or can be activated by specific strains [28]. Among the commonly used microorganisms applied in sourdough, LAB contribute to improving the quality, shelf life and functional properties of fermented foods by converting food components into added-value compounds [29]. They also contribute to modifying the bioaccessibility and bioavailability of polyphenols mainly through the release of free phenolic acids by the esterase activity of sourdough microorganisms [30,31]. Therefore, the combination of sourdough fermentation and the incorporation of a rich source of phenolics such as okra flour could provide an additional functional trait to the resulting bread. At the same time, the addition of a rich source of phenolics to food formulation should consider the associated safety aspects in order to select the optimal percentage of incorporation. Altunkaya et al. [32] evaluated the chemical safety of wheat bread fortified with pomegranate peel powder using the *Artemia salina* assay. The authors observed an increase in toxicity as a function of the replacement of wheat flour with the plant by-product, thus suggesting that the final products should be tested for safety aspects.

During fermentation, LAB activate many metabolic activities, leading to the production of antimicrobials and exopolysaccharides (EPS), acidification and redox potential modification, oligosaccharides and glutamine and glutamate metabolism [33]. The EPS, divided into homopolysaccharides (HoPS) and heteropolysaccharides (HePS), can be used instead of hydrocolloids as texturizing, antistaling or prebiotic additives in bread, improving its textural and nutritional characteristics [34].

To date, okra has mainly been subjected to spontaneous fermentation, resulting in the improvement in nutrient composition, antioxidants and some functional properties [35,36]. Recently, Wang et al. [37] used okra juice as a fermentation substrate using a *Lactiplantibacillus plantarum* strain to evaluate the effect of the process on the physico-chemical properties, antioxidant activity and immunomodulatory ability of polysaccharides. Until now no data have been reported on the fermentation of okra flours in a wheat-based sourdough formulation by using selected starter strains.

Based on the above considerations, the aim of this study was to evaluate the suitability of dehydrated okra pods as functional food ingredients after fermentation. Specifically, okra flours produced by three different drying technologies (sun drying, air drying, freeze drying) were fermented in combination with wheat flour to produce sourdoughs, which were further used for bread preparation. The results are the starting point for expanding the application of okra in food products, allowing its valorization.

2. Materials and Methods

The list of abbreviations used in this research can be found in the Abbreviations Section.

2.1. Raw Materials

The ingredients used in this study were commercial soft wheat flour type 0 (fat 2%, carbohydrates 72%, fiber 2%, protein 11%) (Molino Casillo S.p.A, Corato, Italy) and 3 types of okra flour: sun-dried okra flour (S, direct exposure to solar radiation at ≃30 °C for 24 h), oven-dried okra flour (O, drying in a forced air oven for 48–72 h) and freeze-dried okra flour (F, freezing at −80 °C overnight and followed by freeze drying for 72–96 h). All okra flour samples with particle size less than 300 μm were provided by the University of Gabes, Tunisia. The chemical composition of the okra flours obtained is reported in Table 1.

Table 1. Chemical composition of produced okra flours.

	F	O	S
Moisture (g/100 g)	0.22 ± 0.02	9.85 ± 0.1	0.00 ± 0.0
Protein content (g/100 g db [#])	18.01 ± 0.25	19.04 ± 0.18	18.39 ± 0.15
Total fats (g/100 g db)	18.26 ± 0.36	2.40 ± 0.276	1.18± 0.08
Total carbohydrates (g/100 g db)	26.71 ± 0.26	35.27 ± 0.37	24.88 ± 3.34

[#] db, dry basis. Flour codes are reported in Section 2.1. Data are presented as mean ± standard deviation.

2.2. Microorganisms

For the production of sourdough, three LAB strains belonging to the Culture Collection of the Institute of Sciences of Food Production, National Research Council of Italy, were used as fermentation starters. In particular, the following strains were used: the proteolytic strain *Lactiplantibacillus plantarum* ITM21B, isolated from sourdough [38] and used for the production of bread with reduced salt content [39,40], *Leuconostoc mesenteroides* C43-2M and the dextran producer *Weissella cibaria* C43-11 [41], both isolated from wheat semolina [42], the latter used for the production of focaccia bread with reduced fat content [43,44]. The *W. cibaria* species, although not yet included in the Qualified Presumption of Safety (QPS) list [45] and recognized as a potential bacterial starter in food fermentation, was included in the study because it is a well characterized dextran-producer strain [41].

For long-term storage, stock cultures were prepared by mixing 800 μL of a culture in de Man Rogosa Sharpe (MRS) broth (Conda, Madrid, Spain) with 200 μL of Bacto glycerol (Fisher Scientific UK Ltd., Leicestershire, UK) and freezing this mixture at −80 °C. To obtain fresh cultures, strains were subcultured twice (1% *v*/*v*) in MRS broth for 24 h before use.

2.3. Microbiological Analysis of Raw Materials

A sterile Bacto-peptone (Difco, Becton Dickinson, Co., Sparks, MD, USA) solution (0.1% *w/v*) was added to 20 g of each sample up to 200 g, homogenized in a blender for 2 min, serially diluted and plated on de Man Rogosa Sharpe (MRS) (Biolife Italiana S.r.l., Milan, Italy) agar containing cycloheximide (0.1 g/L) to detect LAB. Petri dishes were then incubated aerobically at 30 °C for 48 h for colony detection.

2.4. Production of Sourdough

The strains *W. cibaria* C43-11, *Leuc. mesenteroides* C43-2M and *L. plantarum* ITM21B were selected as single starters for fermentation. The strains were cultured in MRS broth and incubated at 37 °C (*L. plantarum* ITM21B) or 30 °C (*W. cibaria* and *Leuc. mesenteroides*). In the sourdough formulation, the okra flour/wheat flour ratio and water adjustments were defined using S flour. This preliminary investigation indicated that a level of okra flour higher than 9% (*w/w*) with respect to wheat flour and a dough yield (DY) lower than 300 altered the sourdough's consistency. Therefore, final formulations were prepared by mixing 1 g of okra flour, 10 g of wheat flour and 22 mL of water in order to select the most suitable strain and fermentation conditions for the enrichment of sourdough with bioactive molecules such as amino acids, proteins, EPS, l-glutamic acid and organic acids, and a set of experiments was prepared using O, F and S flours. Mixtures were separately inoculated at 8 log CFU/g with *W. cibaria* C43-11, *Leuc. mesenteroides* C43-2M and *L. plantarum* ITM21B as starters and incubated for 14 h at the optimal growth temperature for each. Not inoculated mixtures before fermentation were used as controls (CTR): S_CTR_T0, F_CTR_T0 and O_CTR_T0. The resulting fermentation products and controls were analyzed. The experiments were performed in duplicate. The following parameters were also registered before and after fermentation: pH, titratable acidity (TTA) and total LAB count.

2.5. Bread-Making Test

Among the sourdough formulations, the preliminary screening did not highly differentiate the samples. Therefore, selection of the sourdough formulation suitable for bread making was based on the type of flour and the starter strain: O flour was chosen since the oven-drying technology can be easily applied in the food sector, and the strain *Leuc. mesenteroides* C43-2M was selected as an EPS-producing strain that has not yet been deeply investigated and belongs to a species included in the QPS list. Furthermore, the sourdough formulation based on O flour and *Leuc. mesenteroides* (O_LeuMes) contained the highest EPS content.

On the basis of these considerations, the O_LeuMes sourdough was selected for the laboratory bread-making test. Bread variants were prepared according to a standard recipe as shown in Table 2. The sourdough was added at 20% and 40% *w/w* (weight of sourdough/weight of total flour) corresponding to 0.6% and 1.2% of fermented O flour on the total flour weight, giving O_20% and O_40% breads. Unfermented O flour at 0.6% and 1.2% was used to produce control breads (CTR_20% and CTR_40%, respectively) in addition to a control bread (CTR_0%) produced with wheat flour as the standard recipe. The samples were prepared using a Princess® Bread Maker Homemade Deluxe, type 152007 bread machine (Princess Household appliance BV, Breda, The Netherland). All ingredients were mixed and the dough was leavened and baked using the "Basic" program. Bread samples were cooled at room temperature for two hours, sliced and stored for physico-chemical analyses.

Table 2. Formulations of bread containing O_LeuMes sourdough at two different percentages in comparison to controls.

	Bread Variant *				
	CTR_0%	O_20%	O_40%	CTR_20%	CTR_40%
Ingredients	Weight (g)				
Wheat flour	350	326.67	303.33	347.88	345.76
Salt	6.3	6.3	6.3	6.3	6.3
Olive oil	10.5	10.5	10.5	10.5	10.5
Dry yeast	5.25	5.25	5.25	5.25	5.25
Tap water	210	163.33	116.67	210	210
O_LeuMes	-	70	140	-	-
O flour	-	-	-	2.12	4.24

* O_20%, bread containing O_LeuMes sourdough at 20% sourdough *w/w* flour; O_40%, bread containing O_LeuMes sourdough at 40% sourdough *w/w* flour; CTR_0%, not containing O flour; CTR_20%, bread containing unfermented O flour (corresponding to okra flour content in O_20%); CTR_40%, bread containing unfermented O flour (corresponding to okra flour content in O_40%).

2.6. Microbiological Analysis of Sourdough

A Sterile Bacto-peptone solution (0.1% *w/v*) was added 1:10 to the sourdough and homogenized in a stomacher for 2 min. Serial decimal dilutions of each sourdough were prepared in sterile NaCl (0.85% *w/v*) + Tween 80 (0.025%), and 100 µL aliquots of each dilution was spread on MRS plates which were incubated for 48 h at the optimal growth temperature of each strain. The total LAB count was expressed as log CFU/g. The presence of starters in each sourdough was monitored as described by Valerio et al. [40]. Bacterial DNA was extracted from each colony of overnight cultures grown in MRS broth at 30 or 37 °C as previously described [46]. Genetic identification of the strains was based on a comparison of the REP-PCR profile of each isolate with the specific pattern obtained from pure cultures of *W. cibaria* C43-11, *Leuc. mesenteroides* C43-2M and *L. plantarum* ITM21B.

2.7. Physico-Chemical Analyses of Sourdough and Bread Samples

2.7.1. pH and TTA Measurement

The pH of the sourdough and bread samples was recorded using a portable pH-meter (type 110, Eutech Instruments, Singapore City, Singapore) equipped with a Double Pore D electrode (Hamilton, Bonaduz, Switzerland). TTA was measured according to AOAC Method No. 981.12 [47].

2.7.2. Protein Content and Protein Profile Analysis

Total proteins were extracted from sourdoughs (before and after fermentation) and bread samples under reducing conditions as reported by Valerio et al. [44]. Protein concentration was determined using the Bradford method [48] with bovine serum albumin as standard and expressed as mg protein per gram sourdough. Total protein profiles were obtained on the 2100Bioanalyzer microchip electrophoresis system (Agilent Technologies, Waldbronn, Germany) using the Lab on-a-chip (Loac) Protein230 reagent kit (14–230 kDa) as reported in Valerio et al. [44]. According to Mw, three different areas were considered in the electropherograms: A1 (14–20 kDa), A2 (21–42 kDa) and A3 (53–63 kDa), and percentages of peaks were obtained for each Mw area. All experiments were performed in duplicate.

2.7.3. Determination of Organic Acids

Samples were prepared as described by Bavaro et al. [43]. The organic acids (lactic, acetic, propionic) were separated on an HPLC system (AKTABasic10, P-900 series pump; Amersham Biosciences AB, Uppsala, Sweden) using Rezex ROA-organic acid H^+ (8%) (7.80 × 300 mm; Phenomenex, Torrance, CA, USA) and a three-channel UV detector (Amersham Biosciences 900) set at 210 nm. The mobile phase was 0.007 mol/L H_2SO_4 (Fluka, Deisenhofen, Germany) pumped through the column heated to 70 °C at a flow rate of 0.6 mL/min. The amounts of organic acids were determined by integrating calibration curves obtained from relevant standards and expressed as mmol/kg sourdough. The limit of detection (LOD) and limit of quantification (LOQ) were calculated considering a signal-to-noise ratio (S/N) of 3 and 6, respectively. LOD values of the organic acids in the extracted samples were the following for sourdough: lactic acid, 0.41 μmol/mL; acetic acid, 0.68 μmol/mL; and propionic acid 0.51 μmol/mL. The LOD values for bread samples were: lactic acid, 0.228 μmol/mL; acetic acid, 5.42 μmol/mL; and propionic acid 0.41 μmol/mL. The LOQ values corresponded to 2 × LOD. The final concentration of each organic acid in the liquid sourdough was calculated taking into account concentration and/or dilution factors and expressed as mmol/kg or μmol/kg product.

2.7.4. Determination of L-Glutamic Acid

Sourdough samples were diluted (1:5) with water, centrifuged (15,000× *g*, 20 min, 4 °C) and the supernatant filtered by centrifugation (7000× *g*, 1 h, 2 °C) through a 3000 Da cut-off micro-concentrator. Bread samples were diluted (1:10) with water, centrifuged (15,000× *g*, 20 min, 4 °C) and the supernatant freeze dried. The freeze-dried samples were resuspended in a volume of sterile water to obtain 10-fold concentrated solutions. The fractions containing molecules with a molecular weight lower than 3000 Da were analysed. The l-glutamic acid content was measured as reported by Valerio et al. [40] by spectrophotometric analysis using a commercially available l-glutamic acid test kit (K-GLUT 07/12; Megazyme International, Bray, Ireland) according to the manufacturer's instructions and using a microtiter plate spectrophotometer BioscreenC (Oy Growth Curves AB Ltd.; Helsinki, Finland).

2.7.5. EPS Quantification

Exopolysaccharides were extracted from sourdough and bread samples as reported by Valerio et al. [44], with slight modifications. Briefly, samples were diluted with distilled water and centrifuged (8000× *g*, 10 min). The supernatants were treated with trichloroacetic acid to a final concentration of 4% and stored at 4 °C for 2 h, to remove proteins that may interfere with EPS quantification. The precipitate was removed by centrifugation (20,000× *g*, 10 min, 4 °C) and the resulting supernatants were treated with three volumes of chilled 96–99% (*v/v*) ethanol to allow EPS precipitation. The solutions were stored overnight at 4 °C. Precipitated EPS were collected by centrifugation (8000× *g*, 20 min, 25 °C), resuspended in the original volume of distilled water, dialyzed (12–14 kDa) against distilled water for 48 h at 4 °C, lyophilized and rehydrated with distilled water to the original volume. The concentration of EPS (g/kg) was determined by the phenol-sulfuric method [49], using glucose as standard (LOD 0.078 g/L or g/kg).

2.7.6. Determination of Total Free Amino Acids

Water/salt-soluble extracts of sourdoughs and breads were prepared according to the method originally described by Osborne [50] and modified by Weiss et al. [51] and used for the analysis of total free amino acids (TFAAs) by the cadmium–ninhydrin method as reported by Doi et al. [52].

2.8. Technological Analyses of Bread Samples

To evaluate the impact of sourdough and unfermented okra flour on bread quality, moisture content, water activity (a_w), color and firmness were assessed on the crumb and specific volume on the loaf.

2.8.1. Moisture Content

The moisture content of the crumb was measured using a gravimetric method [53]. Approximately 2 g of crumb sample were dried in a vacuum oven (1.32 kPa) at 75 °C to a constant weight (12 h).

2.8.2. Water Activity

Crumb water activity was determined using a dew-point measurement system (Aqualab 4TEV, Decagon Devices, Inc., Pullman, WA, USA) at 25 ± 1 °C.

2.8.3. Bread Crumb Firmness

Bread crumb firmness was measured by a puncture test using an Instron 4301 (Instron Ltd., High Wycombe, UK). Instrument settings and operations were performed using Automated Materials Testing System software (version 5, Series IX, Instron Ltd., High Wycombe, UK). Three slices were taken from each loaf. The central part of the slice (4.0 × 4.0 × 0.9 cm) was sampled by manual cutting with a sharp knife. A uniaxial vertical compression test was performed under ambient conditions (20 ± 2 °C, ambient humidity). Samples were penetrated using a 12.7 mm diameter cylindric probe mounted on a 100 N compression head. The crosshead speed was set at 50 mm/min. Force–distance curves were obtained and firmness was taken as the force (N) required to compress the bread crumb by 0.3 cm. Five measurements were taken at different locations on the bread slice.

2.8.4. Crumb Color Evaluation

A tristimulus colorimeter (Chromameter-2 Reflectance, Minolta, Osaka, Japan) equipped with a CR-300 measuring head was used. Petri dishes with a diameter of 90 mm and a height of 16 mm were completely filled with the sample. The instrument was standardized against a white tile. A minimum of 5 measurements were taken for each sample. Color was expressed in CIE L^*, a^* and b^* scale parameters.

Total color difference (ΔE^*) was calculated using Equation (1):

$$\Delta E^* = \sqrt{(L_r^* - L_i^*)^2 + (a_r^* - a_i^*)^2 + (b_r^* - b_i^*)^2} \quad (1)$$

where L_r^*, a_r^*, b_r^* are color parameters of bread without okra flour (CTR_0%) and L_i^*, a_i^*, b_i^* the color parameters of breads containing okra flour. Color differences can be classified as small difference ($\Delta E^* < 1.5$), distinct ($1.5 < \Delta E^* < 3.0$) and very distinct.

2.8.5. Bread Volume

The volume (mL) of the final product was obtained by rapeseed displacement according to Approved Method 10–05 [54].

2.9. Statistical Analysis

Data are presented as mean values ± standard error of the mean. Data relevant to sourdough and bread samples (protein content, EPS, TFAA, organic acids, l-glutamic acid) were analyzed by one-way ANOVA followed by the Fisher's LSD test. Moisture, a_w, color, firmness and specific volume were similarly compared using the Tukey post hoc test. Results were considered statistically significant when the p-value was less than 0.05. Pearson's correlation coefficients were estimated to establish relationships between all the data registered on the sourdough and between the technological and analytical data on the bread samples. Statistical analyses were performed using Statistica 13 software (Dell Inc., Round Rock, TX, USA, 2015).

The data were also analyzed by principal component analysis (PCA) in order to investigate the correlation between the physico-chemical parameters of sourdough. Multivariate analysis was performed using Unscrambler (version 10.1, CAMO, Oslo, Norway).

3. Results and Discussion

3.1. Fermentation Studies

Data from microbiological analysis highlighted the presence of preexisting contamination on the okra flours: the higher microbial load was found in O flour (6.52 ± 0.24 log CFU/g), followed by S flour (5.47 ± 0.26 log CFU/g) and F flour (4.19 ± 0.72 log CFU/g).

However, the starter strains were able to dominate the natural microbiota during sourdough fermentation, reaching values ranging from 9.16 to 9.74 log CFU/g (Table 3). The highest bacterial count was registered for *L. plantarum* in F_LacPla. At the same time, the ITM21B strain reached the lowest pH values in all sourdoughs with the highest TTA values registered in O_LacPla. Acidification occurring in sourdough is one of the key factors influencing gluten protein hydrolysis by activating proteolytic enzymes, degradation of antinutritional compounds, flavor profile, bread staling, bread texture and dough rheological properties, dextran functionality, improved shelf life and other features [16].

Table 3. LAB load, pH and TTA values before (CTR) and after fermentation in the sourdough formulations based on O, F and S flours.

Okra Flour Type	Sourdough	LAB (log CFU/g) ± SE	pH ± SE	TTA (mL NaOH 0.1 N) ± SE
Oven-dried	O_CTR_T0	5.16 ± 0.26 b	5.92 ± 0.02 a	1.00 ± 0.10 a
	O_WeiCib	9.33 ± 0.08 cd	4.03 ± 0.02 c	4.75 ± 0.25 bc
	O_LeuMes	9.43 ± 0.16 cd	3.96 ± 0.02 c	4.45 ± 0.55 b
	O_LacPla	9.26 ± 0.23 cd	3.57 ± 0.01 d	6.50 ± 0.50 d
Freeze-dried	F_CTR_T0	3.15 ± 0.15 a	6.00 ± 0.02 a	0.65 ± 0.15 a
	F_WeiCib	9.45 ± 0.26 cd	4.02 ± 0.01 c	4.65 ± 0.35 bc
	F_LeuMes	9.57 ± 0.33 cd	4.00 ± 0.07 c	4.50 ± 0.50 b
	F_LacPla	9.74 ± 0.03 d	3.70 ± 0.12 d	6.00 ± 1.00 cd
Sun-dried	S_CTR_T0	3.15 ± 0.15 a	5.62 ± 0.04 b	0.95 ± 0.15 a
	S_WeiCib	9.17 ± 0.13 c	3.95 ± 0.09 c	3.45 ± 0.55 b
	S_LeuMes	9.16 ± 0.09 c	3.98 ± 0.06 c	4.75 ± 0.25 bc
	S_LacPla	9.26 ± 0.06 cd	3.55 ± 0.00 d	4.40 ± 0.10 b

Within each column, values (mean ± standard error of the mean) with different lower-case letters are significantly different ($p < 0.05$).

All strains were able to produce the three acids monitored, with some differences in the concentrations (Figure 1). *L. plantarum* ITM21B was the highest producer of lactic acid, as already demonstrated in previous studies [44,55]. The strain *Leuc. mesenteroides* C43-2M produced acetic acid in all sourdough samples at levels significantly higher than the other strains, followed by *W. cibaria*, while strain ITM21B produced a lower amount only in S_LacPla. Propionic acid was mainly produced in ITM21B samples based on F and O flours. The production of short chain organic acids is related to the conversion of flour carbohydrates in sourdough. These compounds contribute to prolonging the bread's shelf life, and the consequent acidification of the environment activates or inhibits many enzymatic activities [16]. Proteolytic enzymes are activated by an acidic environment, while starch-hydrolyzing enzymes are inhibited in the presence of low pH, thus resulting in slower starch digestion and a reduced glycemic response [56]. In addition, acidification enhances proteolytic activity, allowing protein breakdown and the consequent production of free amino acids involved in the Maillard reaction and the formation of volatile compounds in bread [57].

Considering that okra pods contain a large amount of polysaccharides in the form of mucilage [58], which can exert technological and health activities and can be used as a carbohydrate source during fermentation, their inclusion in the form of flour in sourdough formulations can positively influence the fermentation process. At the same time, during fermentation, LAB can produce exopolysaccharides, biopolymers with technological and functional properties very similar to those of okra polysaccharides. The content of these biopolymers was registered in all the sourdough samples, although the formulation did not contain any additional carbohydrate source as a precursor of the bacterial polymers. In fact, the EPS content in each sourdough formulation was maintained at low levels although some differences were observed between samples (Figure 2). A higher level of polysaccharides was observed in O_LeuMes samples compared to the corresponding CTR. The *Leuc. mesenteroides* species is generally associated with EPS production. Recently, Bi et al. [59] prepared an HMw dextran-enriched sourdough using *Leuc. mesenteroides* ATCC 8293 as starter. The authors obtained a final product with improved bread texture and volume due to the inhibitory effect of dextran on gluten depolymerization in the acidic environment. These results confirm the positive role that polysaccharides can exert in combination with LAB fermentation.

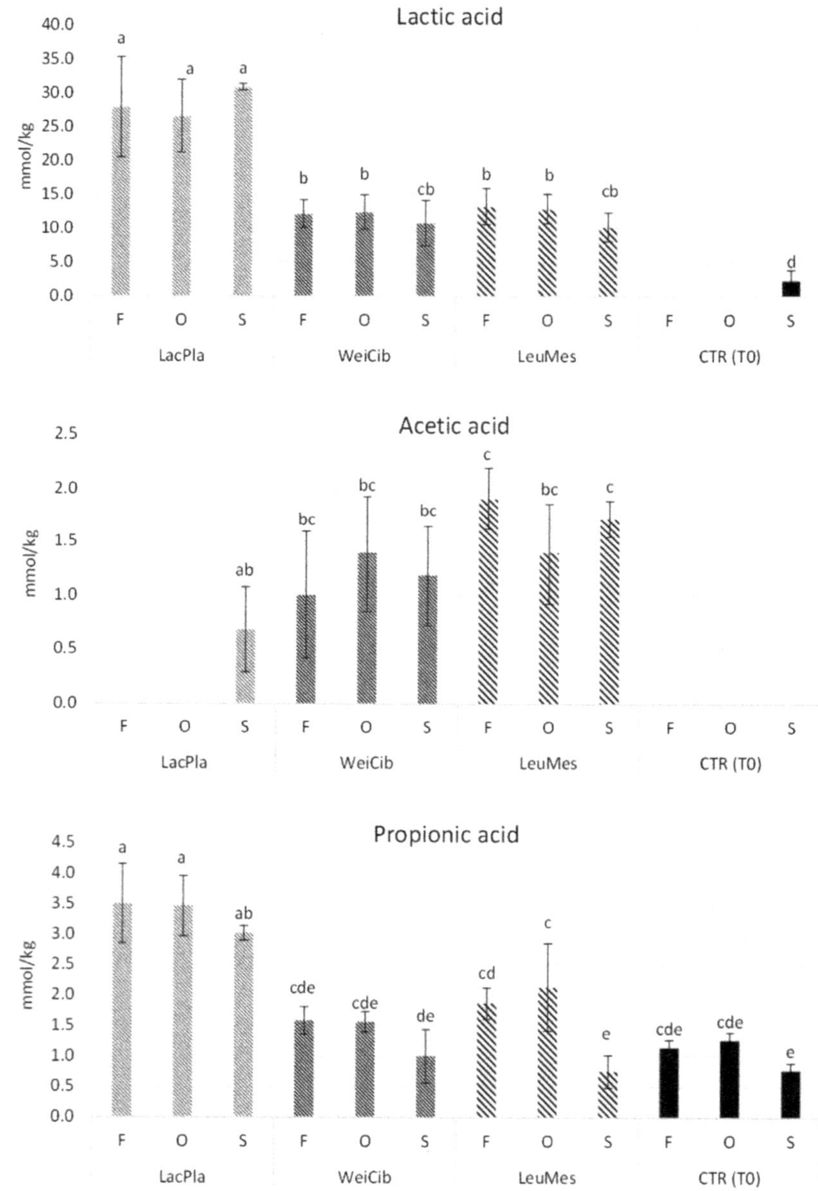

Figure 1. Effect of different okra flours on the content of lactic, acetic and propionic acids registered in sourdough obtained by *L. plantarum* ITM21B, *W. cibaria* C43-11 and *Leuc. mesenteroides* C43-2M after 14 h of fermentation and at start of fermentation (CTR_T0). Values with different lower-case letters are significantly different ($p < 0.05$).

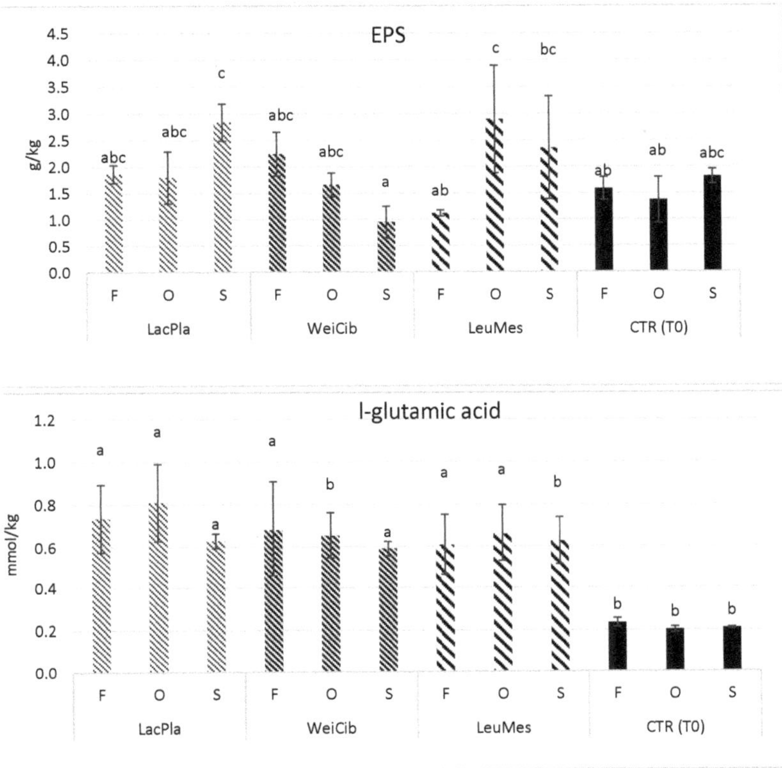

Figure 2. Exopolysaccharides (EPS) and l-glutamic acid content in sourdough formulations, before (CTR_T0) and after fermentation. Values with different lower-case letters are significantly different ($p < 0.05$).

Glutamate or l-glutamic acid, considered as a taste-active compound that can be produced during sourdough fermentation mainly in protein-rich matrices by glutamine deamidation, thus improving the taste of the relevant final product [60,61], was produced in all sourdough samples, while a lower amount ($p < 0.05$) was registered before fermentation (Figure 2).

Slight differences were observed in the content and profile of total proteins (Figures 3 and 4). The total protein content mainly decreased in the presence of the *L. plantarum* strain in S_LacPla, although without a significant difference. The results were consistent with the protein profile observed after fermentation (Figure 4). In fact, the HMw proteins were mainly transformed into smaller proteins (Figure 4) and amino acids (Figure 5).

The electrophoretic patterns of total proteins extracted from sourdoughs based on each okra flour at the beginning (CTRs) and after 14 h of fermentation (T14) are reported in Figure 4. At the beginning of fermentation, the banding patterns of the three okra flours were similar and characterized by about 11 protein bands in the range of 14–64 kDa distributed in Mw A1 (43.9 ± 11.8%), Mw A2 (46.4 ±14.9%) and Mw A3 (9.8 ± 3.1%); major peaks/bands (43.9% of the total) were displayed at 14.8 ± 0.2 and 17 ± 0.2 kDa and at 32.9 kDa only in S_CTR_T0 (17.9% of the total). After 14 h of fermentation, modified protein patterns were observed in all sourdough samples, showing about eight protein peaks/bands in the 14–41 kDa range and the complete disappearance of high molecular weight proteins in Mw A3; the main peaks, accounting for 48.7% of the total, were still present at 14.2 ± 0.1 and 16.5 ± 0.1 kDa. Although the protein patterns of sourdoughs fermented by strains C43-11 and C43-2M were similar, a different increase in the percentage of peaks/bands in Mw areas was observed depending on the strain and flour type: in F_WeiCib and F_LeuMes sourdoughs, a 43.6% and a 27.1% increase in peaks/bands in Mw A1 (14–30 kDa) was observed, whereas in O_WeiCib and O_LeuMes sourdoughs, a 24.7% and 33.5% increase in peaks/bands in Mw A2 (21–42 kDa), respectively, was detected; in sourdoughs based on S flour, slight increases were observed in both areas and with both strains. When O flour was fermented with the *L. plantarum* ITM21B strain, a higher proteolysis occurred than in the corresponding CTR, as six protein peaks/bands were distributed in the 14–36 kDa range and a 24.9% increase in percentage of peaks/bands in Mw A2 was observed; in F_LacPla and S_LacPla, eight protein peaks/bands ranging from 14 to 41 kDa were visualized with an overall 25% increase in the percentage of peaks/bands in Mw A1.

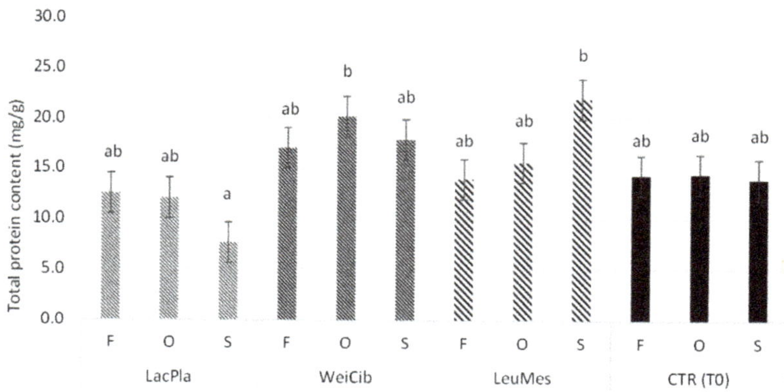

Figure 3. Total protein content in sourdough formulations, before (CTR_T0) and after fermentation. Values with different lower-case letters are significantly different among all samples ($p < 0.05$).

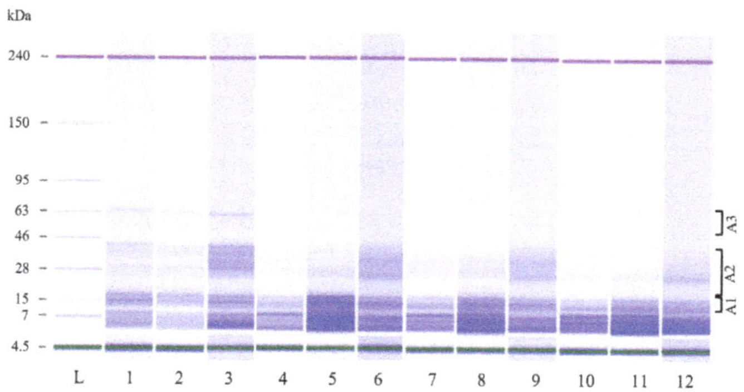

Figure 4. Protein profile of total proteins extracted from sourdough formulations before (T0) (CTR_T0) and after (T14) fermentation (C43-11, C43-2M and ITM21B). Lane L, sizing ladder; CTRs (T0): O (1), F (2), S (3); O_C43-11 (4), F_C43-11 (5) and S_C43-11 (6); O_C43-2M (7), F_C43-2M (8) and S_C43-2M (9); O_ITM21B (10), F_ITM21B (11) and S_ITM21B (12); brackets indicate molecular weight areas as reported in paragraph 2.7.2.

Samples fermented by the *L. plantarum* ITM21B strain, mainly sourdoughs containing O and S flours, showed the most significant increase in TFAA content when compared to CTR samples (Figure 5). In fact, fermentation induced the increase in TFAA mainly in S_LacPla, O_LacPla, F_LeuMes and S_LeuMes. These compounds are fundamental for the nutritional value of the product and for its flavor [57].

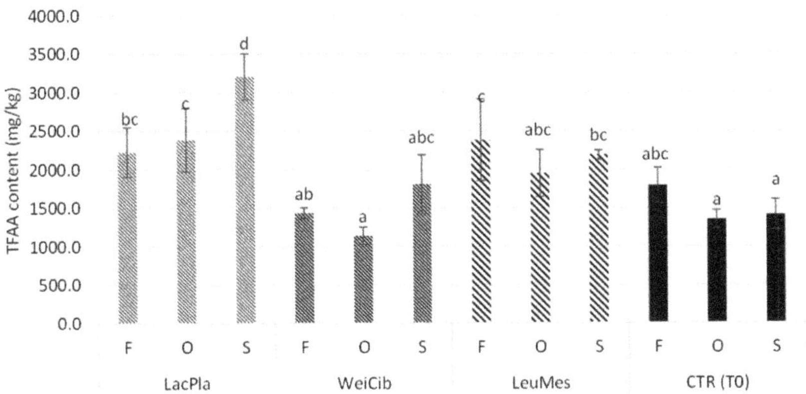

Figure 5. Total free amino acids content in sourdough formulations, before (CTR_T0) and after fermentation. Values with different lower-case letters are significantly different among all samples ($p < 0.05$).

Considering the small amount of okra flour used to prepare sourdough, the changes observed in the metabolite composition after fermentation suggested the potential impact that this vegetable could exert on the quality of the food product. Data from the fermentation test highlighted that organic acids, which play an important role in dough acidification and in the subsequent activation of cereal enzymes and antimicrobial properties, were produced differently in the sourdough formulations. In particular, lactic and propionic acid production was mainly registered in sourdough started with *L. plantarum* with all okra flours, whereas acetic acid was only produced by *Leuc. mesenteroides* and *W. cibaria* in all sourdough formulations. EPS, metabolites that could have a role in the technological properties of the final product, were mainly produced after fermentation in O_LeuMes sourdough, while the l-glutamate, a taste-active compound that can be used in the substitution of salt, was produced in almost all sourdough samples. The total protein content indicated the occurrence of proteolysis during fermentation, mainly related to the conversion of HMw proteins into smaller ones and into amino acids. Although no significant decrease in total protein content was observed after fermentation, the electrophoretic pattern highlighted a modification of the protein profile with the production of small proteins mainly in *L. plantarum* sourdough. Finally, in the case of TFAA, associated to the flavor characteristics of bread, a higher production after fermentation was observed in the O_LacPla, S_LacPla and S_WeiCib sourdough samples. To detect any effect of okra flour addition, the nature of the starter and dehydration process on the quality parameters monitored in the current study, Principal Component Analysis was performed.

3.2. Principal Component Analysis

To better discriminate the sourdough samples, PCA was performed on all data (pH, TTA, EPS, organic acids, total protein content, TFAA, l-glutamic acid) (Figure 6). Figure 6a (loading plot) and the correlation analysis (Table 4) clearly show the negative correlation between the content of total protein and that of small molecules related to the secondary metabolism, such as TFAA. Conversely, the higher levels of TTA were positively correlated with the higher protein content and then to the absence of proteolysis. A positive correlation was observed between lactic acid and all parameters except for total protein content and pH, indicating the key role of this compound in sourdough metabolism. As shown in the score plot (Figure 6b), all control samples (before fermentation) were located in the lower left part of the graph, corresponding to the higher pH values as initial conditions. Fermented okra flours were discriminated mainly on the basis of the starter strain. Sourdoughs fermented by *W. cibaria* C43-11 and *Leuc. mesenteroides* C43-2M were more similar in composition with only minor differences between samples. The F_WeiCib, F_LeuMes and O_LeuMes sourdoughs were grouped in the right part of the graph because they were characterized by higher LAB counts, l-glutamic acid contents and TTA values. On the other hand, the acetic acid content mainly distinguished S_LeuMes and O_WeiCib from the other samples. In the lower right part of the graph, the sourdoughs inoculated with *L. plantarum* ITM21B strain were located and distinguished by the lower protein content, higher TFAA, EPS, and propionic and lactic acids.

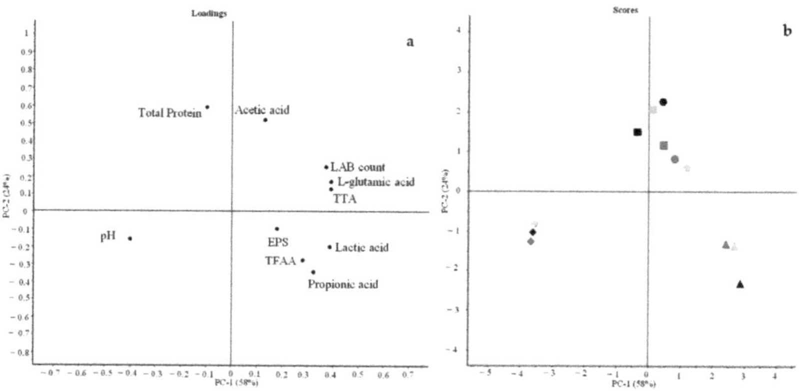

Figure 6. Principal component analysis (PCA) of all data registered on sourdough. (**a**) Loading plot indicating the correlation between variables (pH; TTA; LAB count; EPS; l-glutamic acid; organic acids—lactic; acetic; propionic; total proteins; TFAA content). (**b**) Score plot indicating sample distribution basing on the inoculated strains (triangle LacPla; circle LeuMes; square WeiCib; diamond CTRs) used to ferment F (red symbol), O (green symbol) and S (black symbol) okra flours.

Table 4. Pearson correlation coefficients among analytical data (EPS, organic acid content, l-glutamic acid, total proteins, TFAA, pH and TTA) registered on sourdough samples.

	pH	**TTA**	**LAB Count**	**EPS**	**Lactic Acid**	**Acetic Acid**	**Propionic Acid**	**L-Glutamic Acid**	**Total Proteins**	**TFAA**
pH	-	*−0.94*	*−0.95*	−0.26	*−0.76*	*−0.44*	−0.52	*−0.70*	0.00	*−0.49*
TTA	*−0.94*	-	*0.91*	0.17	*0.75*	*0.33*	0.57	*0.72*	0.04	*0.40*
LAB count	*−0.95*	*0.91*	-	0.17	*0.65*	*0.51*	0.44	*0.70*	0.09	*0.37*
EPS	−0.26	0.17	0.17	-	*0.29*	0.15	*0.33*	0.07	−0.16	*0.32*
lactic acid	*−0.76*	*0.75*	*0.65*	*0.29*	-	0.11	*0.76*	*0.32*	*−0.41*	*0.68*
acetic acid	*−0.44*	*0.33*	*0.51*	0.15	0.11	-	−0.01	*0.30*	0.17	0.16
propionic acid	*−0.52*	*0.57*	*0.44*	*0.33*	*0.76*	−0.01	-	0.19	*−0.40*	*0.58*
l-glutamic acid	*−0.70*	*0.72*	*0.70*	0.07	*0.32*	*0.30*	0.19	-	0.22	0.11
total proteins	0.00	0.04	0.09	−0.16	*−0.41*	0.17	*−0.40*	0.22	-	*−0.37*
TFAA	*−0.49*	*0.40*	*0.37*	*0.32*	*0.68*	0.16	*0.58*	0.11	*−0.37*	-

Values in italic character indicate significant correlation ($p < 0.05$).

Among the sourdough formulations, since only small differences were observed between the samples and the oven-drying technology is easily applicable in the food sector, the sourdough based on the starter *Leuc. mesenteroides* C43-2M and O flour was selected for the baking test, to further investigate the potential of this formulation in the production of bakery products.

3.3. Bread Properties

The fermented sourdough prepared with *Leuc. mesenteroides* C43-2M, the O flour and the unfermented O flour were used to prepare bread prototypes. The general appearance of the bread samples is shown in Figure 7. The images show the suitability of sourdough as an ingredient for yeast-leavened bread making. Table 5 reports the specific volume of the bread loaf and the moisture content, a_w, color and firmness of the crumb. The inclusion of O_LeuMes sourdough in the bread formulation did not significantly modify the moisture content of the crumb, with only minor changes in a_w. The latter was probably due to differences in batch preparation rather than a specific effect of the presence of okra flour. As expected, despite the small amount of okra flour used in the formulation (0.6% and 1.21% O flour replacement), its presence led to significant color changes in the crumb, with a decrease in the whiteness/lightness L^* value and an increase in redness a^* values, as also shown by the ΔE values. This darkening effect was independent from the occurrence of fermentation and was attributed to the brown color of the O flour. Previous studies have reported the strong impact of oven-dried okra flour on bread color [5,18,20,60].

CTR_0% O_20% O_40% CTR_20% CTR_40%

Figure 7. Bread samples prepared using fermented or unfermented oven-dried okra flour (O). CTR_0%, standard recipe; bread O_20%, bread O_40% (*w/w* flour weight), both containing O_LeuMes corresponding to 0.6% and 1.21% (*w/w*) of O flour replacement; breads CTR_20% and CTR_40% (containing not fermented O flour at 0.6% and 1.21% *w/w*, respectively).

Table 5. Moisture, a_w, color and firmness of crumb and loaf specific volume of bread samples prepared using unfermented O flour (CTR_20% and CTR_40%) or O_LeuMes sourdough (O_20% and O_40%). CTR_0% is the standard bread without okra flour. Values with different lower-case letters are significantly different among all samples ($p < 0.05$).

Bread Samples	Specific Volume (cm^3/g)	Moisture	a_w	L^*	a^*	b^*	ΔE	Firmness (N)
CTR_0%	4.08 ± 0.81 a	66.87 ± 0.90 a	0.64 ± 0.01 a	71.48 ± 3.16 a	−1.44 ± 0.15 d	15.21 ± 0.81 a		1.13 ± 0.81 c
O_20%	2.30 ± 0.81 b	68.86 ± 0.71 a	0.63 ± 0.01 ab	62.29 ± 2.71 b	−1.18 ± 0.20 c	15.48 ± 0.99 a	9.92	1.87 ± 0.07 b
O_40%	2.44 ± 0.81 b	65.87 ± 1.46 a	0.61 ± 0.02 bca	57.65 ± 2.17 c	−0.69 ± 0.13 a	15.23 ± 0.92 a	13.85	1.75 ± 0.03 b
CTR_20%	2.38 ± 0.81 b	65.66 ± 1.13a	0.60 ± 0.01 bc	62.58 ± 2.52 b	−1.00 ± 0.11 bc	15.73 ± 0.74 a	8.92	1.98 ± 0.14 b
CTR_40%	1.75 ± 0.81 c	68.74 ± 1.78 a	0.58 ± 0.02 c	60.77 ± 2.13 bc	−0.85 ± 0.11 ab	16.10 ± 0.54 a	10.76	3.01 ± 0.11 a

It is noteworthy that the presence of okra induced a reduction in specific volume with a concomitant increase in firmness. This was particularly evident in the bread sample containing unfermented okra flour at the higher content (1.21%). This behaviour can be attributed to the presence of okra components with high water binding capacity (e.g., pectin and other polysaccharides). These components may compete with other ingredients for water, reducing the availability of water for gluten–starch network formation and thus reducing dough elasticity, gas entrapment and leavening capacity. This is known from the literature for other plant fibers [21,62–66]. Our data on volume and firmness are consistent with the literature. Alamri [64] tested the effect of freeze-dried okra extract incorporated into bread at 4, 7, 10 and 13%: firmness increased and color became darker as a function of okra flour replacement. Recently, Bai-Ngew et al. [20] investigated the effect of two okra powders obtained by microwave vacuum or a hot air-drying process, incorporated in bread formulation at 1, 3 and 5%. The results indicated that the specific volume of the bread was inversely correlated with the addition of okra. The exception was the 1% replacement, which resulted in an improvement in specific volume, the bread's hardness and bioactive compound content.

As a naturally gluten-free (GF) material, okra flour can also be a valuable ingredient in GF bread. Interesting results were obtained by Tufaro et al. [18] who produced oven-dried okra powder (whole and fine) and incorporated it into the GF bread recipe at 1.5% of dough weight in combination with hydrocolloids. The authors observed that okra powders mainly affected the amount of water needed to achieve the desired dough consistency and, to a lesser extent, the color of the final product, while the technological traits were improved by the type of hydrocolloid used (hydroxypropyl methylcellulose rather than *Psyllium* fiber).

In our study, the partial replacement of wheat flour with unfermented O flour at 0.6% and 1.21% had some negative effects on bread quality as compared to the CTR_0% not containing okra flour, while the fermentation process improved the specific volume and reduced the firmness of bread containing the higher amount of sourdough (40% corresponding to the 1.21% flour replacement) with respect to the CTR bread containing unfermented O flour at the same level (Table 5). Actually, even if these are preliminary results and a better understanding of the effects of okra flour type and percentage replacement on bread characteristics there is needed, it seems that fermentation partially reduced these detrimental effects, especially considering the O_40% bread This effect could be due to the microbial activity leading to acidification of the dough and hydrolysis of okra components as well as to the slightly higher content of EPS (Figures 8 and 9).

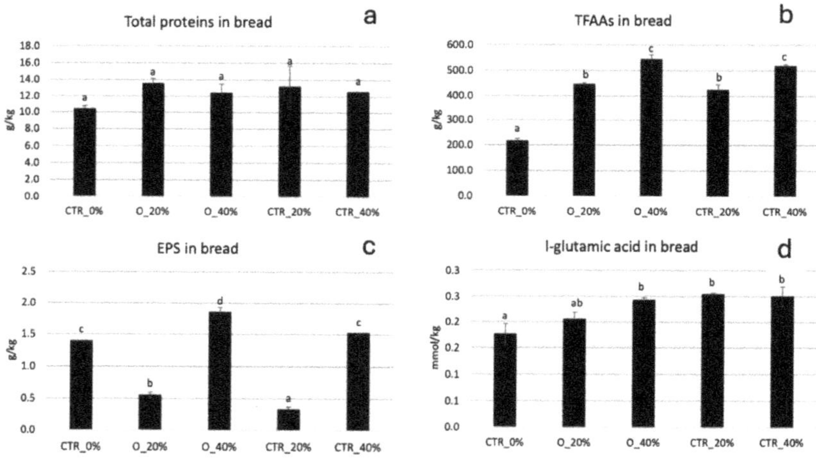

Figure 8. Total proteins (**a**), TFAAs (**b**), EPS (**c**) and l-glutamic acid (**d**) content in bread samples prepared using unfermented (CTR_20% and CTR_40%) O flour or O_LeuMes sourdough (O_20% and O_40%). CTR_0% is the standard bread without okra flour. Values with different lower-case letters are significantly different among all samples ($p < 0.05$).

As reported by Nawab et al. [67], a polysaccharide such as dextran, which has a high molecular weight and linear chain structure binds moisture, then acting as a plasticizer that can limit starch swelling and gelatinization, thereby delaying the staling process and prolonging shelf life. Dextran, a homopolysaccharide belonging to the exopolysaccharide compounds produced by LAB, is frequently recovered in sourdough and is characterized by several technological and health properties [29,68,69]. In particular, it can act as a hydrocolloid in bread formulation, contributing to the improvement in the product's texture and sensory qualities, as well as meeting consumer demands for clean-label and healthy foods [70]. Dough quality is highly dependent on gluten solubility, which increases in an acidic environment. The presence of polysaccharides with a high water holding capacity leads to increased formation of β-structures, aggregation and abnormal folding of gluten proteins [71]. Bi et al. [59] demonstrated the influence of the acidic environment on the interaction between HMw dextran and gluten structure as the depolymerization of gluten aggregates caused by HMw dextran was weakened under acidic conditions (pH = 3.6), while it was promoted in a weakly acidic environment (pH = 4.4–4.6). These results confirm the positive role that polysaccharides can play in combination with a controlled LAB fermentation.

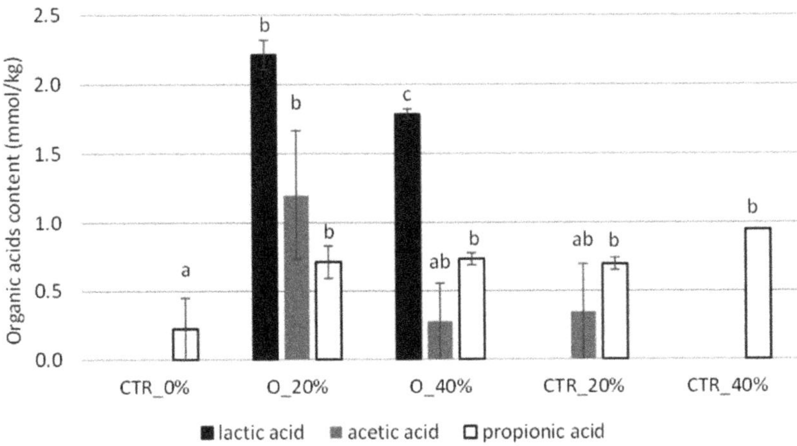

Figure 9. Organic acids (lactic, acid and propionic) content in bread samples prepared using unfermented (CTR_20% and CTR_40%) or O_LeuMes sourdough (O_20% and O_40%). CTR_0% is the standard bread without okra flour. Values of each acid with different lower-case letters are significantly different among all samples ($p < 0.05$).

The EPS content in bread was in agreement with other published data [72]. In our study, the main aim of quantifying EPS was to compare breads containing sourdough with those containing unfermented okra flour. As shown in Figure 8, significant differences in the content of EPS were registered among the samples, with the highest value observed in bread containing the fermented O_LeuMes sourdough at 40%. The microbial contribution to the polysaccharidic fraction in bread can be hypothesized since the amount of these compounds in both sourdough-containing breads (O_20% and O_40%) was higher than controls containing unfermented okra flour (CTR_20% and CTR_40%). The higher amount of polysaccharides in CTR_40% than CTR_20% could be due to the higher concentration of okra flour. The differences between control samples could be related to the presence of starch, mainly in CTR_0%, which led to an increase in the carbohydrate content as quantified by the phenol-sulfuric acid method.

The content of l-glutamic acid was significantly higher in samples containing sourdough and unfermented O flour, suggesting the contribution of okra flour to the content of this amino acid. Actually, glutamic acid is one of the most abundant amino acids in okra pods and particularly in seeds [73]. The protein content did not vary between samples since okra pods are not particularly rich in these compounds while containing many amino acids [3]. In fact, the amount of TFAAs was significantly higher in sourdough-based bread samples. Among the organic acids, a higher level was observed in the samples produced with O flour compared to the standard bread not containing okra flour (CTR_0%) (Figure 9). In particular, the lactic acid level was significantly higher in bread produced with the sourdough, while the other acids were found to be at very low levels in all samples.

In order to better understand the correlations between technological and analytical measures registered on bread, Pearson correlation coefficients were calculated. The results are presented in Table 6 and clearly show the significant correlation between total protein content, TFAA, pH, TTA and almost all the technological parameters. All analytical parameters, except for pH, were negatively correlated with specific volume, a_w and lightness L^*, and positively correlated with firmness, moisture, redness *a** and yellowness *b**, suggesting the influence of the microbial metabolism characterizing sourdough on bread quality attributes. In detail, specific volume was positively correlated with pH and negatively correlated with all the other parameters. Actually, the acidification of bread (low pH and high TTA) can have a negative effect on the specific volume and firmness of bread due to the breakdown of the gluten protein [16].

Table 6. Pearson correlation coefficients between technological (color, moisture, a_w, firmness of crumb and loaf specific volume) and analytical (EPS, organic acids content, l-glutamic acid, total proteins, TFAA, pH and TTA) data relevant to bread.

		Analytical Data								
		EPS	Lactic Acid	Acetic Acid	Propionic Acid	L-Glutamic Acid	Total Proteins	TFAA	pH	TTA
Technological data	*L**	−0.07	−0.18	−0.11	*−0.93*	*−0.79*	*−0.58*	*−0.88*	*0.60*	*−0.30*
	*a**	*0.34*	*0.28*	0.05	*0.87*	*0.64*	*0.43*	*0.87*	*−0.50*	*0.41*
	*b**	0.09	*0.42*	0.18	0.09	0.14	*0.30*	*0.28*	*−0.44*	*0.50*
	moisture	0.07	−0.06	*−0.27*	−0.09	*0.40*	0.00	0.13	*−0.54*	0.17
	a_w	*−0.47*	*−0.74*	−0.24	*−0.54*	*−0.36*	*−0.37*	*−0.79*	*0.71*	*−0.89*
	firmness	*0.33*	*0.65*	0.19	*0.48*	*0.54*	*0.41*	*0.79*	*−0.90*	*0.85*
	specific volume	−0.01	*−0.52*	*−0.30*	*−0.80*	*−0.79*	*−0.72*	*−0.94*	*0.95*	*−0.66*

Values in italic character indicate significant correlation ($p < 0.05$).

4. Conclusions

The data obtained in the current study suggest that the valorization of okra pod flours by fermentation in the sourdough process represents a biotechnological strategy that can ameliorate the technological disadvantages caused by the use of unprocessed okra flour in bread making, paving the way for the production of innovative fermented food ingredients based on this precious vegetable. The fermentation process allowed the content of exopolysaccharides (EPS), total free amino acids (TFAA), l-glutamic acid and organic acids to be increased in the sourdough and the incorporation of okra flour into the bread-making process to be simplified. Bread containing the *Leuconostoc mesenteroides* wheat-based sourdough with 9% oven-dried okra flour added at 40% *w/w* in its formulation, corresponding to a replacement level of okra flour of 1.21%, was improved in nutritional and technological properties, mainly for the content of TFAA, EPS and l-glutamic acid, and showed an improved specific volume and firmness with respect to the unfermented oven-dried okra flour bread.

Author Contributions: Conceptualization: F.V. and H.M.; methodology: F.V., S.C. and H.M.; Validation: F.V., S.C., L.M., M.D.B., V.C., S.L.L. and S.P.; formal analysis: M.D.B., V.C., F.V. and S.P.; investigation: M.D.B., S.C., L.M., V.C., S.L.L., A.M. and S.P.; writing—original draft: F.V., V.C., M.D.B., S.C., H.M. and A.M.; writing—review and editing: F.V., V.C., M.D.B., S.C., H.M., A.M., S.L.L. and L.M.; resources: H.M., S.L.L., M.D.B. and A.M.; visualization: F.V.; supervision: F.V. and H.M.; project administration: F.V. and H.M.; funding acquisition: F.V. and H.M. All authors have read and agreed to the published version of the manuscript.

Funding: This research was funded by ExplOkra Project—"Exploitation of Okra (*Abelmoschus esculentus* L.) through sustainable approaches for food applications" CNR/MHESR Tunisia 2023—2024—AD002-186.

Data Availability Statement: The original contributions presented in the study are included in the article, further inquiries can be directed to the corresponding author.

Acknowledgments: The authors thank Marinella Cavallo, Simonetta Martena and Filomena Epifani for their technical support in the management of the ExplOkra Project.

Conflicts of Interest: The authors declare no conflicts of interest.

Abbreviations

S.No	Notation	Description
1	O	oven-dried okra flour
2	F	freeze-dried okra flour
3	S	sun-dried okra flour
4	DY	dough yield
5	CTR	control
6	O_CTR_T0	not inoculated mixtures, containing O flour, before fermentation
7	O_WeiCib	sourdough inoculated with *W. cibaria* C43-11 strain, containing O flour, after 14 h of fermentation
8	O_LeuMes	sourdough inoculated with *Leuc. mesenteroides* C43-2M strain, containing O flour, after 14 h of fermentation
9	O_LacPla	sourdough inoculated with *L. plantarum* ITM21B strain, containing O flour, after 14 h of fermentation
10	F_CTR_T0	not inoculated mixtures, containing F flour, before fermentation
11	F_LeuMes	sourdough inoculated with *W. cibaria* C43-11 strain, containing F flour, after 14 h of fermentation
12	F_LeuMes	sourdough inoculated with *Leuc. mesenteroides* C43-2M strain, containing F flour, after 14 h of fermentation
13	F_LacPla	sourdough inoculated with *L. plantarum* ITM21B strain, containing F flour, after 14 h of fermentation
14	S_CTR_T0	not inoculated mixtures, containing S flour, before fermentation
15	S_WeiCib	sourdough inoculated with *W. cibaria* C43-11 strain, containing S flour, after 14 h of fermentation
16	S_LeuMes	sourdough inoculated with *Leuc. mesenteroides* C43-2M strain, containing S flour, after 14 h of fermentation
17	S_LacPla	sourdough inoculated with *L. plantarum* ITM21B strain, containing S flour, after 14 h of fermentation
18	Mw	molecular weight
19	HMw	high molecular weight
20	GF	gluten-free
21	CTR_0%	bread not containing okra flour, produced with standard recipe
22	O_20%	bread containing 20% of O_LeuMes (*w/w* flour weight), corresponding to 0.6% (*w/w*) of O flour replacement
23	O_40%	bread containing 40% of O_LeuMes (*w/w* flour weight), corresponding to 1.21% (*w/w*) of O flour replacement
24	CTR_20%	bread containing not fermented O flour at 0.6% *w/w*
25	CTR_40%	bread containing not fermented O flour at 1.21% *w/w*

References

1. Islam, M.T. Phytochemical information and pharmacological activities of Okra (*Abelmoschus esculentus*): A literature-based review. *Phytother. Res.* **2019**, *33*, 72–80. [CrossRef] [PubMed]
2. Wu, D.T.; Nie, X.R.; Shen, D.D.; Li, H.Y.; Zhao, L.; Zhang, Q.; Lin, D.R.; Qin, W. Phenolic compounds, antioxidant activities, and inhibitory effects on digestive enzymes of different cultivars of okra (*Abelmoschus esculentus*). *Molecules* **2020**, *25*, 1276. [CrossRef] [PubMed]
3. Romdhane, M.H.; Chahdoura, H.; Barros, L.; Dias, M.I.; Corrêa, R.C.G.; Morales, P.; Ciudad-Mulero, M.; Flamini, G.; Majdoub, H.; Ferreira, I.C. Chemical composition, nutritional value, and biological evaluation of Tunisian okra pods (*Abelmoschus esculentus* L. Moench). *Molecules* **2020**, *25*, 4739. [CrossRef] [PubMed]
4. Adelakun, O.E.; Oyelade, O.J. Potential use of okra seed (*Abelmoschus esculentus* Moench) flour for food fortification and effects of processing. In *Flour and Breads and Their Fortification in Health and Disease Prevention*, 1st ed.; Preedy, V.R., Watson, R.R., Patel, V.B., Eds.; Academic Press: London, UK, 2011; pp. 205–212.
5. Machine, A.P.; Massingue, A.A.; Asaam, S. Physicochemical and sensory properties of bread with sweet potato flour (*Ipomea batatas* L.) as partial replacer of wheat flour supplemented with okra hydrocolloids. *Afr. J. Food Sci.* **2020**, *14*, 385–394.
6. Gemede, H.F.; Ratta, N.; Haki, G.D.; Woldegiorgis, A.Z.; Beyene, F. Nutritional quality and health benefits of okra (*Abelmoschus esculentus*): A review. *Int. J. Nutr. Food Sci.* **2015**, *4*, 208–215. [CrossRef]
7. Nie, X.R.; Fu, Y.; Wu, D.T.; Huang, T.T.; Jiang, Q.; Zhao, L.; Zhang, Q.; Lin, D.R.; Chen, H.; Qin, W. Ultrasonic-assisted extraction, structural characterization, chain conformation, and biological activities of a pectic-polysaccharide from okra (*Abelmoschus esculentus*). *Molecules* **2020**, *25*, 1155. [CrossRef]

8. Nie, X.R.; Li, H.Y.; Du, G.; Lin, S.; Hu, R.; Li, H.Y.; Zhao, L.; Zhang, Q.; Chen, H.; Wu, D.T.; et al. Structural characteristics, rheological properties, and biological activities of polysaccharides from different cultivars of okra (*Abelmoschus esculentus*) collected in China. *Int. J. Biol. Macromol.* **2019**, *139*, 459–467. [CrossRef]
9. Maalej, H.; Maalej, A.; Bayach, A.; Zykwinska, A.; Colliec-Jouault, S.; Sinquin, C.; Marchand, L.; Ktari, N.; Bardaa, S.; Salah, R.B.; et al. A novel pectic polysaccharide-based hydrogel derived from okra (*Abelmoschus esculentus* L. Moench) for chronic diabetic wound healing. *Eur. Polym. J.* **2023**, *183*, 111763. [CrossRef]
10. Liu, H.; Gong, F.; Wei, F.; Wu, H. Artificial simulation of salivary and gastrointestinal digestion, and fermentation by human fecal microbiota, of polysaccharides from *Dendrobium aphyllum*. *RSC Adv.* **2018**, *8*, 13954–13963. [CrossRef]
11. Xia, F.; Zhong, Y.; Li, M.; Chang, Q.; Liao, Y.; Liu, X.; Pan, R. Antioxidant and anti-fatigue constituents of okra. *Nutrients* **2015**, *7*, 8846–8858. [CrossRef]
12. Yuan, Q.; Lin, S.; Fu, Y.; Nie, X.R.; Liu, W.; Su, Y.; Han, Q.H.; Zhao, L.; Zhang, Q.; Lin, D.R.; et al. Effects of extraction methods on the physicochemical characteristics and biological activities of polysaccharides from okra (*Abelmoschus esculentus*). *Int. J. Biol. Macromol.* **2019**, *127*, 178–186. [CrossRef] [PubMed]
13. Wu, D.T.; Fu, Y.; Guo, H.; Yuan, Q.; Nie, X.R.; Wang, S.P.; Gan, R.Y. In vitro simulated digestion and fecal fermentation of polysaccharides from loquat leaves: Dynamic changes in physicochemical properties and impacts on human gut microbiota. *Int. J. Biol. Macromol.* **2021**, *168*, 733–742. [CrossRef]
14. Sengkhamparn, N.; Bakx, E.J.; Verhoef, R.; Schols, H.A.; Sajjaanantakul, T.; Voragen, A.G. Okra pectin contains an unusual substitution of its rhamnosyl residues with acetyl and alpha-linked galactosyl groups. *Carbohydr. Res.* **2009**, *344*, 1842–1851. [CrossRef] [PubMed]
15. Manzocco, L.; Basso, F.; Nicoli, M.C. Effect of hyperbaric storage at room temperature on the activity of polyphenoloxidase in model systems and fresh apple juice. *Food Bioprocess Technol.* **2023**, *16*, 2247–2256. [CrossRef]
16. Islam, M.A.; Islam, S. Sourdough bread quality: Facts and Factors. *Foods* **2024**, *13*, 2132. [CrossRef]
17. Padayachee, A.; Netzel, G.; Netzel, M.; Day, L.; Zabaras, D.; Mikkelsen, D.; Gidley, M.J. Binding of polyphenols to plant cell wall analogues–Part 2: Phenolic acids. *Food Chem.* **2012**, *135*, 2287–2292. [CrossRef]
18. Tufaro, D.; Bassoli, A.; Cappa, C. Okra (*Abelmoschus esculentus*) powder production and application in gluten-free bread: Effect of particle size. *Food Bioprocess Technol.* **2022**, *15*, 904–914. [CrossRef]
19. Machine, A.P.; Massingue, A.A.; Balane, S.G.; Armando, E.J. Enhancing bread quality: Investigating the physicochemical and sensory properties of bread with sweet potato flour and okra hydrocolloids as substitutes for wheat flour. In *Current Perspectives in Agriculture and Food Science*, 1st ed.; Son, L., Ed.; B P International: Hooghly, India, 2024; Volume 8, pp. 28–48.
20. Bai-Ngew, S.; Phimolsiripol, Y.; Walter, P. Characterization of hot air drying and microwave vacuum drying of okra powder and application in bread. *Food Appl. Biosci. J.* **2024**, *12*, 11–28.
21. Xu, K.; Guo, M.; Roman, L.; Pico, J.; Martinez, M.M. Okra seed and seedless pod: Comparative study of their phenolics and carbohydrate fractions and their impact on bread-making. *Food Chem.* **2020**, *317*, 126387. [CrossRef]
22. Ou, J.; Wang, M.; Zheng, J.; Ou, S. Positive and negative effects of polyphenol incorporation in baked foods. *Food Chem.* **2019**, *284*, 90–99. [CrossRef]
23. Li, M.; Koecher, K.; Hansen, L.; Ferruzzi, M.G. Phenolics from whole grain oat products as modifiers of starch digestion and intestinal glucose transport. *J. Agric. Food Chem.* **2017**, *65*, 6831–6839. [CrossRef] [PubMed]
24. Pico, J.; Martinez, M.M. Unraveling the inhibition of intestinal glucose transport by dietary phenolics: A review. *Curr. Pharm. Des.* **2019**, *25*, 3418–3433. [CrossRef] [PubMed]
25. Pico, J.; Corbin, S.; Ferruzzi, M.G.; Martinez, M.M. Banana flour phenolics inhibit trans-epithelial glucose transport from wheat cakes in a coupled in vitro digestion/Caco-2 cell intestinal model. *Food Funct.* **2019**, *10*, 6300–6311. [CrossRef] [PubMed]
26. Alamri, M.S. Okra-gum fortified bread: Formulation and quality. *J. Food Sci. Technol.* **2014**, *51*, 2370–2381. [CrossRef]
27. Buczkowski, B.K. Sourdough Bread Enriched with Soluble Fibres: Development, Characterisation and Nutritional Aspects of a Functional Food Product. Doctoral Dissertation, Manchester Metropolitan University, Manchester, UK, 2013.
28. De Vuyst, L.; Van Kerrebroeck, S.; Harth, H.; Huys, G.; Daniel, H.-M.; Weckx, S. Microbial ecology of sourdough fermentations: Diverse or uniform? *Food Microbiol.* **2014**, *37*, 11–29. [CrossRef]
29. Arendt, E.K.; Ryan, L.A.; Dal Bello, F. Impact of sourdough on the texture of bread. *Food Microbiol.* **2007**, *24*, 165–174. [CrossRef]
30. Ripari, V.; Bai, Y.; Gänzle, M.G. Metabolism of phenolic acids in whole wheat and rye malt sourdoughs. *Food Microbiol.* **2019**, *77*, 43–51. [CrossRef]
31. Gabriele, M.; Sparvoli, F.; Bollini, R.; Lubrano, V.; Longo, V.; Pucci, L. The impact of sourdough fermentation on non-nutritive compounds and antioxidant activities of flours from different *Phaseolus vulgaris* L. genotypes. *J. Food Sci.* **2019**, *84*, 1929–1936. [CrossRef]
32. Altunkaya, A.; Hedegaard, R.V.; Brimer, L.; Gökmen, V.; Skibsted, L.H. Antioxidant capacity versus chemical safety of wheat bread enriched with pomegranate peel powder. *Food Funct.* **2013**, *4*, 722–727. [CrossRef]
33. Gänzle, M.G. From gene to function: Metabolic traits of starter cultures for improved quality of cereal foods. *Int. J. Food Microbiol.* **2009**, *134*, 29–36. [CrossRef]
34. Korakli, M.; Rossmann, A.; Gänzle, M.G.; Vogel, R.F. Sucrose metabolism and exopolysaccharide production in wheat and rye sourdoughs by *Lactobacillus sanfranciscensis*. *J. Agr. Food Chem.* **2001**, *49*, 5194–5200. [CrossRef] [PubMed]

35. Adelakun, O.E.; Olanipekun, B.F.; Akingbaso, O.; Adhikar, B.M. Effect of fermentation and variety on quality attributes of okra seed (*Abelmoschus esculentus* (L.) Moench) flour. *Donn. J. Food Sci. Tech.* **2017**, *3*, 001–006.
36. Adetuyi, F.O.; Ibrahim, T.A. Effect of fermentation time on the phenolic, flavonoid and vitamin C contents and antioxidant activities of okra (*Abelmoschus esculentus*) seeds. *Niger. Food J.* **2014**, *32*, 128–137. [CrossRef]
37. Wang, X.; Hu, K.; Chen, Y.; Lai, J.; Zhang, M.; Li, J.; Li, Q.; Zhao, N.; Liu, S. Effect of *Lactiplantibacillus plantarum* fermentation on the physicochemical, antioxidant activity and immunomodulatory ability of polysaccharides from Lvjian okra. *Int. J. Biol. Macromol.* **2024**, *257*, 128649. [CrossRef] [PubMed]
38. Corsetti, A.; Lavermicocca, P.; Morea, M.; Baruzzi, F.; Tosti, N.; Gobbetti, M. Phenotypic and molecular identification and clustering of lactic acid bacteria and yeasts from wheat (species *Triticum durum* and *Triticum aestivum*) sourdoughs of southern Italy. *Int. J. Food Microbiol.* **2001**, *64*, 95–104. [CrossRef]
39. Di Biase, M.; Bavaro, A.R.; Lonigro, S.L.; Pontonio, E.; Conte, A.; Padalino, L.; Minisci, A.; Lavermicocca, P.; Valerio, F. *Lactobacillus plantarum* ITM21B fermentation product and chickpea flour enhance the nutritional profile of salt reduced bakery products. *Int. J. Food Sci. Nutr.* **2019**, *70*, 701–713. [CrossRef]
40. Valerio, F.; Conte, A.; Di Biase, M.; Lattanzio, V.M.; Lonigro, S.L.; Padalino, L.; Pontonio, E.; Lavermicocca, P. Formulation of yeast-leavened bread with reduced salt content by using a *Lactobacillus plantarum* fermentation product. *Food Chem.* **2017**, *221*, 582–589. [CrossRef]
41. De Bellis, P.; Ferrara, M.; Bavaro, A.R.; Linsalata, V.; Di Biase, M.; Musio, B.; Gallo, V.; Valerio, F. Characterization of dextran produced by the food-related strain *Weissella cibaria* C43-11 and of the relevant dextransucrase gene. *Foods* **2022**, *11*, 2819. [CrossRef]
42. Valerio, F.; Favilla, M.; De Bellis, P.; Sisto, A.; de Candia, S.; Lavermicocca, P. Antifungal activity of strains of lactic acid bacteria isolated from a semolina ecosystem against *Penicillium roqueforti*, *Aspergillus niger* and *Endomyces fibuliger* contaminating bakery products. *Syst. Appl. Microbiol.* **2009**, *32*, 438–448. [CrossRef]
43. Bavaro, A.R.; Di Biase, M.; Conte, A.; Lonigro, S.L.; Caputo, L.; Cedola, A.; Del Nobile, A.; Logrieco, A.F.; Lavermicocca, P.; Valerio, F. *Weissella cibaria* short-fermented liquid sourdoughs based on quinoa or amaranth flours as fat replacer in focaccia bread formulation. *Int. J. Food Sci. Technol.* **2021**, *56*, 3197–3208. [CrossRef]
44. Valerio, F.; Bavaro, A.R.; Di Biase, M.; Lonigro, S.L.; Logrieco, A.F.; Lavermicocca, P. Effect of amaranth and quinoa flours on exopolysaccharide production and protein profile of liquid sourdough fermented by *Weissella cibaria* and *Lactobacillus plantarum*. *Front. Microbiol.* **2020**, *11*, 967. [CrossRef]
45. EFSA Panel on Biological Hazards (BIOHAZ); Koutsoumanis, K.; Allende, A.; Álvarez-Ordóñez, A.; Bolton, D.; Bover-Cid, S.; Chemaly, M.; De Cesare, A.; Hilbert, F.; Lindqvist, R.; et al. Scientific Opinion on the update of the list of qualified presumption of safety (QPS) recommended microorganisms intentionally added to food or feed as notified to EFSA. *EFSA J.* **2023**, *21*, 7747.
46. De Bellis, P.; Valerio, F.; Sisto, A.; Lonigro, S.L.; Lavermicocca, P. Probiotic table olives: Microbial populations adhering on olive surface in fermentation sets inoculated with the probiotic strain *Lactobacillus paracasei* IMPC2.1 in an industrial plant. *Int. J. Food Microbiol.* **2010**, *140*, 6–13. [CrossRef]
47. AOAC. *Official Methods of Analysis*; Association of Official Analytical Chemists: Washington, DC, USA, 1990.
48. Bradford, M.M. A rapid and sensitive method for the quantitation of microgram quantities of protein utilizing the principle of protein-dye binding. *Anal. Biochem.* **1976**, *72*, 248–254. [CrossRef]
49. Dubois, M.; Gilles, K.A.; Hamilton, J.K.; Rebers, P.T.; Smith, F. Colorimetric method for determination of sugars and related substances. *Anal. Chem.* **1956**, *28*, 350–356. [CrossRef]
50. Osborne, T.B. *The Proteins of the Wheat Kernel (Monograph)*; Carnegie Inst.: Washington, DC, USA, 1907.
51. Weiss, W.; Vogelmeier, C.; Gorg, A. Electrophoretic characterization of wheat grain allergens from different cultivars involved in bakers' asthma. *Electrophoresis* **1993**, *14*, 805–816. [CrossRef]
52. Doi, E.; Shibata, D.; Matoba, T. Modified colorimetric ninhydrin methods for peptidase assay. *Anal Biochem.* **1981**, *118*, 173–184. [CrossRef]
53. AOAC. *Official Methods of Analysis*; Association of Official Analytical Chemists: Gaithersburg, MD, USA, 2000.
54. American Association of Cereal Chemists (AACC). *Approved Methods of the AACC*, 10th ed.; St. Paul: Eagan, MN, USA, 2000.
55. Lavermicocca, P.; Valerio, F.; Evidente, A.; Lazzaroni, S.; Corsetti, A.; Gobbetti, M. Purification and characterization of novel antifungal compounds from the sourdough *Lactobacillus plantarum* strain 21B. *Appl. Environ. Microbiol.* **2000**, *66*, 4084–4090. [CrossRef]
56. Montemurro, M.; Pontonio, E.; Gobbetti, M.; Rizzello, C.G. Investigation of the nutritional, functional and technological effects of the sourdough fermentation of sprouted flours. *Int. J. Food Microbiol.* **2019**, *302*, 47–58. [CrossRef]
57. Pétel, C.; Onno, B.; Prost, C. Sourdough volatile compounds and their contribution to bread: A review. *Trends Food Sci. Technol.* **2017**, *59*, 105–123. [CrossRef]
58. Dantas, T.L.; Alonso Buriti, F.C.; Florentino, E.R. Okra (*Abelmoschus esculentus* L.) as a potential functional food source of mucilage and bioactive compounds with technological applications and health benefits. *Plants* **2021**, *10*, 1683. [CrossRef]
59. Bi, Q.; Hong, T.; Mei, X.; Xu, X.; Xu, D. Effect of high-molecular weight dextran-enriched sourdough fermented using *Leuconostoc mesenteroides* ATCC 8293 on bread quality and gluten. *Food Biosci.* **2023**, *53*, 102777. [CrossRef]
60. Zhao, C.J.; Kinner, M.; Wismer, W.; Gänzle, M.G. Effect of glutamate accumulation during sourdough fermentation with *Lactobacillus reuteri* on the taste of bread and sodium-reduced bread. *Cereal Chem.* **2015**, *92*, 224–230. [CrossRef]

61. Vermeulen, N.; Gänzle, M.G.; Vogel, R.F. Glutamine deamidation by cereal-associated lactic acid bacteria. *J. Appl. Microbiol.* **2007**, *103*, 1197–1205. [CrossRef]
62. Peressini, D.; Sensidoni, A. Effect of soluble dietary fibre addition on rheological and breadmaking properties of wheat doughs. *J. Cereal Sci.* **2009**, *49*, 190–201. [CrossRef]
63. Foschia, M.; Peressini, D.; Sensidoni, A.; Brennan, C.S. The effects of dietary fibre addition on the quality of common cereal products. *J. Cereal Sci.* **2013**, *58*, 216–227. [CrossRef]
64. Ameh, M.O.; Gernah, D.I.; Igbabul, B.D. Physico-chemical and sensory evaluation of wheat bread supplemented with stabilized undefatted rice bran. *Food Nutr. Sci.* **2013**, *4*, 43–48. [CrossRef]
65. Chareonthaikij, P.; Uan-On, T.; Prinyawiwatkul, W. Effects of pineapple pomace fibre on physicochemical properties of composite flour and dough, and consumer acceptance of fibre-enriched wheat bread. *Int. J. Food Sci. Technol.* **2016**, *51*, 1120–1129. [CrossRef]
66. Plazzotta, S.; Sillani, S.; Manzocco, L. Exploitation of lettuce waste flour to increase bread functionality: Effect on physical, nutritional, sensory properties and on consumer response. *Int. J. Food Sci. Technol.* **2018**, *53*, 2290–2297. [CrossRef]
67. Nawab, A.; Alam, F.; Haq, M.A.; Hasnain, A. Effect of guar and xanthan gums on functional properties of mango (*Mangifera indica*) kernel starch. *Int. J. Biol. Macromol.* **2016**, *93*, 630–635. [CrossRef]
68. Di Cagno, R.; De Angelis, M.; Limitone, A.; Minervini, F.; Carnevali, P.; Corsetti, A.; Gänzle, M.; Ciati, R.; Gobbetti, M. Glucan and fructan production by sourdough *Weissella cibaria* and *Lactobacillus plantarum*. *J. Agric. Food Chem.* **2006**, *54*, 9873–9881. [CrossRef]
69. Dong, Y.; Ronholm, J.; Fliss, I.; Karboune, S. Screening of lactic acid bacteria strains for potential sourdough and bread applications: Enzyme expression and exopolysaccharide production. *Probiotics Antimicrob. Proteins* **2024**, 1–23. [CrossRef]
70. Zhang, Y.; Wang, D.; Zhang, Z.; Guan, H.; Zhang, Y.; Xu, D.; Xu, X.; Li, D. Improvement on wheat bread quality by in situ produced dextran—A comprehensive review from the viewpoint of starch and gluten. *Compr. Rev. Food Sci. Food Saf.* **2024**, *23*, 13353. [CrossRef]
71. Nawrocka, A.; Krekora, M.; Niewiadomski, Z.; Miś, A. Characteristics of the chemical processes induced by celluloses in the model and gluten dough studied with application of FTIR spectroscopy. *Food Hydrocoll.* **2018**, *85*, 176–184. [CrossRef]
72. Pepe, O.; Ventorino, V.; Cavella, S.; Fagnano, M.; Brugno, R. Prebiotic content of bread prepared with flour from immature wheat grain and selected dextran-producing lactic acid bacteria. *Appl. Environ. Microbiol.* **2013**, *79*, 3779–3785. [CrossRef]
73. Sami, R.; Lianzhou, J.; Yang, L.; Ma, Y.; Jing, J. Evaluation of fatty acid and amino acid compositions in okra (*Abelmoschus esculentus*) grown in different geographical locations. *Biomed Res. Int.* **2013**, *1*, 574283. [CrossRef]

foods

MDPI

Article

Fermented *Gastrodia elata* Bl. Alleviates Cognitive Deficits by Regulating Neurotransmitters and Gut Microbiota in D-Gal/$AlCl_3$-Induced Alzheimer's Disease-like Mice

Yu Wang [1,2], Min Zhao [1,2], Chunzhi Xie [3], Lilang Li [1,2], Ling Lin [1,2], Qiji Li [1,2], Liangqun Li [1,2], Faju Chen [1,2], Xiaosheng Yang [1,2], Juan Yang [1,2] and Ming Gao [1,2,*]

1 State Key Laboratory for Functions and Applications of Medicinal Plants, Guizhou Medical University, Guiyang 550014, China; wang_yu@gmc.edu.cn (Y.W.); luckluckzm@163.com (M.Z.); lililang@gmc.edu.cn (L.L.); 17685222422@163.com (L.L.); leeqiji@126.com (Q.L.); liliangqun2010@163.com (L.L.); chenfaju202212@126.com (F.C.); gzcnp@sina.cn (X.Y.); yangxz2002@126.com (J.Y.)
2 Natural Products Research Center of Guizhou Province, Guiyang 550014, China
3 College of Food and Biotechnology Engineering, Xuzhou University of Technology, Xuzhou 221018, China; xcz0611@xzit.edu.cn
* Correspondence: gaoming@gmc.edu.cn; Tel.: +86-18345701431

Abstract: Alzheimer's disease (AD) is a common neurological disease with recognition ability loss symptoms and a major contributor to dementia cases worldwide. *Gastrodia elata* Bl. (GE), a food of medicine–food homology, has been reported to have a mitigating effect on memory and learning ability decline. However, the effect of GE fermented by *Lactobacillus plantarum*, *Acetobacter pasteurianus*, and *Saccharomyces* (FGE) on alleviating cognitive deficits in AD was not studied. Mice were randomly divided into six groups, control, model, donepezil, low, medium, and high doses of FGE, and D-Galactose/Aluminum chloride (D-Gal/$AlCl_3$) was used to establish an AD-like mouse model. The results indicated that FGE could improve the production of neurotransmitters and relieve oxidative stress damage in AD-like mice, which was evidenced by the declined levels of amyloid-β (Aβ), Tau, P-Tau, acetylcholinesterase (AchE), and malondialdehyde (MDA), and increased acetylcholine (Ach), choline acetyltransferase (ChAT), and superoxide dismutase (SOD) levels in brain tissue. Notably, FGE could enhance the richness of the gut microbiota, especially for beneficial bacteria such as *Lachnospira* and *Lactobacillus*. Non-target metabolomics results indicated that FGE could affect neurotransmitter levels by regulating amino acid metabolic pathways to improve AD symptoms. The FGE possessed an ameliorative effect on AD by regulating neurotransmitters, oxidative stress levels, and gut microbiota and could be considered a good candidate for ameliorating AD.

Keywords: *Gastrodia elata* Bl.; Alzheimer's disease; fermentation; neurotransmitter; gut microbiota

1. Introduction

Alzheimer's disease (AD), a type of neurodegenerative disease with insidious onset, is characterized by memory impairment, aphasia, apraxia, agnosia, and executive and spatial recognition dysfunction [1–3]. There are over 55 million patients with dementia worldwide, according to the "*World Alzheimer's Disease 2023 Report*" released by the Alzheimer's Association International (ADI). Even more concerning, the number of patients with dementia is expected to reach 139 million by 2050, with the greatest increases in low- and middle-income countries, of which about 60–70% will be patients with AD [4]. It seriously affects the quality of patient life and is a threat to human health. Unfortunately, despite several decades of research, the pathogenesis of AD is still not fully clear. Recent studies revealed that AD is caused by various pathogenesis factors and involves many pathological mechanisms [5]. The imbalanced production and clearance of amyloid-β (Aβ) is the initial factor of neuronal degeneration and dementia. The excessive Aβ oligomers between brain neurons may accelerate neurodegeneration and then induce Tau hyperphosphorylation, which is one of the main pathogens for the occurrence of AD [6–8]. Tau protein, a microtubule-related protein, is mainly distributed in the axons of neurons in the central nervous system. Excessive phosphorylation of Tau protein causes the loss of the ability to bind to microtubules, resulting in microtubule deaggregation and axon transport obstruction, forming Tau aggregates, which then cause the apoptosis of nerve cells and lead to the occurrence of AD [9].

Current studies have confirmed that the gut microbiota, immune system, intestinal mucosal barrier, blood–brain barrier, and the "microbial–gut–brain" axis of the human body all change during aging, which further influences the composition of intestinal flora, metabolites, and the associated inflammatory state of the body [10,11]. The activated neuroimmune–inflammatory response further promotes Aβ production, which then induces and aggravates AD. The composition of intestinal flora in AD patients was obviously different from that of non-AD elderly people. At the phylum level, the abundance of *Bacteroidetes* increased while that of *actinomyces* decreased. At the genus level, the number of *Bacteroides* and *Ruminococcus* increased while that of *Clostridium trichoides* decreased. These results indicated that an increase in pro-inflammatory bacteria in AD patients, and the change in cognitive function may be closely linked to the peripheral systemic inflammation caused by intestinal pro-inflammatory bacteria [12]. Xu et al. found that yeast β-glucans, a type of polysaccharide, prominently shape the intestinal flora, produce SCFAs, and could be a novel dietary supplement to prevent early-stage pathologies associated with AD [13]. Reshaping the balance of intestinal flora and reducing the production of abnormal metabolites of intestinal flora, Aβ deposition, and Tau hyperphosphorylation could be a potential effective strategy to relieve AD symptoms.

Gastrodia elata Bl. (GE), an edible and medicinal plant, is used as a traditional functional food to improve sleep and relieve headaches, with a long history according to the *Compendium of Materia Medica*. People living in the Guizhou and Yunnan provinces of China treated GE as a food to cook soup. In 2019, the Chinese government officially announced that GE can be used as both medicine and food [14,15]. Previous studies have shown that the main components of GE, phenols and glycosides, could effectively improve the central nervous system damage and ameliorate the decline of memory and learning ability caused by AD, depression, and Parkinson's disease [16]. Notably, GE could positively modulate the gut microbiota and pathological changes in AD-related proteins to protect against cognitive impairment [17,18].

Microbial fermentation has been used to process food and traditional Chinese medicine for centuries in China with high efficiency, low toxicity, and safety features, and has played a vital role in the production and development of human society [19]. However, studies about the mitigative effect of fermented GE on AD have not been performed. Our previous study confirmed that FGE was associated with better improvement in an animal behavior test than the GE group in AD-like mice (Figure S1). Thus, this study aimed to clarify the anti-AD effects of FGE via regulating neurotransmitters and the gut microbiota in D-Gal/$AlCl_3$-induced AD-like mice.

2. Materials and Methods

2.1. Preparation of FGE and Experimental Animals

2.1.1. Preparation of FGE

GE powder was mixed with sterile water at a ratio of 1:20 (*m*/*v*), and then, 8% white sugar, 0.02% cellulose, starter (0.3%, *Lactobacillus plantarum* PL15A (Thankcome Bio, Suzhou, China), *Acetobacter pasteurianus* AS141 (Yishui Jinrun Bio, Yishui, China), and *Saccharomyces cerevisiae* BV818 (Angel, Shenzhen, China) at a ratio of 1:1:1 (*w*/*w*/*w*)) were added. The mixture was cultured at 37 °C in a fermenter for 30 days, and the liquid supernatant was boiled and centrifuged at 4000 rpm for 5 min to remove the residue.

2.1.2. Animal Experiments

SPF-grade KM mice (male, 25 ± 2 g, 5–6 weeks) were provided by Liaoning Changsheng Biotechnology Co., Ltd. (Changchun, China, SCXK-2015-0001). The mice were housed in a standard animal feeding center at 23–25 °C with 55–60% humidity and were fed adaptively for one week. All experimental procedures were approved by the Animal Care and Use Committee of the Guizhou Medical University (No. 2304037). The mice were randomly divided into six groups (n = 10): control, model, donepezil (1 mg/kg·bw), FGE-L (120 mg/kg·bw), FGE-M (240 mg/kg·bw), and FGE-H (720 mg/kg·bw) groups. The intragastric administration of D-galactose (D-Gal, 120 mg/kg·bw) and aluminum trichloride ($AlCl_3$, 40 mg/kg·bw) was conducted in the morning each day for seven weeks to establish an AD-like animal model. The donepezil and FGE groups were administered with the related samples (0.2 mL/d) in the afternoon. The control and model groups received an equal volume of normal saline. Intragastric drug administration was adopted, and the experiments were conducted for 42 days.

2.2. Behavioral Tests

2.2.1. Shuttle Box Test

The mice in each group were allowed to move freely in the shuttle box and given 2 min to adapt to the environment. Then, the mice were given a certain degree of light, electrical, and sound stimulation, with each cycle interval being 15 s, and trained for 30 cycles daily for five days. In the conditioned stimulus period, the mice escaped to the other side through the middle channel, which was recorded as an active escape. In the unconditioned stimulus period, the mice shuttled to the opposite atrial ventricle, which was a passive escape, and the numbers of active escapes were recorded. During the memory test, each mouse was placed in the left ventricle to face the wall and received 10 s of beeping sound and 10 s of foot stimulation, with a 15 s interval for each cycle, for a total of 30 cycles. After each mouse completed the test, 75% alcohol was used to clean the surfaces of the instrument to remove the odor of the previous mouse.

2.2.2. Open Field Test

The motion of mice was recorded with a camera for 5 min in an open-field device box, and the total movement distance of the mice was recorded. Similarly, the residual odor of mice in the box was cleaned with 75% alcohol to prevent interference by residual odors.

2.3. Determination of Neurotransmitters and Oxidative Stress Indexes

The hippocampus area of brain tissue was mixed with PBS buffer solution (w/v = 1:9), and then the mixture was homogenated by a tissue homogenizer and centrifuged for 15 min at 10,000 rpm at 4 °C to obtain the supernatant. The contents of Aβ, Tau, P-Tau, acetylcholine (Ach), acetylcholinesterase (AchE), acetylcholintransferase (ChAT), and malondialdehyde (MDA) and the activity of superoxide dismutase (SOD) in brain tissue were investigated using correlated assay kits according to the manufacturer's instructions.

2.4. H&E Staining

The fixed brain tissue was infiltrated with paraffin and sliced with a slicer, and then the paraffin sections were dewaxed with xylene solution. The following procedures were conducted for staining and dehydration: cleaning solution I for 10 min, cleaning solution II for 10 min, cleaning solution III for 10 min; gradient alcohol dewaxing hydration: dehydration with 100% ethanol for 15 min, dehydration with 95% ethanol for 5 min, dehydration with 80% ethanol for 5 min, rinsing with double steaming water twice, 5 min/times, staining with hematoxylin for 6 min, rinsing with pure water, differentiating liquid for 30 s, returning to blue water for 10 min, staining with eosin for 2 min, and rinsing with pure water for 10 min. And then, ethanol dehydration, xylene transparency, and the sealed sheet were assessed.

2.5. Immunohistochemistry Analysis

The brain tissue was fixed in 4% paraformaldehyde for 3 h and dehydrated with alcohol and xylene. Xylene I, II, and III solutions were used for dewaxing for 10 min. Antigen repair solution (200 mL) was heated in a microwave oven, paraffin sections were added for 5 min, and the sections were washed with PBS thrice. The sample was allowed to inactivate after adding peroxidase and a serum sealer for 10 min at room temperature; the primary antibody was added and incubated overnight at 4 °C. Secondary antibodies were added for 10 min at room temperature after the samples were rewarmed. Antibiotin–peroxidase solution was added and incubated at room temperature for 10 min. Then, the DAB reaction was performed at room temperature for 5 min, and hematoxylin was added for 3–4 min, followed by PBS and washing with 95% ethanol for 10 min. The samples were immersed in xylene I, II, and III solutions for 5 min each and sealed.

2.6. Gut Microbiota DNA Extraction and Analysis of Gut Microbiota

Cecal contents were collected from mice for DNA extraction using an E.Z.N.A.® Soil DNA Kit (Omega Bio-tek, Norcross, GA, USA) according to the manufacturer's protocols. TBS-380 and NanoDrop2000 were applied to determine the concentration and purity of the extracted DNA from the intestinal canal, respectively. Then, Covaris M220 (Gene Company Limited, Hong Kong, China) was used to clip the DNA extract to an average size of approximately 400 bp for paired-end library construction. Paired-end sequencing was conducted on an Illumina Novaseq 6000 (Illumina Inc., San Diego, CA, USA) at Majorbio Bio-Pharm Technology Co., Ltd. (Shanghai, China).

2.7. UPLC-MS Analysis

The FGE was dissolved in HPLC-grade methanol and filtered through a 0.22 μm microporous membrane to perform LC-MS analysis. The LC-MS analysis was conducted in a UPLC-Q-Exactive MS (Thermo Scientific, Waltham, MA, USA) with a Waters Aaquity UPLC HSS T3 column (100 mm × 2.1 mm, 1.8 μm) by using gradient elution at a flow rate of 0.3 mL/min (5% B (0 min)—5% B (2 min)—95% B (42 min)—95% B (47 min)—5% B (47.1 min)—5% B (50 min); A: H_2O + 0.1% formic acid; B: acetonitrile + 0.1% formic acid). The temperature of the column oven was set at 60 °C and the injection volume at 10 μL. The fractions were analyzed in data-dependent acquisition (DDA) mode: The full MS survey scan was performed for 100 ms in the range of 100–1500 Da, and the gradient of collision energy was set as 35 eV. Compounds were identified by comparison with the MS spectral data from previously isolated compounds and the mzValut (Thermo Fisher Scientific, Waltham, MA, USA) database.

2.8. Statistical Analysis

Robust regression analysis was conducted using GraphPad Prism 8 (GraphPad Prism, La Jolla, CA, USA). One-way analysis of variance was used to analyze the differences between groups. All data were presented as mean ± SD. Tukey's post hoc tests were used to analyze the differences between groups. Statistical significance was set at $p < 0.05$.

3. Results

3.1. Identification of Main Components in FGE

UPLC-HR/MS was applied to identify the main components of FGE, and 67 compounds were identified as shown in Figure 1 and Table S1. Among them, 42 compounds were identified as phenolic compounds, including gastrodin, p-hydroxybenzyl alcohol, p-hydroxybenzoic acid, p-hydroxybenzaldehyde, parishin D, etc. Meanwhile, 16 organic acids were confirmed, including fumaric acid, citric acid, glucuronic acid, gluconic acid, etc.

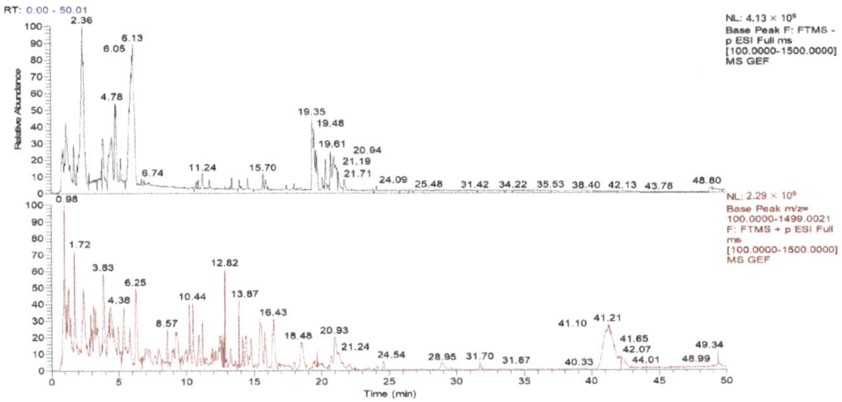

Figure 1. UPLC/Q-TOF-MS/MS chromatogram of the FGE in positive ionization mode.

3.2. The Effect of FGE on Cognitive Function in AD-like Mice

The shuttle box test is generally used to train the active conditioned reflex of the mouse brain, which could represent changes in the learning and memory ability of animals. As illustrated in Figure 2, the number of active escapes in the model group was significantly reduced compared with the control group ($p < 0.01$), indicating that an AD-like animal model was successfully established. After FGE administration, the number of active escapes was significantly higher than that in the model group in a dose-dependent manner ($p < 0.01$). These results indicated that the FGE groups could effectively improve the learning ability, memory loss, and inactive behavior of AD-like mice.

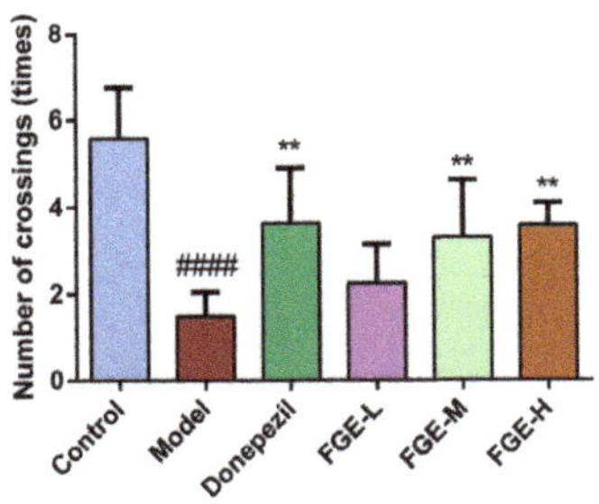

Figure 2. Effects of FGE on number of crossings of AD-like mice. FGE-L (120 mg/kg b. w.), FGE-M (240 mg/kg b. w.), and FGE-H (720 mg/kg b. w.). Data are expressed as means ± SD (n = 10) of six independent experiments. #### $p < 0.0001$ compared with control group, ** $p < 0.05$ compared with model group.

3.3. The Effect of FGE on Anxiety-like Behavior in AD-like Mice

The total distance of movement in an open-field experiment is an important factor for evaluating the memory ability and behavior activity of animals [20]. As demonstrated in Figure 3, the total movement distance was significantly reduced in the model group compared with that in the control group ($p < 0.01$), indicating that an AD-like animal model was successfully established. Notably, the total movement distance of mice in the donepezil and FGE groups was significantly increased compared with the model group ($p < 0.01$). Accordingly, administrating FGE could improve the behavioral activity and cognition of AD-like mice.

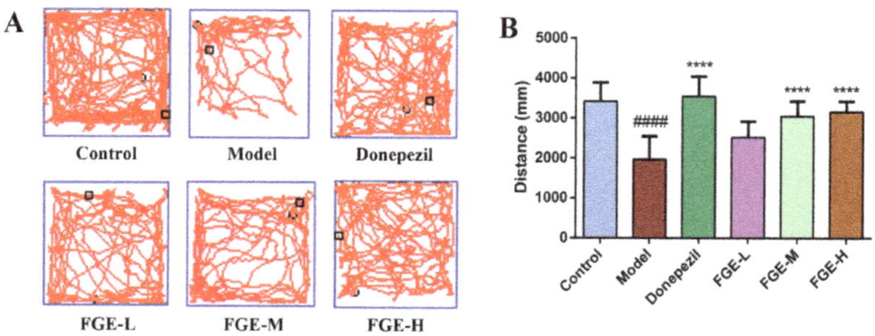

Figure 3. (**A**,**B**) Effects of FGE on total movement distance of AD-like mice. FGE-L (120 mg/kg b. w.), FGE-M (240 mg/kg b. w.), and FGE-H (720 mg/kg b. w.). Data are expressed as means ± SD (n = 10) of six independent experiments. #### $p < 0.0001$ compared with control group, **** $p < 0.0001$ compared with model group.

3.4. The Effect of FGE on the Levels of Neurotransmitters in Brain Tissue of AD-like Mice

The damage of neuronal cells in brain tissue is generally caused by an abnormal increase in Aβ, which leads to the misfolding and assembly of Tau proteins throughout the cortex, eventually leading to nervous system failure, neurodegeneration, and cognitive decline. As displayed in Figure 4, the contents of Aβ, Tau, P-Tau, and AchE were significantly increased, while Ach and ChAT sharply decreased in the model group compared to the control group ($p < 0.05$). Interestingly, treatment with FGE could significantly improve the secretion of Aβ, Tau, P-Tau, and AchE and increase Ach and ChAT contents in the brain tissue. FGE could positively regulate neurotransmitter contents in AD-like mice.

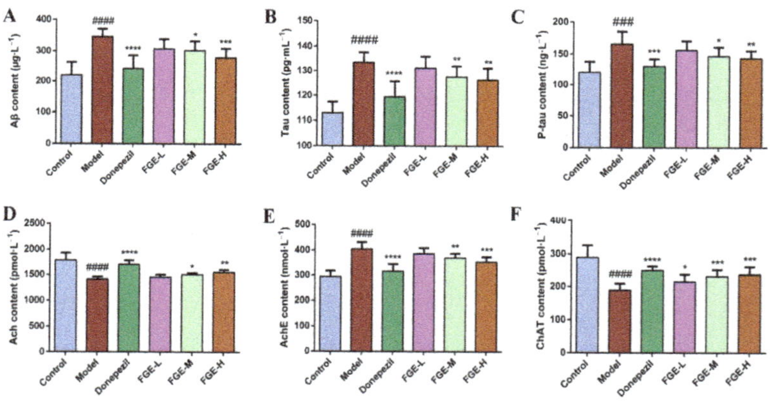

Figure 4. Effects of FGE on the production of neurotransmitters in the brain tissue of AD-like mice. (**A**) Aβ; (**B**) Tau; (**C**) P-Tau; (**D**) Ach; (**E**) AchE; (**F**) ChAT; FGE-L (120 mg/kg b. w.), FGE-M (240 mg/kg b. w.), and FGE-H (720 mg/kg b. w.). Data are expressed as means ± SD (n = 10) of six independent experiments. ### $p < 0.001$, #### $p < 0.0001$ compared with control group; * $p < 0.01$, ** $p < 0.05$, *** $p < 0.001$, **** $p < 0.0001$ compared with model group.

3.5. The Effect of FGE on the Content of Antioxidants in AD-like Mice

Oxidative stress (OS) is a series of adaptive reactions caused by an imbalance between reactive oxygen species and the antioxidant system of the body [21]. It interferes with the normal redox state of cells and damages proteins, lipids, and nucleic acids. The imbalance of OS drives AD, PD, atherosclerosis, heart failure, and cancer. As depicted in Figure 5, the MDA content was significantly increased ($p < 0.01$) and the SOD content was decreased in the model group compared to the control group. Compared with the model group, the levels of MDA and SOD were decreased and increased in the brain tissue in a dose-dependent manner in the donepezil and FGE groups.

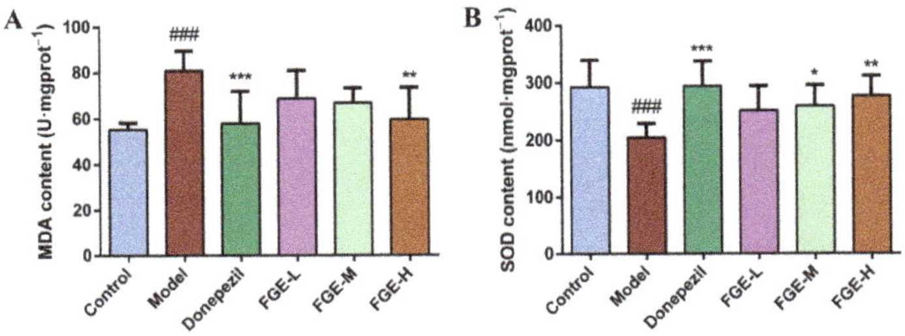

Figure 5. Effects of FGE on the levels of antioxidants in the brain tissue of AD-like mice. (**A**) MDA; (**B**) SOD; FGE-L (120 mg/kg b. w.), FGE-M (240 mg/kg b. w.), and FGE-H (720 mg/kg b. w.). Data are expressed as means ± SD (n = 10) of six independent experiments. ### $p < 0.001$ compared with control group; * $p < 0.05$, ** $p < 0.05$, *** $p < 0.001$ compared with model group.

3.6. The Effect of FGE on the Damage of Neuronal Cells in the Hippocampus of AD-like Mice

As displayed in Figure 6, the number of nerve cells was reduced, the arrangement was irregular, the integrity of the pyramidal cell layer was lost, and the cell body and nucleus were shrunken in the model group compared with the control group. Treatment with FGE improved neuronal cell damage, decreased cell shrinkage, improved cell morphology, increased cell layers and neat cell arrangement, and significantly increased cytoplasm and nucleus clarity compared with the model group. Accordingly, FGE could improve the symptoms of AD-like mice by protecting nerve cells in the brain tissue.

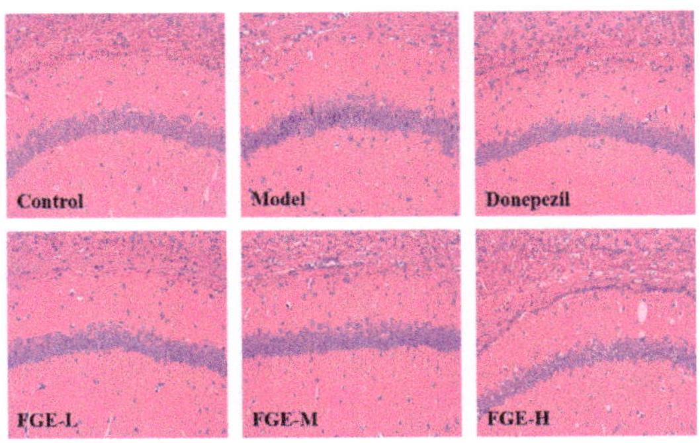

Figure 6. Effects of FGE on hippocampal neurons in CA1 of AD-like mice. Sections were stained with hematoxylin and eosin and observed using Leica microscope (n = 3). FGE-L (120 mg/kg b. w.), FGE-M (240 mg/kg b. w.), and FGE-H (720 mg/kg b. w.).

3.7. The Effect of FGE on the Protein Expression of Aβ in the Hippocampus of AD-like Mice

Aβ, a type of amyloid protein in the body, has an important effect on cognition, learning, and memory, which is the main contributing factor to AD pathogenesis. Aβ deposition was significantly increased in the model group compared with the control group, indicating that the production and degradation system of Aβ in the brain was disturbed, causing the brain Aβ not to be degraded. Remarkably, Aβ accumulation in the hippocampus of the brain tissue was significantly reduced in FGE-treated groups compared with the model group, indicating that FGE could improve AD symptoms by degrading the content of Aβ in brain tissue (Figure 7).

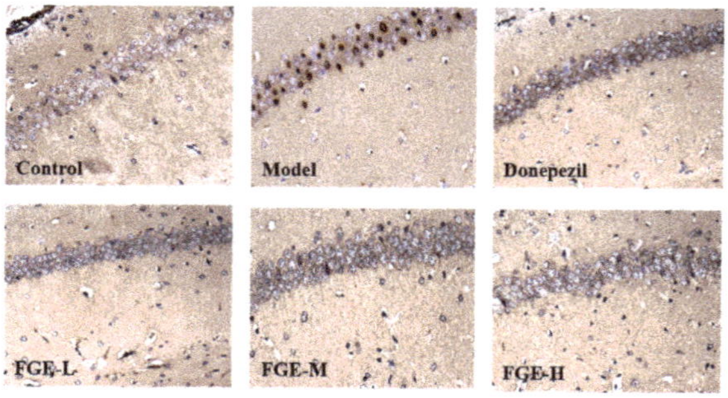

Figure 7. Effects of FGE on the content of Aβ in the brain tissue of AD-like mice (n = 3). FGE-L (120 mg/kg b. w.), FGE-M (240 mg/kg b. w.), FGE-H (720 mg/kg b. w.).

3.8. The Effect of FGE on the Abundance of Intestinal Flora in AD-like Mice

To investigate the effect of FGE on the gut microbiota composition, the 16S rRNA gene sequences of 18 samples from six groups were sequenced using Majorbio HiSeq (Shanghai Majorbio Biomedical Technology Co., Ltd., Shanghai, China). The taxonomic composition at the genus level was characterized by operational taxonomic units (OTUs). At the genus level, the Venn diagram indicated that there were 2504 shared OTUs among the six groups and 134, 22, 24, 16, 9, and 15 unique OTUs in the control, model, donepezil, FGE-L, FGE-M, and FGE-H groups, respectively (Figure 8A). Alpha diversity estimations and PCA analyses indicated that the intestinal microbiota was separated in these six groups (Figure 8B–F). After treatment with FGE, the gut microbiota was similar to that in the control group.

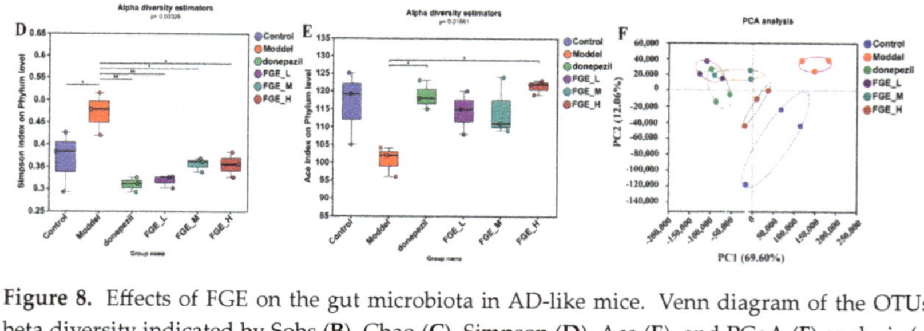

Figure 8. Effects of FGE on the gut microbiota in AD-like mice. Venn diagram of the OTUs (**A**); beta diversity indicated by Sobs (**B**), Chao (**C**), Simpson (**D**), Ace (E), and PCoA (**F**) analysis (n = 3); * $p < 0.05$, ** $p < 0.05$, compared with model group.

At the phylum level, Bacteroidetes, Firmicutes, Bacteria, Proteobacteria, and Parabasalia were the most abundant, whereas the abundance of Firmicutes and Bacteroidetes was markedly decreased and increased in the model group compared with the control group (Figure 9A,B). Intriguingly, FGE treatment could increase the abundance of Firmicutes, and decrease the abundance of Bacteroidetes. The imbalance of the two phyla can affect glucose metabolism in the brain tissue, and abnormal glucose metabolism can lead to increased Aβ deposition, which ultimately increases the risk of AD. At the species level, the abundance of *Muribaculaceae*, *Rikenellaceae*, and *Bacteroides* was obviously increased, and the abundance of *Lachnospiraceae* and *Prevotella* sp. was decreased in the model group compared with the control group (Figure 10). Prominently, treatment with FGE could improve the abundance of *Lachnospiraceae* and decrease the abundance of *Muribaculaceae* and Bacteroidaceae compared with the model group. The present study confirmed that the abundance of *Lachnospiraceae* was decreased in the intestinal canal of patients with AD, indicating that treatment with FGE could ameliorate the symptoms of AD.

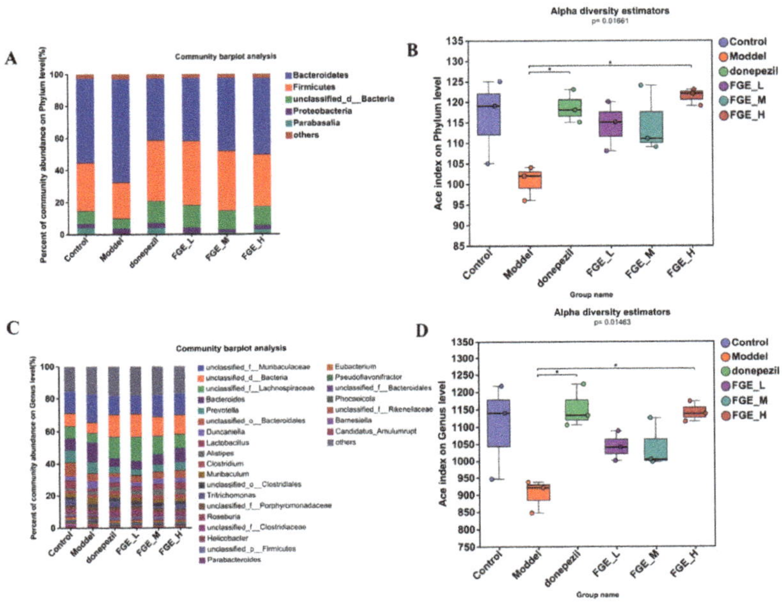

Figure 9. Relative abundance of bacteria at the phylum and genus levels. Histogram of relative abundance at the phylum level (**A**); Alpha diversity determined by Ace diversity index at the phylum level (**B**); Histogram of relative abundance at the genus level (**C**); Alpha diversity determined by Ace diversity index at the genus level (**D**). One-way ANOVA with LSD *t*-test was applied to evaluate significant differences between the groups; * $p < 0.05$ ($n = 3$).

3.9. FGE Regulates the Imbalance of Amino Acid Metabolism

The intestinal flora is regulated by host genes, immune response, and dietary factors, and its metabolic and immune potential determines its importance in host health and disease [22]. Therefore, targeting the gut microbiota and its associated metabolic pathways is an effective strategy for treating diseases. In this study, the metabolic pathways in the six groups were analyzed using the KEGG database. As demonstrated in Figure 11A, metabolism, genetic information processing, and environmental information processing contributed to the top three level-1 pathways in these six groups. Further analysis of the KEGG secondary metabolic map indicated that the global and overview maps, carbohydrate metabolism, and amino acid metabolism are the top 3 metabolic functions among the 38 metabolic functions. As demonstrated in Figure 11B, the greatest difference in a metabolic pathway is amino acid metabolism among the six groups. AD is a multi-mechanism metabolic disorder closely linked to energy metabolism, in which amino acid metabolism plays an important role. Moreover, studies have found that amino acid levels have been identified as potential biomarkers for patients with AD. Consequently, we speculate that FGE may improve AD by intervening in the amino acid metabolism.

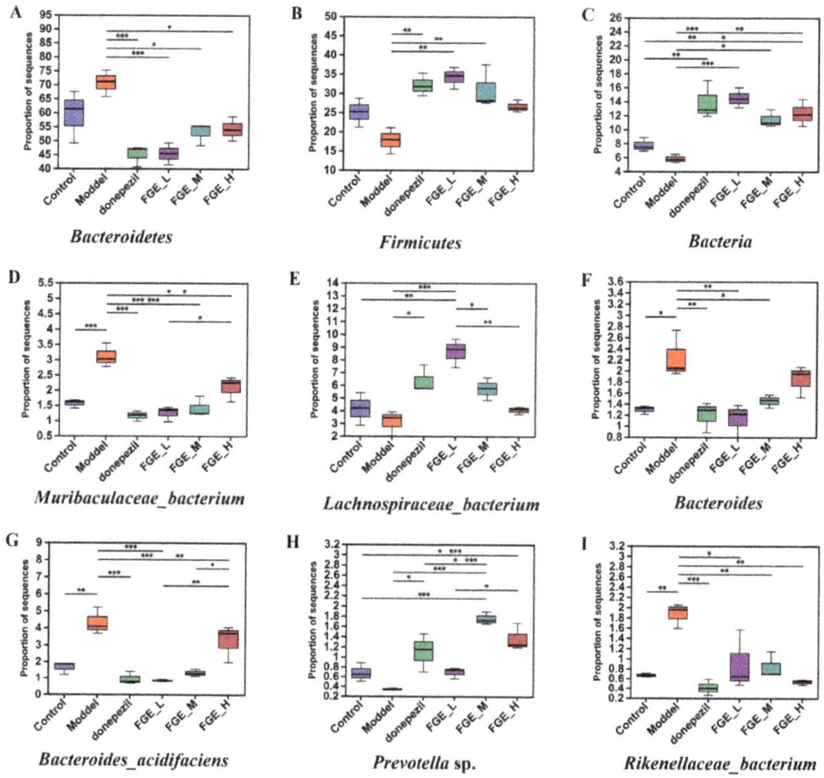

Figure 10. Abundance of bacterial taxa at different taxonomic levels (n = 3). (**A**) Phylum Bacteroidetes; (**B**) phylum Firmicutes; (**C**) phylum Bacteria; (**D**) species *Muribaculaceae_bacterium;* (**E**) species *Lachnospiraceae_bacterium;* (**F**) species *Bacteroides;* (**G**) species *Bacteroides_acidifaciens;* (**H**) species *Prevotella* sp.; (**I**) species *Rikenellaceae_bacterium*. * $p < 0.05$, ** $p < 0.05$, *** $p < 0.001$ compared with model group.

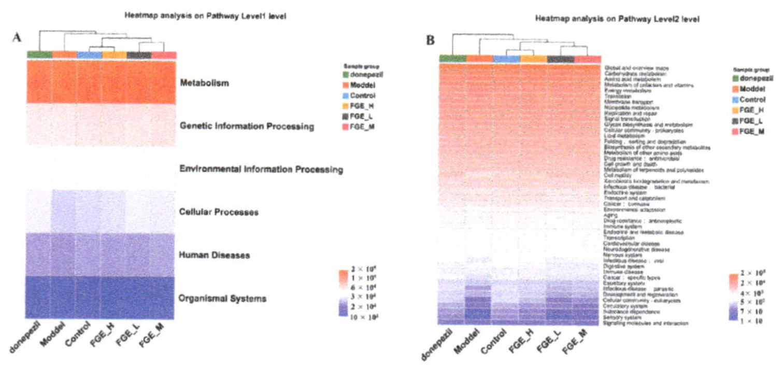

Figure 11. Heat map describing the primary and secondary metabolic function of intestinal microorganisms in AD-like mice (n = 3).

3.10. The Effect of FGE Treatment on Neurotransmitters by Regulating Gut Microbiota

Amino acid metabolism, the most varied metabolic pathway, is linked to neurotransmitter and oxidative stress levels. As illustrated in Figure 12A, the intestinal flora of mice in different groups was significantly different at the genus level. The red and blue colors indicate positive and negative correlations with the relevant indicators, respectively. Lactobacillus, *Rikenellaceae*, Parabacteroides, Eubacterium, Parasutterella, and *Rikenellaceae* were significantly correlated with Aβ, Tau, P-Tau, Ach, AchE, ChAT, SOD, and MDA contents. Consequently, the structure of the intestinal flora has an impact on the level of neurotransmitters and oxidative stress, indicating that treatment with FGE could regulate the brain–nervous system by maintaining the intestinal flora structure of the body to improve the effect of AD.

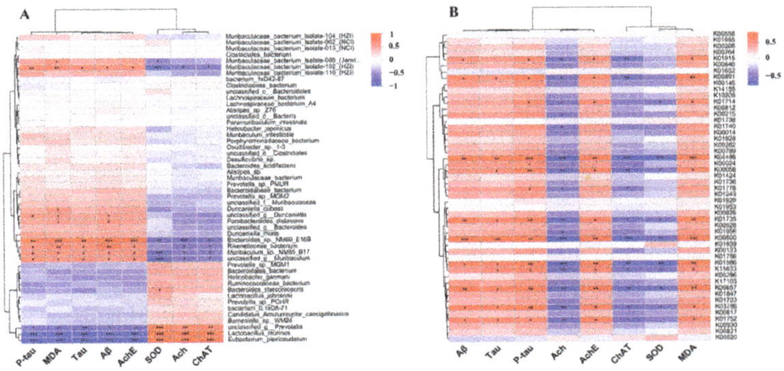

Figure 12. Heat map describing the correlation of the abundances of key bacterial genera at species levels and AD-related parameters (**A**); the correlation of the amino acid metabolic pathways and AD-related parameters (**B**). * $p < 0.05$, ** $p < 0.01$, *** $p < 0.001$.

3.11. Correlation Analysis between Amino Acid Metabolism and Oxidative Stress Level in AD-like Mice

Amino acid metabolism is the most abundant differential metabolic pathway involved in the synthesis of proteins, including neurotransmitters. Our results indicated that the metabolic pathways of K00891, K04486, K00058, K01735, K00800, K01586, K00657, K03786, and K01752 are crucial for the synthesis of neurotransmitters and OS-related indicators (Figure 12B). Further studies have indicated that the overactive biosynthesis pathway of phenylalanine, tyrosine, and tryptophan could lead to AD development and exacerbation. Phenylalanine could be converted into hippuric acid (HA) by an imbalance of intestinal flora, and excessive HA causes neuroinflammation and affects the level of neurotransmitters and oxidative stress in the body. Remarkably, FGE could improve AD by interfering with the HA synthesis by regulating the amino acid pathway.

4. Discussion

Phenolic compounds are the main active ingredients in GE, especially gastrodin, which has been widely used and studied, and has obvious application effects in neurological diseases such as stroke, epilepsy, and AD [23]. In our study, only parishin D was identified from FGE, indicating that the microbial fermentation could degrade the parishin to gastrodin and then increase its clinical efficacy for AD [24]. In addition, microbial fermentation can improve the taste and flavor of products, such as ethyl succinate, a flavoring agent, which was identified from FGE and is mainly produced by *Saccharomyces cerevisiae*. The dynamic changes in flavor components, including organic acids, alcohol, lipid, and ether organic acids, during the fermentation have been studied previously [25].

The unbalanced production and clearance of Aβ are the initial factors in neuronal degeneration and dementia. The plaques formed between brain neurons due to abnormal levels of Aβ protein are neurotoxic and lead to neuronal degeneration. Aβ, a large molecule of protein from the fatty membrane of nerve cells, has sticky chemical properties and is a major component of amyloid plaques [26]. Certain types of Aβ, such as Aβ38/40/42, coagulate to form small clumps of protein that accumulate to form plaques. This sick Aβ can infect normal Aβ, thereby increasing plaque formation. Tau protein is a microtubule-related protein, and one of its main functions is to maintain the stability of axonal microtubules and ensure normal brain function. In our study, FGE treatment could significantly inhibit Aβ deposition in the hippocampal region by protecting the mechanism of Aβ movement in neurons.

The defective and hyperphosphorylated Tau protein can bind to microtubules and cause neurofibrillary tangles. In summary, the formation of Aβ plaques and neurofibrillary tangles of Tau proteins accelerates damage to healthy neurons. The deposition of Aβ plaque can trigger a series of chain reactions, which lead to the misfolding and assembly of Tau protein in the cell and then spread the disease to the entire nerve circuit and cortex, eventually leading to nervous system failure and cognitive decline [27].

The cholinergic hypothesis is the first hypothesis that attempts to explain AD pathogenesis. It is believed that AD occurrence is due to defects in neurotransmitters in the brains of patients with AD, resulting in damage to cholinergic neurons. This hypothesis suggests that a decrease in AChE and ChAT activity is the main reason for the decrease in Ach concentration and cholinergic activity [28]. The acetylcholine receptor antagonist scopolamine could cause cognitive dysfunction, increase the production of Aβ, and inhibit the activity of α-secreting enzyme [29]. Interestingly, we found that FGE positively regulated neurotransmitter levels in the brain tissue of AD-like mice, such as Aβ, Tau, P-Tau, Ach, AchE, and ChAT [30].

Studies have found that many diseases, such as cerebral ischemia, epilepsy, schizophrenia, PD, and AD neuronal degeneration, are closely associated with excessive free radicals [31]. An increased degree of OS results from the production of a large number of lipid peroxides, such as MDA, originating from the oxidation of intracellular lipids [32]. SOD is an important antioxidant enzyme that removes superoxide anion free radicals in organisms, which can protect cells from the damage caused by oxygen free radicals. SOD and MDA, as important indexes to evaluate the antioxidant capacity and oxidative capacity of oxidative stress, respectively, are jointly applied in the field of disease research [33]. Amin et al. reported that SOD and MDA could be markers of OS to study the association of benzene exposure and insulin resistance in children [34]. In this study, we found that FGE treatment could reduce the MDA levels and increase the SOD content, indicating that FGE could regulate the imbalance of OS in the body. The "microbial–gut–brain" axis is a two-way communication pathway between the gut and the brain conducting neural, immune, endocrine, and metabolic signals [35,36]. Abnormal composition of the gut microbiota has been found in several neuropsychiatric disorders, such as autism spectrum disorders, depression, PD, and AD. Recent studies have disclosed that intestinal flora plays a key role in regulating intestinal and brain functions [37]. Exploring the relationship between intestinal flora and AD can provide new strategies and ideas for preventing and treating AD. Ferreiro et al. found that disturbances in the gut microbiota occur in the preclinical phase of AD and are closely linked to the formation of pathological markers of Aβ and Tau proteins [38]. *Firmicutes* play an important role in host intestinal homeostasis through short-chain fatty acid synthesis, normalizing intestinal permeability, and participating in brain–gut axis regulation [39]. *Lactobacillus*, a Gram-positive bacterium belonging to Firmicutes, could produce acetate, and lactic acid, gamma-aminobutyric acid and convert tryptophan metabolism into neurotransmitters such as 5-HT [40]. Studies have found that low levels of 5-HT can affect the signaling of other neurotransmitters in brain tissue and increase the accumulation of Aβ plaques in the brain [41]. Interestingly, the proportion of *Lactobacillus* in intestinal flora sharply increased after treatment with FGE compared with the model group. In addition, our study found that the abundance of *Lachnospira* markedly increased in the FGE groups. Studies have found that *Lachnospira* is a potentially beneficial bacterium in the human gut, which is involved in the metabolism of various carbohydrates and is the main source of energy for the host [42]. Moreover, the increased abundance of *Lachnospira* could promote the growth of *Bacteroides* in the intestinal canal. Chen et al. found that the levels of *Bacteroides* and *Lachnospira* were reduced, and those of *Prevotella* were increased in patients with AD compared with healthy controls, indicating that FGE could restore intestinal homeostasis [43].

Changes in the composition of gut microbes are associated with host physiology and human diseases, such as diabetes, heart disease, and depression, through excessive production of short-chain fatty acids such as butyric acid, propionic acid, succinic acid, and acetic acid which affect the host metabolism [44,45]. Cui et al. reported that the glutamine/glutamate and glycine/serine/threonine metabolism pathways were abnormally altered in the serum of patients with AD [46].

The phenylalanine, tyrosine, and tryptophan metabolic pathways have a positive contribution in regulating neurotransmitters in serum, such as Aβ, Tau, P-Tau, Ach, AchE, ChAT, SOD, and MDA, which were confirmed in this study. Moreover, abnormal levels of phenylalanine could be converted into hippuric acid (HA) by intestinal microorganisms, and excessive HA causes neuroinflammation and affects the level of neurotransmitters and oxidative stress in the body. Tryptophan metabolism is regulated by intestinal microorganisms, and its metabolites possess immune, metabolic, and neuroregulatory functions, which have become therapeutic targets for various diseases. Trp metabolism has been implicated in several neurodegenerative diseases, including Huntington's disease (HD), AD, amyotrophic lateral sclerosis (ALS), and PD. Trp metabolism, a potential mechanism of neurodegeneration, includes excitatory toxicity through TRP-dependent protein toxicity, accumulation of Trp metabolites, and energy imbalance due to NAD+ depletion [47].

5. Conclusions

In this study, we found that FGE improved behavioral cognitive deficits in AD-like mice treated with GE powder and could ameliorate the damage to neuronal cells in the hippocampus by positively regulating neurotransmitters such as Aβ, Tau, P-Tau, AchE, Ach, and ChAT. Moreover, analysis of the gut microbiota composition found that FGE could enhance the abundance of probiotics, such as *Lachnospira* and *Lactobacillus*, which were correlated with the production of neurotransmitters. Furthermore, we confirmed that the amino acid metabolism pathway is the most differential metabolic pathway that is closely correlated with AD development. As a result, microbial-based fermentation can be a potential strategy for treating GE to achieve greater medicinal value. In addition, FGE could be a new potential source for treating AD by regulating the gut microbiota.

Supplementary Materials: The following supporting information can be downloaded at: https://www.mdpi.com/article/10.3390/foods13132154/s1, Figure S1: Effects of GE powder and FGE on number of crossings of AD-like mice; Figure S2: Effects of GE powder and FGE on total movement distance of AD-like mice; Figure S3: LDA Effect Size analysis of community differences between groups; Table S1: Chemical composition identification of FGE using HR-LC/MS in positive mode.

Author Contributions: Conceptualization, Y.W. and C.X.; methodology, M.Z. and J.Y.; software, M.Z. and F.C.; validation, L.L. (Ling Lin), Q.L. and L.L. (Liangqun Li); formal analysis, Y.W.; investigation, M.Z. and L.L. (Ling Lin); resources, Y.W.; writing—original draft preparation, M.G.; writing—review and editing, J.Y. and M.G.; supervision, X.Y. and C.X.; funding acquisition, M.G., L.L. (Lilang Li) and X.Y. All authors have read and agreed to the published version of the manuscript.

Funding: This research was funded by the Science and Technology Projects of Guizhou Province (QKHJC-ZK [2024]YB639, QKHCG [2024]YB087), the National Key Research and Development Program of China (No. 2021YFD1601001), the Science and Technology Projects of Guizhou Province (No. QKHCG [2020]4013), and the Guizhou Province Innovation Capability Improvement Platform (No. 2020-018).

Institutional Review Board Statement: All experimental procedures were approved by the Animal Care and Use Committee of the Guizhou Medical University (No. 2304037).

Informed Consent Statement: Not applicable.

Data Availability Statement: The original contributions presented in the study are included in the article/Supplementary Materials, further inquiries can be directed to the corresponding author.

Conflicts of Interest: The authors declare no conflicts of interest.

References

1. Scheltens, P.; Strooper, B.D.; Kivipelto, M.; Holstege, H.; Chételat, G.; Teunissen, C.E.; Cummings, J.; van der Flier, W.M. Alzheimer's disease. *Lancet* **2021**, *397*, 1577–1590. [CrossRef] [PubMed]
2. Breijyeh, Z.; Karaman, R. Comprehensive Review on Alzheimer's Disease: Causes and Treatment. *Molecules* **2020**, *25*, 5789–5816. [CrossRef]
3. Ren, R.; Qi, J.; Lin, S.; Liu, X.; Yin, P.; Wang, Z.; Tang, R.; Wang, J.; Huang, Q.; Li, J.; et al. The China Alzheimer Report 2022. *Gen. Psychiatry* **2022**, *35*, 100751. [CrossRef]
4. Sola-Sevilla, N.; Puerta, E. SIRT2 as a potential new therapeutic target for Alzheimer's disease. *Neural Regen. Res.* **2024**, *19*, 124–131. [CrossRef] [PubMed]
5. Kim, Y.; Cho, M.; Jang, C.H.; Lee, J.S.; Kim, J.-S.; Oh, J.; Lim, J. Oral Administration of Euonymus alatus Leaf Extract Ameliorates Alzheimer's Disease Phenotypes in 5xFAD Transgenic Mice. *Foods* **2024**, *13*, 682–693. [CrossRef] [PubMed]
6. Johnson, E.C.B.; Dammer, E.B.; Duong, D.M.; Ping, L.; Zhou, M.; Yin, L.; Higginbotham, L.A.; Guajardo, A.; White, B.; Troncoso, J.C.; et al. Large-scale proteomic analysis of Alzheimer's disease brain and cerebrospinal fluid reveals early changes in energy metabolism associated with microglia and astrocyte activation. *Nat. Med.* **2020**, *26*, 769–780. [CrossRef] [PubMed]
7. Jin, M.; Wang, S.; Gao, X.; Zou, Z.; Hirotsune, S.; Sun, L. Pathological and physiological functional cross-talks of α-synuclein and tau in the central nervous system. *Neural Regen. Res.* **2024**, *19*, 855–862. [CrossRef]
8. Kuragano, M.; Yoshinari, W.; Lin, X.; Shimamori, K.; Uwai, K.; Tokuraku, K. Evaluation of Amyloid β42 Aggregation Inhibitory Activity of Commercial Dressings by A Microliter-Scale High-Throughput Screening System Using Quantum-Dot Nanoprobes. *Foods* **2020**, *9*, 825–836. [CrossRef]
9. Tracy, T.E.; Madero-Pérez, J.; Swaney, D.L.; Chang, T.S.; Moritz, M.; Konrad, C.; Ward, M.E.; Stevenson, E.; Hüttenhain, R.; Kauwe, G.; et al. Tau interactome maps synaptic and mitochondrial processes associated with neurodegeneration. *Cell* **2022**, *185*, 712–728. [CrossRef] [PubMed]
10. Cheng, L.; Chou, P.; Hou, A.; Huang, C.; Shiu, W.; Wang, S. Lactobacillus paracasei PS23 improves cognitive deficits via modulating the hippocampal gene expression and the gut microbiota in d-galactose-induced aging mice. *Food Funct.* **2022**, *13*, 5240–5251. [CrossRef] [PubMed]
11. Verhaar, B.J.H.; Hendriksen, H.M.A.; de Leeuw, F.A.; Doorduijn, A.S.; van Leeuwenstijn, M.; Teunissen, C.E.; Barkhof, F.; Scheltens, P.; Kraaij, R.; van Duijn, C.M.; et al. Gut Microbiota Composition Is Related to AD Pathology. *Front. Immunol.* **2022**, *12*, 794519. [CrossRef] [PubMed]
12. He, Y.; Li, B.; Sun, D.; Chen, S. Gut Microbiota: Implications in Alzheimer's Disease. *J. Clin. Med.* **2020**, *9*, 2042–2059. [CrossRef] [PubMed]
13. Xu, M.; Mo, X.; Huang, H.; Chen, X.; Liu, H.; Peng, Z.; Chen, L.; Rong, S.; Yang, W.; Xu, S.; et al. Yeast β-glucan alleviates cognitive deficit by regulating gut microbiota and metabolites in Aβ1–42-induced AD-like mice. *Int. J. Biol. Macromol.* **2020**, *161*, 258–270. [CrossRef] [PubMed]
14. Chen, P.; Sheen, L. Gastrodiae Rhizoma (天麻 tiān má): A review of biological activity and antidepressant mechanisms. *J. Tradit. Complement. Med.* **2011**, *1*, 31–40. [CrossRef]
15. Guan, J.; Chen, Z.; Guo, L.; Wang, C.; Xu, T.; Wan, F.; Zhou, T.; Cui, X.; Yang, Y. Evaluation of how sulfur-fumigation reduces the edible quality and flavor of *Gastrodia elata* Blume Rhizoma. *LWT-Food Sci. Technol.* **2023**, *187*, 115296. [CrossRef]
16. Fasina, O.B.; Wang, J.; Mo, J.; Osada, H.; Ohno, H.; Pan, W.; Xiang, L.; Qi, J. Gastrodin from *Gastrodia elata* Enhances Cognitive Function and Neuroprotection of AD Mice via the Regulation of Gut Microbiota Composition and Inhibition of Neuron Inflammation. *Front. Pharmacol.* **2022**, *13*, 814271. [CrossRef] [PubMed]
17. Zhao, W.; Wang, J.; Latta, M.; Wang, C.; Liu, Y.; Ma, W.; Zhou, Z.; Hu, S.; Chen, P.; Liu, Y. Rhizoma Gastrodiae Water Extract Modulates the Gut Microbiota and Pathological Changes of P-TauThr231 to Protect against Cognitive Impairment in Mice. *Front. Pharmacol.* **2022**, *13*, 903659. [CrossRef] [PubMed]
18. Lin, H.; Wu, S.; Weng, Z.; Wang, H.; Shi, R.; Tian, M.; Wang, Y.; He, H.; Wang, Y.; Liu, X.; et al. Tianma Formula Alleviates Dementia via ACER2-Mediated Sphingolipid Signaling Pathway Involving Aβ. *Evid. -Based Complement. Altern. Med.* **2021**, *2021*, 6029237. [CrossRef] [PubMed]
19. Zheng, X.; Qiu, J.; Zhang, H.; Gao, N.; Jiang, T.; Gong, Y.; Zhang, W.; Li, Z.; Feng, X.; Hong, Z. PD184352 exerts anti-inflammatory and antioxidant effects by promoting activation of the Nrf2/HO-1 axis. *Biochem Pharmacol* **2023**, *211*, 115542. [CrossRef] [PubMed]
20. Yang, Y.; Chen, W.; Wang, X.; Ge, W. Impact of mitochondrial aldehyde dehydrogenase 2 on cognitive impairment in the AD model mouse. *Acta Biochim. Biophys. Sin.* **2021**, *53*, 837–847. [CrossRef] [PubMed]
21. Davalli, P.; Mitic, T.; Caporali, A.; Lauriola, A.; D'Arca, D. ROS, Cell Senescence, and Novel Molecular Mechanisms in Aging and Age-Related Diseases. *Oxidative Med. Cell. Longev.* **2016**, *2016*, 3565127. [CrossRef]
22. Wu, J.; Wang, K.; Wang, X.; Pang, Y.; Jiang, C. The role of the gut microbiome and its metabolites in metabolic diseases. *Protein Cell* **2020**, *12*, 360–373. [CrossRef] [PubMed]

23. Zhou, H.; Lu, S.; Yu, Z.; Zhang, J.; Mei, Z. Mechanisms for the biological activity of *Gastrodia elata* Blume and its constituents: A comprehensive review on sedative-hypnotic, and antidepressant properties. *Phytomedicine* **2024**, *123*, 155251. [CrossRef] [PubMed]
24. Li, Y.; Liu, X.; Liu, S.; Liu, D.; Wang, X.; Wang, Z. Transformation Mechanisms of Chemical Ingredients in Steaming Process of *Gastrodia elata* Blume. *Molecules* **2019**, *24*, 3159–3172. [CrossRef]
25. Zhao, M.; Wang, Y.; Li, L.; An, Z.; Xie, C.; Lin, L.; Yang, X.; Yang, J. Dynamic study on flavor components and antioxidant activity in the fermentation of *Gastrodia elata*. *Food Ferment. Ind.* **2021**, *47*, 92–98.
26. Cacciamani, F.; Sambati, L.; Houot, M.; Habert, M.O.; Dubois, B.; Epelbaum, S. Reduced awareness of cognitive decline is associated with brain amyloid load in asymptomatic elderly individuals at risk for AD: A longitudinal study. *Alzheimer's Dement.* **2020**, *16*, 037138. [CrossRef]
27. Roda, A.R.; Montoliu-Gaya, L.; Serra-Mir, G.; Villegas, S. Both Amyloid-β Peptide and Tau Protein Are Affected by an Anti-Amyloid-β Antibody Fragment in Elderly 3xTg-AD Mice. *Int. J. Mol. Sci.* **2020**, *21*, 6630–6652. [CrossRef] [PubMed]
28. Hampel, H.; Mesulam, M.M.; Cuello, A.C.; Khachaturian, A.S.; Vergallo, A.; Farlow, M.R.; Snyder, P.J.; Giacobini, E.; Khachaturian, Z.S. Revisiting the Cholinergic Hypothesis in Alzheimer's Disease: Emerging Evidence from Translational and Clinical Research. *J. Prev. Alzheimer's Dis.* **2018**, *6*, 1–14. [CrossRef]
29. Liskowsky, W.; Schliebs, R. Muscarinic acetylcholine receptor inhibition in transgenic Alzheimer-like Tg2576 mice by scopolamine favours the amyloidogenic route of processing of amyloid precursor protein. *Int. J. Dev. Neurosci.* **2006**, *24*, 149–156. [CrossRef] [PubMed]
30. Snowden, S.G.; Ebshiana, A.A.; Hye, A.; Pletnikova, O.; O'Brien, R.; Yang, A.; Troncoso, J.; Legido-Quigley, C.; Thambisetty, M. Neurotransmitter Imbalance in the Brain and Alzheimer's Disease Pathology. *J. Alzheimer's Dis.* **2019**, *72*, 35–43. [CrossRef]
31. Reddy, V.P. Oxidative Stress in Health and Disease. *Biomedicines* **2023**, *11*, 2925–2941. [CrossRef]
32. Ohira, H.; Oikawa, D.; Kurokawa, Y.; Aoki, Y.; Omura, A.; Kiyomoto, K.; Nakagawa, W.; Mamoto, R.; Fujioka, Y.; Nakayama, T. Suppression of colonic oxidative stress caused by chronic ethanol administration and attenuation of ethanol-induced colitis and gut leakiness by oral administration of sesaminol in mice. *Food Funct.* **2022**, *13*, 9285–9298. [CrossRef] [PubMed]
33. Liguori, I.; Russo, G.; Curcio, F.; Bulli, G.; Aran, L.; Della-Morte, D.; Gargiulo, G.; Testa, G.; Cacciatore, F.; Bonaduce, D.; et al. Oxidative stress, aging, and diseases. *Clin. Interv. Aging* **2018**, *13*, 757–772. [CrossRef]
34. Amin, M.M.; Rafiei, N.; Poursafa, P.; Ebrahimpour, K.; Mozafarian, N.; Shoshtari-Yeganeh, B.; Hashemi, M.; Kelishadi, R. Association of benzene exposure with insulin resistance, SOD, and MDA as markers of oxidative stress in children and adolescents. *Environ. Sci. Pollut. Res.* **2018**, *25*, 34046–34052. [CrossRef]
35. Mohajeri, M. Brain Aging and Gut–Brain Axis. *Nutrients* **2019**, *11*, 424–426. [CrossRef] [PubMed]
36. Zhao, L.; Li, D.; Chitrakar, B.; Li, C.; Zhang, N.; Zhang, S.; Wang, X.; Wang, M.; Tian, H.; Luo, Y. Study on Lactiplantibacillus plantarum R6-3 from Sayram Ketteki to prevent chronic unpredictable mild stress-induced depression in mice through the microbiota–gut–brain axis. *Food Funct.* **2023**, *14*, 3304–3318. [CrossRef]
37. Banc, R.; Rusu, M.E.; Filip, L.; Popa, D.-S. The Impact of Ellagitannins and Their Metabolites through Gut Microbiome on the Gut Health and Brain Wellness within the Gut–Brain Axis. *Foods* **2023**, *12*, 270–310. [CrossRef]
38. Feyissa, A.M.; Chandran, A.; Stockmeier, C.A.; Karolewicz, B. Reduced levels of NR2A and NR2B subunits of NMDA receptor and PSD-95 in the prefrontal cortex in major depression. *Prog. Neuro-Psychopharmacol. Biol. Psychiatry* **2009**, *33*, 70–75. [CrossRef] [PubMed]
39. Jandhyala, S.M. Role of the normal gut microbiota. *World J. Gastroenterol.* **2015**, *21*, 8787–8803. [CrossRef]
40. Riezzo, G.; Chimienti, G.; Orlando, A.; D'Attoma, B.; Clemente, C.; Russo, F. Effects of long-term administration of Lactobacillus reuteri DSM-17938 on circulating levels of 5-HT and BDNF in adults with functional constipation. *Benef. Microbes* **2019**, *10*, 137–147. [CrossRef] [PubMed]
41. Baranger, K.; Giannoni, P.; Girard, S.D.; Girot, S.; Gaven, F.; Stephan, D.; Migliorati, M.; Khrestchatisky, M.; Bockaert, J.; Marchetti-Gauthier, E.; et al. Chronic treatments with a 5-HT 4 receptor agonist decrease amyloid pathology in the entorhinal cortex and learning and memory deficits in the 5xFAD mouse model of Alzheimer's disease. *Neuropharmacology* **2017**, *126*, 128–141. [CrossRef]
42. Jalanka, J.; Major, G.; Murray, K.; Singh, G.; Nowak, A.; Kurtz, C.; Silos-Santiago, I.; Johnston, J.; de Vos, W.; Spiller, R. The Effect of Psyllium Husk on Intestinal Microbiota in Constipated Patients and Healthy Controls. *Int. J. Mol. Sci.* **2019**, *20*, 433–444. [CrossRef]
43. Chen, W.; Bi, Z.; Zhu, Q.; Gao, H.; Fan, Y.; Zhang, C.; Liu, X.; Ye, M. An analysis of the characteristics of the intestinal flora in patients with Parkinson's disease complicated with constipation. *Am. J. Transl. Res.* **2021**, *13*, 13710–13722.
44. Yuan, X.; Chen, B.; Duan, Z.; Xia, Z.; Ding, Y.; Chen, T.; Liu, H.; Wang, B.; Yang, B.; Wang, X.; et al. Depression and anxiety in patients with active ulcerative colitis: Crosstalk of gut microbiota, metabolomics and proteomics. *Gut Microbes* **2021**, *13*, 1987779. [CrossRef] [PubMed]
45. Fan, Y.; Pedersen, O. Gut microbiota in human metabolic health and disease. *Nat. Rev. Microbiol.* **2020**, *19*, 55–71. [CrossRef]

46. Cui, M.; Jiang, Y.; Zhao, Q.; Zhu, Z.; Liang, X.; Zhang, K.; Wu, W.; Dong, Q.; An, Y.; Tang, H.; et al. Metabolomics and incident dementia in older Chinese adults: The Shanghai Aging Study. *Alzheimer's Dement.* **2020**, *16*, 779–788. [CrossRef] [PubMed]
47. Platten, M.; Nollen, E.A.A.; Röhrig, U.F.; Fallarino, F.; Opitz, C.A. Tryptophan metabolism as a common therapeutic target in cancer, neurodegeneration and beyond. *Nat. Rev. Drug Discov.* **2019**, *18*, 379–401. [CrossRef] [PubMed]

foods

MDPI

Article

An Exploration of the Antioxidative and Anti-Inflammatory Role of *Lactiplantibacillus plantarum* 106 via Improving Mitochondrial Function

Mengchun Qin [1], Yinfei Xing [2], Maocheng Sun [1], Lin Ma [1], Xiaolei Li [1], Fumin Ma [1], Dan Li [1] and Cuicui Duan [1,*]

[1] Key Laboratory of Agro-Products Processing Technology, Education Department of Jilin Province, Changchun University, 6543 Weixing Road, Changchun 130022, China
[2] College of Veterinary Medicine, Jilin University, Changchun 130062, China; xingyf23@mails.jlu.edu.cn
* Correspondence: duancui19841111@163.com; Tel.: +86-431-85250229

Abstract: In this present study, bioinformatics analysis and the experimental validation method were used to systematically explore the antioxidant activity and anti-inflammatory effect of *Lactiplantibacillus plantarum* A106, which was isolated from traditional Chinese pickles, on lipopolysaccharide (LPS)-induced RAW264.7 macrophages. *L. plantarum* A106 had a good scavenging ability for DPPH, ABTS, and hydroxyl radicals. Furthermore, *L. plantarum* A106 could increase the activity of RAW264.7 macrophages; raise the SOD and GSH levels, with or without LPS sensitization; or decrease the MDA, TNF–α, and IL–6 levels. In order to deeply seek the antioxidant and anti-inflammatory role and mechanism, bioinformatic analysis, including GO, KEGG, and GSEA analysis, was used to conduct an in-depth analysis, and the results showed that the LPS treatment of RAW264.7 macrophages significantly upregulated inflammatory-related genes and revealed an enrichment in the inflammatory signaling pathways. Additionally, a network analysis via the Cytoscape software (version 3.9.1) identified key central genes and found that LPS also disturbed apoptosis and mitochondrial function. Based on the above bioinformatics analysis, the effects of *L. plantarum* A106 on inflammation-related gene expression, mitochondrial function, apoptosis, etc., were detected. The results indicated that *L. plantarum* A106 restored the declined expression levels of crucial genes like TNF–α and IL–6; mitochondrial membrane potential; and apoptosis and the expression of apoptosis-related genes, Bcl–2, Caspase–3, and Bax. These results suggest that *L. plantarum* A106 exerts antioxidant activity and anti-inflammatory effects through regulating inflammatory and apoptosis-related gene expression, restoring the mitochondrial membrane potential.

Keywords: *Lactiplantibacillus plantarum*; antioxidant; bioinformatics analysis; mitochondrial function; apoptosis

1. Introduction

Lactic acid bacteria (LAB), a class of probiotics, not only maintain the ecological balance of the intestinal microbiota but also exhibit biologically active properties that have antitumor, immunomodulatory, and antioxidant effects. These properties have contributed to extensive research and application in the food and health product industries. Recently, LAB have been widely recognized for their crucial role in modulating inflammatory responses, including the regulation of cytokine production, the improvement of the intestinal barrier function, and the increase of immune cell activity [1,2]. In one study, *L. reuteri* LM1071 inhibited the production of inflammatory mediators like NO, pro-inflammatory factors, and COX proteins, thus demonstrating anti-inflammatory potential [3]. In another, *L. plantarum* DP189 downregulated pro-inflammatory cytokines such as TNF–α, IL–6, and IL–1β in neurodegenerative diseases, thereby inhibiting inflammatory responses and modulating the gut microbiota [4]. Another study, which used a D-galactose-induced senescent mouse model to explore the antioxidant effects of lactobacillus on hepatotoxicity, found that *L. plantarum* AR113 and AR501 attenuated liver injury and reduced the abnormal activities of SOD, GSH-Px, and catalase to normal levels in mice [5]. Oxidative stress was generated during the inflammatory response process, which could reduce the cellular immune function and increase and aggregate the inflammatory factors in cells, thereby promoting cell damage. Consequently, the level of inflammatory factors in the cells was a direct indication of the level of oxidative stress in the cells [6].

The RAW264.7 cell model, derived from murine macrophages, has been extensively utilized to study immune responses, particularly in the context of inflammation and infection. Lipopolysaccharide (LPS)-treated RAW264.7 macrophages were commonly used to explore the immune response mechanisms. It was reported that LPS could activate macrophages and induce the secretion of several pro-inflammatory cytokines, including tumor necrosis factor-alpha (TNF–α), interleukin–1β (IL–1β), IL–6, etc. [7,8]. A study investigated the anti-inflammatory potential and underlying mechanisms of andrographolide in LPS-induced RAW264.7 macrophages, revealing that andrographolide inhibited the activation of NF-κB and MAPK, thereby inhibiting autophagy and alleviating the inflammatory response [9]. Additionally, the phenotype and function of the macrophages were regulated by their surrounding microenvironment, and they were susceptible to reactive oxygen species (ROS), leading to cell death [10]. Thus, macrophages can also serve as a cellular model for oxidative stress. Under exogenous stimuli, macrophages could phagocytose, process, and present foreign substances, a process accompanied by electron transfer, which was a primary source of oxidative stress [11]. Tang et al. (2019) found that both quercetin and quercitrin could suppress the LPS-induced inflammatory response and oxidative stress in macrophages [12]. *L. paracasei* KW3110 inhibited the mitochondrial dysfunction induced by LPS in RAW264.7 macrophages, which, in turn, alleviated inflammatory stress [13]. However, the potential targets, signaling pathways, organelles, etc. still needed deep analysis before the anti-inflammatory and antioxidative mechanisms could be explored. Bioinformatics analysis provided a comprehensive and precise research direction. For example, some crucial bioactive ingredients, signaling pathways, and potential targets have been acquired via bioinformatics analysis, including the Gene Ontology (GO) and Kyoto Encyclopedia of Genes and Genomes (KEGG) analyses, as well as accounts of protein–protein interaction (PPI) [14]. However, there are few reports on how LAB modulate the LPS-induced inflammatory process and oxidative stress response in RAW264.7 macrophages using bioinformatics analysis combined with in vitro validation.

This study employed bioinformatics analysis and experimental validation methods to investigate the role of *L. plantarum* A106 in modulating inflammatory processes and relieving oxidative stress responses utilizing the LPS-induced oxidative stress in RAW264.7 macrophages model. Specifically, this study explored the protective effects of *L. plantarum* A106 against LPS-induced inflammation and oxidative stress in RAW264.7 macrophages. This study will enhance our understanding of the anti-inflammatory and antioxidative mechanisms of probiotics and may provide a scientific basis for developing new strategies to treat inflammation and oxidative stress-related diseases.

2. Materials and Methods

2.1. Strain and Cell Line

L. plantarum A106 was isolated from the naturally fermented pickles bought from the Yanbian area of Jilin Province. It was cultured in a De Man–Rogosa–Sharp (MRS) broth. The RAW264.7 macrophages were gifted by Jilin University.

2.2. Reagents and Instruments

1,1-Diphenyl-2-trinitrophenylhydrazine (DPPH) and 2,2′-Azido-bis-(3-ethylbenzodihy drothiazoline-6-sulfonic acid) (ABTS) were purchased from Phygene Life Sciences Co., Ltd. (Fuzhou, China). The DEME/high glucose medium was purchased from HyClone (HyClone Laboratories, Inc., Omaha, NE, USA). The serum was bought from Inner Mongolia OPCEL Biotechnology Co., Ltd. (Inner Mongolia, China). The cell counting kit-8 (CCK-8), superoxide dismutase (SOD), reactive oxygen species (ROS), mitochondrial membrane potential, and apoptosis kits were purchased from Beyotime Biotechnology (Beijing, China).

2.3. Culture of L. plantarum A106 and Preparation of Different Components

The *L. plantarum* A106 that was stored at −80 °C was quickly dissolved in a 37 °C water bath and inoculated into the MRS liquid medium with a 3% inoculum. It was incubated at 37 °C for 24 h and subsequently subcultured to 2–3 generations for further experiments.

With respect to the preparation of the culture supernatant and bacterial cells, the *L. plantarum* A106 was put in a shaking incubator (37 °C, 150 rpm) for 18–24 h, then a centrifuge at 5610× *g* for 20 min at 4 °C, before the supernatant was collected and filtered through a 0.22 μm filter membrane. The bacterial cells were washed three times with PBS (0.01 mol/L at a pH of 7.4) and adjusted to 1×10^9 CFU/mL.

With respect to the preparation of the cell-free extract, the lysozyme (1.5 mg/mL) was added to intact *L. plantarum* A106 with a concentration of 1×10^9 CFU/mL, then put in a water bath at 37 °C for 30 min, accompanied by an ultrasonic cell crusher in an ice bath (250 W, 5 s of operation, 10 s of interval, and a duration of 30 min). Afterward, the mixture was centrifuged at 5610× *g*/min for 10 min to remove any bacterial debris and to obtain the cell-free extracts.

2.4. Effectiveness of Different Fractions of L. plantarum A106 on Free Radical Scavenging

2.4.1. DPPH Scavenging Rate

The DPPH scavenging rate was tested according to the method reported by Xu et al. (2019), making appropriate modifications [15]. Specifically, 1 mg of DPPH was dissolved in 25 mL of anhydrous ethanol. Subsequently, 500 μL of different components were thoroughly mixed with 500 μL of DPPH-anhydrous ethanol solution. There was centrifugation at 4909× *g* for 10 min, before the supernatant was discarded, and the absorbance was tested at 517 nm. The DPPH scavenging rate was calculated according to the following formula:

$$\text{DPPH clearance (\%)} = [1 - (A_2 - A_1)/A_0] \times 100$$

where A_0 is the absorbance of the ultrapure water and DPPH; A_1 is the absorbance of the sample and ultrapure water; and A_2 is the absorbance of the sample and DPPH.

2.4.2. ABTS$^+$ Scavenging Rate

The ABTS$^+$ scavenging rate was tested according to the method reported by Li et al. (2015), with minor modifications [16]. The ABTS reaction solution was prepared according to the instructions. In total, 300 μL of various fractions were separately mixed with 300 μL of the ABTS reaction solution. The mixture was shook for 30 s and then kept in dark conditions for 6 min, followed by centrifuging at 5610× *g* for 5 min. The supernatant was transferred to a 96-well plate to read at 734 nm. The ABTS$^+$ scavenging rate was calculated according to the following formula:

$$\text{ABTS}^+ \text{ scavenging rate (\%)} = (1 - A_1/A_0) \times 100$$

where A_0 is the absorbance of the ABTS working solution and A_1 is the absorbance of the sample.

2.4.3. Hydroxyl Radical Scavenging Rate

The hydroxyl radical scavenging rate was detected according to the method described by Yang et al. (2017), with appropriate modifications [17]. Firstly, 0.5 mL of 1,10-o-diazophenanthrene (1.875 mmol/L), 1 mL of PBS (0.01 mmol/L at a pH of 7.4), and 0.5 mL of ferrous sulphate (1.875 mmol/L) were added to a 5 mL Eppendorf tube and mixed thoroughly. Then, 0.5 mL of the samples and 0.5 mL of H_2O_2 (0.1%) were added, and the mixture was vortexed thoroughly and incubated in a water bath at 37 °C for 1 h. Following this, the samples were centrifuged for 10 min at 5610× *g* and the supernatant was transferred to a 96-well plate to read at 536 nm. The hydroxyl radical scavenging rate was calculated according to the following formula:

$$\text{Hydroxyl radical scavenging rate (\%)} = [(A_s - A_n)/(A_b - A_n)] \times 100$$

where A_s is the absorbance of the samples and H_2O_2; A_n is the absorbance of the PBS; and A_b is the absorbance of the H_2O_2.

2.5. Cell Viability Assay

The RAW264.6 macrophages were cultivated in a cell culture incubator at 37 °C with 5% CO_2 using the DMEM/high glucose medium supplemented with 10% fetal bovine serum. After subculturing for 2–3 generations, the cells were digested with trypsin, centrifuged, resuspended in the complete medium, and adjusted to a concentration of 1×10^6 cells/mL, before being inoculated in a 96-well plate with a 200 µL cell suspension per well and cultured for 24 h. In the experimental group, 50 µL of bacterial suspension was added to each well at a different ratio: 1:100 (1×10^4 CFU/mL) or 1:1000 (1×10^3 CFU/mL) relative to the cell concentration using the complete medium. The control group only used the complete medium. Afterwards, 1 µg/mL of LPS was added to the experimental group 4 h before the end of the cultivation; meanwhile, 20 µL of CCK-8 solution was added to each well. The cells were then incubated at 37 °C for 1.5, 2, 2.5, and 3 h. The absorbance values at 450 nm were measured at various incubation periods. The cell viability was determined using the following formula:

$$\text{Cell viability} = [(A_2 - A_0)/(A_1 - A_0)] \times 100\%$$

where A_0 is the absorbance of the positive group, A_1 is the absorbance of the control group, and A_2 is the absorbance of the experimental group.

2.6. Measurement of the Levels of the Cytokines, TNF–α and IL–4

The cells were harvested and centrifuged, and the culture supernatant was collected. The cytokines TNF–α and IL–4 in the supernatant of the culture that underwent different treatments were detected using commercial ELISA kits (BioLegend, San Diego, CA, USA) according to the manufacturer's instruction.

2.7. Download of Microarray Datasets and Extraction of Differentially Expressed Genes (DEGs) in Microarray Datasets

The dataset GSE186886, downloaded from the GEO database (http://www.ncbi.nlm.nih.gov/geo/), included the control group and LPS-induced RAW264.7 group.

The "limma" (version 3.58.1) R package was employed to identify the differentially expressed genes 1 (DEGs1) between the LPS-induced group and the control group in the GSE186886 dataset, where $p < 0.05$ and $|\log2$ (fold change, FC)$| > 1$ were set as thresholds. Subsequently, "ggplot2" (version 3.5.0) was utilized to construct a volcano plot illustrating the expression of the DEGs.

2.8. Functional Enrichment Analysis

Both the Gene Ontology (GO) and Kyoto Encyclopedia of Genes and Genomes (KEGG) pathways analysis of the DEGs2 (differentially expressed genes in DEGs1 with $|\log2$ (fold change, FC)$| > 2$) were conducted using the DAVID (https://davidbioinformatics.nih.gov/) online platform, with p-values < 0.05 and q-values < 0. The results were visualized using the "ggplot2" R package.

In our study, we conducted an enrichment analysis of the DEGs2 targeting the KEGG pathways, GO terms, and Reactome pathways using the "clusterProfiler" (version 4.10.1) in R. The gene set enrichment analysis (GSEA) method was employed to associate the DEGs2 with KEGG pathways, GO terms, and Reactome pathways, considering $p < 0.05$ as significant. For the visualization of these enrichment outcomes, the "ggplot2" package in R was utilized.

2.9. Protein–Protein Interaction (PPI) Network Analysis

To elucidate the functional interactions among the screened potential proteins, the identified DEGs2 were imported into the STRING database (https://string-db.org/) to construct a protein–protein interaction (PPI) network, where the confidence score threshold was set to 0. Subsequently, the PPI network was imported into the Cytoscape software (version 3.9.1), where the significant modules were depicted using the CytoNCA (version 2.1.6) and CytoHubba (version 0.1) plugin.

2.10. Gene Ontology and Pathway Enrichment Analysis Using Metascape

A Metascape analysis offers the advantage of integrating multiple gene list analysis and enrichment tools within a single platform, thereby providing comprehensive, interactive visualizations for a deeper understanding of biological functions and pathways. To deeply explore the biological functions and pathways associated with the genes related to the mitochondrial pathways identified in the Gene Ontology Cellular Component (GO CC) analysis, an enrichment analysis was conducted using Metascape (https://metascape.org). Finally, Metascape's interactive visualization tools were used to present the results.

2.11. Quantitative Real-Time Polymerase Chain Reaction (qPCR)

The total RNA was extracted using Trizol (Sigma-Aldrich Chemical Co. Ltd., St. Louis, MO, USA) and subsequently transcribed into cDNA. A qPCR was performed to assess the expression levels of various genes, according to the procedures as previously described [18]. Quantitative analyses were conducted using the $2^{-\Delta\Delta CT}$ method, and all the data were normalized to the GADPH expression. The primer sequences used were as follows: TNF-α: 5′-CAGGCGGTGCCTATGTCTC-3′ and 5′-CGATCACCCCGAAGTTCAGTAG-3′; IL-1β: 5′-GCAACTGTTCCTGAACTCAACT-3′ and 5′-ATCTTTTGGGGTCCGTCAACT-3′; IL-6: 5′-TCTATACCACTTCACAAGTCGGA-3′ and 5′-GAATTGCCATTGCACAACTCTTT-3′; Caspase-3: 5′-CTGGACTGTGGCATTGAGAC-3′ and 5′-GCAAAGGGACTGGATGAACC-3′; Bax: 5′-CCGGCGAATTGGAGATGAACT-3′ and 5′-CCAGCCCATGATGGTTCTGAT-3′; Bcl-2: 5′-TCAGAGCGAGAAGGTAGGGA3′ and 5′-CTGTGGGGTAACAAGAAGGTC-3′; GAPDH: 5′-AGAAACCTGCCAAGTATGATGAC-3′ and 5′-AGAAACCTGCCAAGTATGA TGAC-3′; CXCL10: 5′-TGAATCCGGAATCTAAGACCATCAA-3′ and 5′-AGGACTAGCCA TCCACTGGGTAAAG-3′; CSF3: 5′-AAGCTGGTGAGTGAGTGTGC-3′ and 5′-GGCCATTC CCAGTTCCA-3′; CCL2: 5′-AGCAGCAGGTGTCCCAAAGA-3′ and 5′-GATCTCATTTGGT TCCGATCCA-3′; and CCL5: 5′-CCTGCTGCTTTGCCTACATTGC-3′ and 5′-ACACACTT GGCGGTTCTTTCGG-3′.

2.12. Determination of SOD, GSH, and MDA Content

The RAW264.7 macrophages were digested, centrifuged, and resuspended. The cell concentration was adjusted to 1×10^6 cells/mL and then inoculated into a 6-well plate with 2 mL per well. In the experimental group, 50 μL of bacterial suspension was added to each well (see the treatment of the *L. plantarum* A106 in Section 2.5). The control group only used the complete medium. The LPS was added at a final concentration of 1 μg/mL to the experimental group 4 h before the end of the cultivation. The cells were lysed using a cell lysis buffer (Beyotime) and then collected and centrifuged, and the supernatant was taken for the determination of the SOD, GSH, and MDA contents using a total SOD assay kit with WST-8, a GSH assay kit with DTNB, and an MDA assay kit (Beyotime), according to the instructions of each kit.

2.13. Determination of ROS Level

The cells cultivated in the 6-well plate underwent the same treatment as described in Section 2.12. The levels of ROS kit (Beyotime) following the provided instructions. The cells were collected, centrifuged, and resuspended before the levels of ROS were analyzed using flow cytometry (CytoFLEX, Beckman Coulter, Indiana, IN, USA).

2.14. Determination of Mitochondrial Membrane Potential (MMP)

The evaluation of the MMP is an important indicator to assess the normal function of the mitochondria. The MMP was measured using an enhanced mitochondrial membrane potential assay kit with JC-1 according to the instructions of operation (Beyotime). Firstly, the cellular states and staining conditions were observed under the fluorescence microscope. Then, the cells were collected, centrifuged, and resuspended, and the intensity of the fluorescence was assayed through a flow cytometer (Beckman Coulter, CytoFLEX).

2.15. Apoptosis Assay

Generally, the loss of MPP is a sign of cellular apoptosis. In this study, cellular apoptosis was also determined using an Annexin V-FITC apoptosis detection kit, according to the protocol of the manufacturer (Beyotime). Briefly, the cells were seeded in a 6-well plate, then collected, centrifuged, and resuspended in 195 μL of Annexin V-FITC binding buffer at the end of the different treatments. This was followed by the addition of 5 μL of Annexin V-FITC and 10 μL of PI (propidium iodide) staining solution, before incubation at room temperature for 10–20 min in the dark and subsequent storage in an ice water bath for analysis in a flow cytometer.

2.16. Statistical Analysis

The statistical analysis was conducted using Origen 2019b and SPSS Statistics 25 software using a one-way ANOVA, where $p < 0.05$ was considered as statistical significance. All the experiments were expressed as the mean $\pm$ SEM.

3. Results

3.1. Effect of Culture Supernatant, Bacterial Cells, and Cell-Free Extracts on the Scavenging Rate of Different Radicals In Vitro

The supernatant, bacterial cells, and cell-free extracts of the *L. plantarum* A106 culture were prepared separately and tested for the scavenging rates of DPPH, $ABTS^+$ and hydroxyl radicals. It can be seen from Figure 1 that the DPPH scavenging rate of the *L. plantarum* A106 culture supernatant was significantly higher than those of the bacterial suspension and the cell-free extract group ($p < 0.01$). Furthermore, the DPPH scavenging rate of the bacterial suspension was also obviously higher than that of the cell-free extract group ($p < 0.01$). Vitamin C (Vc) was the positive group (Figure 1A). Moreover, the results of the $ABTS^+$ radical scavenging rate also indicated a similar trend to the DPPH scavenging rate. (Figure 1B). Interestingly, the hydroxyl radical scavenging rate showed the same trend as the DPPH, with the only difference being that the cell-free extract group had a higher hydroxyl radical scavenging rate than the bacterial suspension group (Figure 1C).

3.2. Effect of L. plantarum A106 on the Cell Viability of LPS-Induced RAW264.7 Macrophages

The cell viability was determined with the CCK-8 assay kit. As shown in Figure 1D, the *L. plantarum* A106 demonstrated no cytotoxicity towards RAW264.7 macrophages. Additionally, the cell viability of the *L. plantarum* A106-treated group (1:100 or 1:1000) was significantly higher ($p < 0.05$) than that of the control group at 1.5 h. Furthermore, there were no significant differences among all groups after cultivating for 2 h; as a result, 2 h was determined to be the ideal duration for the CCK-8 treatment. After a 2 h CCK-8 treatment, there was a noteworthy reduction in the cell viability of the LPS-treated group when compared to the control group. However, the treatment with *L. plantarum* A106 (1:100 or 1:1000) significantly impeded the decrease in cell viability ($p < 0.05$) (Figure 1E).

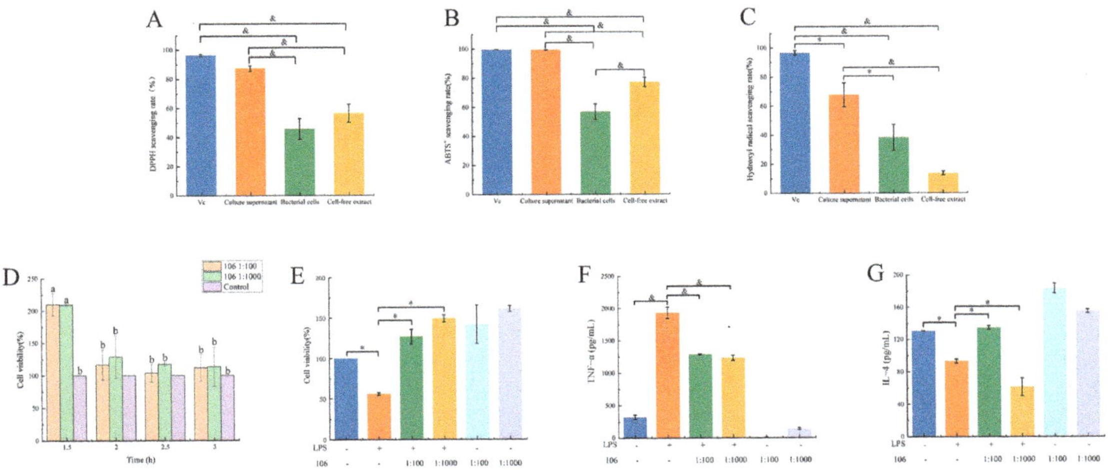

Figure 1. Scavenging rates of DPPH (**A**), ABTS+ (**B**), and hydroxyl radicals (**C**) of different fractions derived from *L. plantarum* A106 (**D**) Changes of cell viability depending on different treatment time (**E**). The effect of *L. plantarum* A106 treatment on the cell viability of LPS-induced RAW264.7 macrophages; and the effect of *L. plantarum* A106 treatment on the levels of TNF–α (**F**) and IL–4 (**G**) in LPS-induced RAW264.7 macrophages. The data are expressed as the mean ± SEM; and the p value was determined by a one-way ANOVA followed by multiple comparisons. * denotes a significant difference at $p < 0.05$, and & denotes a significant difference at $p < 0.01$. Different letters indicated significant differences ($p < 0.05$).

3.3. Effect of L. plantarum *A106 Treatment on the LPS-Induced Secretion of Inflammatory Factors*

As shown in Figure 1F,G, compared with the control group, the levels of the pro-inflammatory factor TNF–α were significantly increased ($p < 0.01$) and of the anti-inflammatory factor IL–4 were significantly reduced in the RAW264.7 macrophages that were stimulated by LPS ($p < 0.05$). In contrast, the treatment with the *L. plantarum* A106 significantly reduced the TNF–α levels ($p < 0.01$) but obviously increased the IL–4 levels ($p < 0.05$).

3.4. Selection and Functional Enrichment Analysis of DEGs

Among the 2465 differentially expressed genes (DEGs1) identified from the GSE186886 dataset, 519 genes were upregulated and 1946 genes were downregulated. The selection criteria were $p < 0.05$ and |log2 (fold change, FC)| > 1. All DEGs1 are presented in a volcano plot (Figure 2A), with red indicating the upregulated genes, blue indicating the downregulated genes, and gray representing the genes that did not show significant expression changes in the dataset.

To further investigate the functions of the DEGs, the GO and KEGG analyses were performed on the DEGs2 (differentially expressed genes from DEGs1 with |log2 (fold change, FC)| > 2) using DAVID, and the top 20 enriched entries were visualized using bubble charts. As shown in Figure 2D, the KEGG enrichment analysis presented several pathways that were significantly enriched with inflammatory responses, including the cytokine—cytokine receptor interaction; the TNF signaling pathway and IL–17 signaling pathway; the NOD-like receptor signaling pathway; and the NF-kappa B signaling pathway. As depicted in Figure 2C, the BP (biological process) category was predominantly composed of the innate immune response, the positive regulation of transcription from the RNA polymerase II promoter, the inflammatory response, and the immune system process. The MF (molecular function) category (Figure 2B) comprised protein binding, identical protein binding, hydrolase activity, protein homodimerization activity, and cytokine activity.

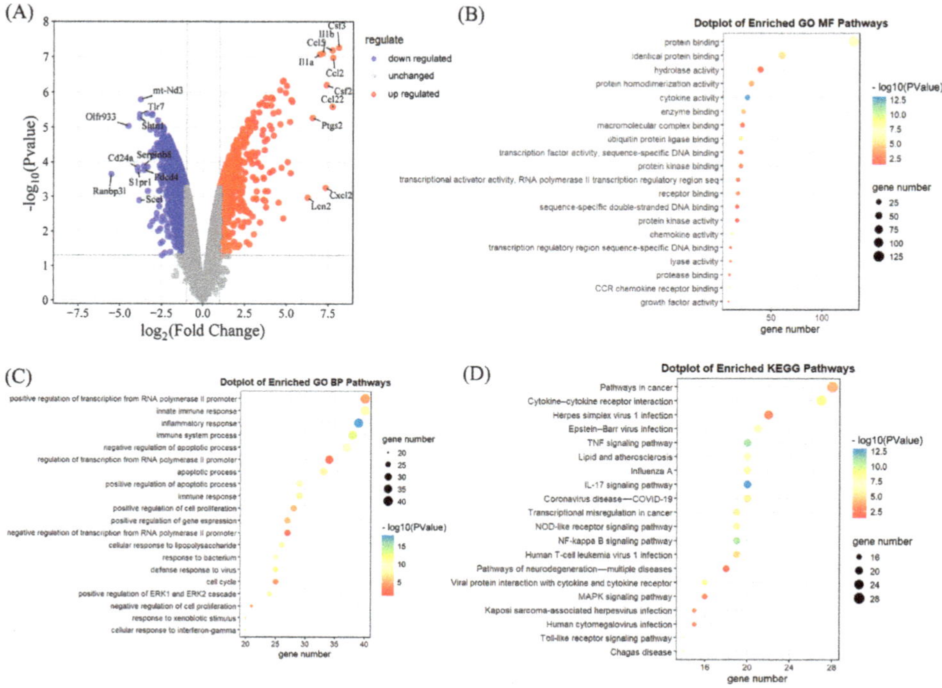

Figure 2. DEGs and enrichment analysis. (**A**) Volcano plot showing DEGs in the LPS-treated group vs. the control group; (**B**) GO:MF enrichment results (top 20); (**C**) GO:BP enrichment results (top 20); and (**D**) KEGG enrichment results (top 20).

3.5. GSEA Analysis of Biological Processes and Functions

The gene set enrichment analysis (GSEA) was conducted to delve deeper into the biological functions and pathways associated with DEGs2, because the results of the GO analysis showed that DEGs2 were mainly enriched in extracellular space and the extracellular region (Figure 3C). In terms of biological processes (Figure 3A), the main enrichment was related to the response to IL–1, monocyte chemotaxis, and granulocyte migration. Similarly, at the molecular function level (Figure 3B), there was found to be high enrichment of the cytokine activity, cytokine receptor binding, and chemokine activity. The pathway analysis based on KEGG and Reactome highlighted the role of related pro-inflammatory cytokines (Figure 3D,E). Therefore, these results suggest that the LPS treatment of RAW264.7 macrophages may contribute to the development of inflammatory responses.

3.6. Construction of Protein–Protein Interaction Network, Selection of Hub Genes, and Metascape Enrichment Analysis

After uploading the selected DEGs2 to the STRING online database with a confidence threshold of >0.7 and visualizing the results in Cytoscape software (version 3.9.1), any isolated nodes were removed to establish a protein–protein interaction (PPI) network consisting of 164 nodes and 667 edges. Subsequently, the betweenness centrality of the nodes was assessed using the CytoNCA plugin, where darker node colors indicated higher betweenness centrality values (Figure 4A). In addition, the top 10 genes were identified in Cytohubba using the maximal clique centrality (MCC) method (Figure 4B), where the node color represented connectivity (red for higher degrees and yellow for lower degrees). Integrating the outcomes of the betweenness centrality and maximal clique centrality analyses, we propose that TNF, IL–6, Cxcl10, and IL–1β potentially play pivotal roles in modulating LPS-induced infections.

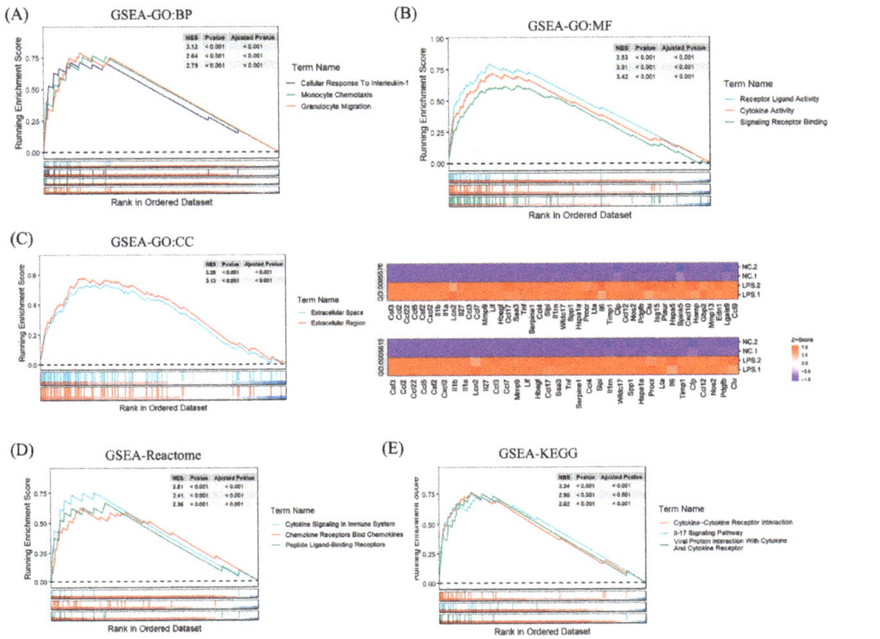

Figure 3. GSEA enrichment analysis. (**A**) The significant GSEA sets for the GO-BP pathways; (**B**) the significant GSEA sets for the GO-MF pathways; (**C**) the significant GSEA sets and heatmap for the GO-CC pathways; (**D**) the significant GSEA sets for the Reactome pathways; and (**E**) the significant GSEA sets for the KEGG pathways.

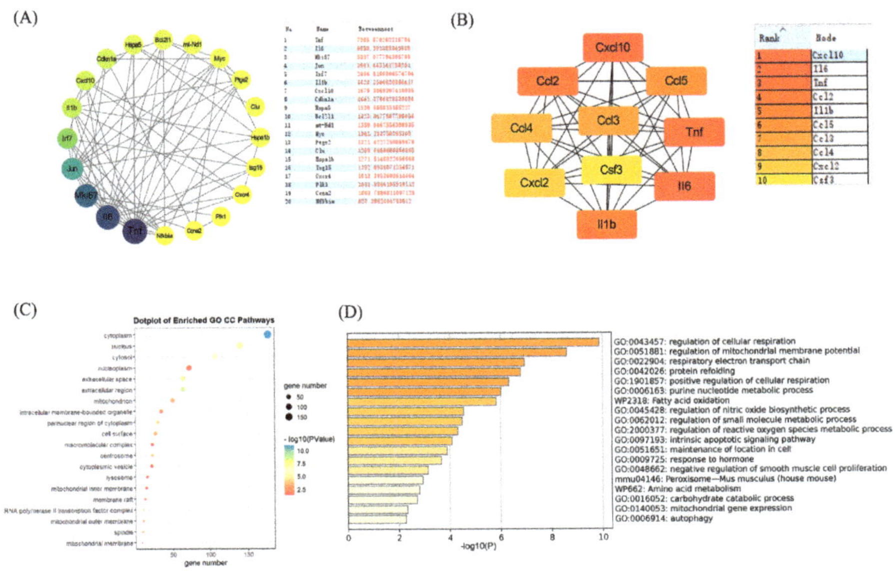

Figure 4. PPI and mitochondrial Metascape analysis. (**A**) Top 20 key genes featured by CytoNCA; (**B**) a key module with 10 genes, the key gene screened by the MCC method; (**C**) the GO:CC enrichment results (top 20); and (**D**) the Metascape results of mitochondrial DEGs.

According to the CC (cellular component) analysis (Figure 4C), the above findings were mainly associated with organelles such as the nucleus and mitochondrion. Given that the role of mitochondria in cellular functionality is critical, because they serve as the cell's powerhouses by producing ATP via respiration and regulating cellular metabolism and apoptosis, we conducted an enrichment analysis on 50 DEGs associated with mitochondria identified in the CC (cellular component) using the Metascape online platform (https://metascape.org). The results primarily indicate enrichment in the regulation of cellular respiration, the regulation of mitochondrial membrane potential, and the respiratory electron transport chain (Figure 4D).

3.7. Effect of *L. plantarum A106* on Inflammatory-Related Gene Expression Levels

TNF-α, IL-1β, IL-6, Cxcl10, Csf3, Ccl2, and Ccl5 were identified as the hub genes and the significantly upregulated genes. As demonstrated in Figure 5, the expression of TNF-α, IL-1β, IL-6, Cxcl10, Csf3, Ccl2, and Ccl5 genes increased following the LPS stimulation ($p < 0.05$). Compared to the LPS-treated group, the addition of *L. plantarum* A106 significantly downregulated the expression of TNF-α, IL-1β, IL-6, Cxcl10, Csf3, Ccl2, and Ccl5 genes, indicating that *L. plantarum* A106 can mitigate the inflammatory levels caused by LPS treatment of RAW264.7 macrophages and exert anti-inflammatory effects.

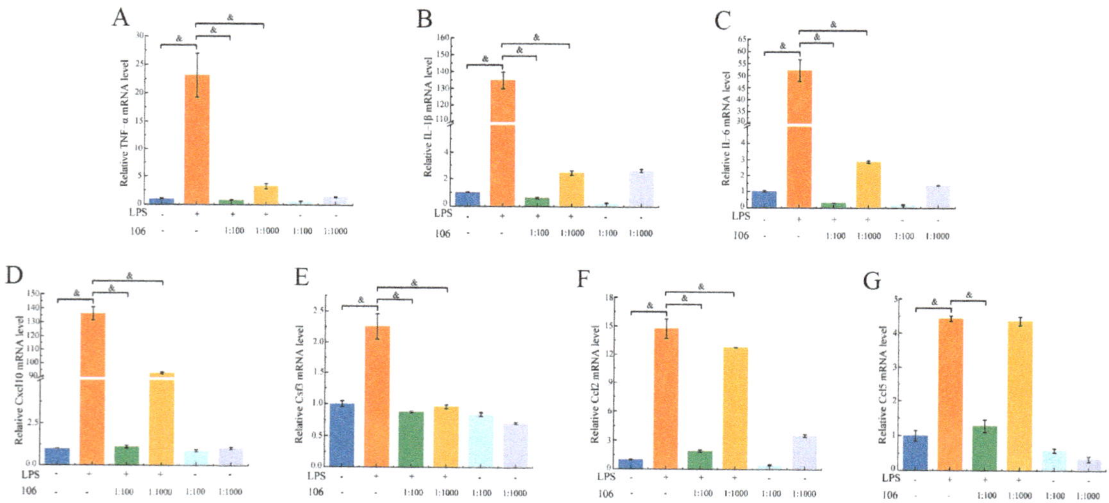

Figure 5. Effect of the *L. plantarum* A106 treatment on the relative levels of the expression of TNF-α mRNA (**A**), IL-1β mRNA (**B**), IL-6 mRNA (**C**), Cxcl10 mRNA (**D**), Csf3 mRNA (**E**), Ccl2 mRNA (**F**), and Ccl5 mRNA (**G**) in the LPS-induced oxidative stress in RAW264.7 macrophages. The data are expressed as the mean ± SEM; and the *p* value was determined by a one-way ANOVA followed by multiple comparisons. & denotes significant difference at $p < 0.01$; and 106 represents *L. plantarum* A106.

3.8. Effect of *L. plantarum A106* on LPS-Induced SOD, GSH, and MDA Activities

RAW264.7 macrophages were used to establish a LPS-induced oxidative stress model to further explore the antioxidant capacity of *L. plantarum* A 106. The results are shown in Figure 6. Under the stimulation of LPS, the SOD and GSH viability of RAW264.7 macrophages was significantly reduced, and the MDA viability was significantly increased compared to the control ($p < 0.01$) (Figure 6A–C). However, the addition of *L. plantarum* A106 significantly increased the SOD and GSH viability ($p < 0.01$) and significantly reduced the MDA viability ($p < 0.05$). The above results indicated that treatment with *L. plantarum* A106 might slow down the LPS-induced oxidative stress in RAW264.7 macrophages.

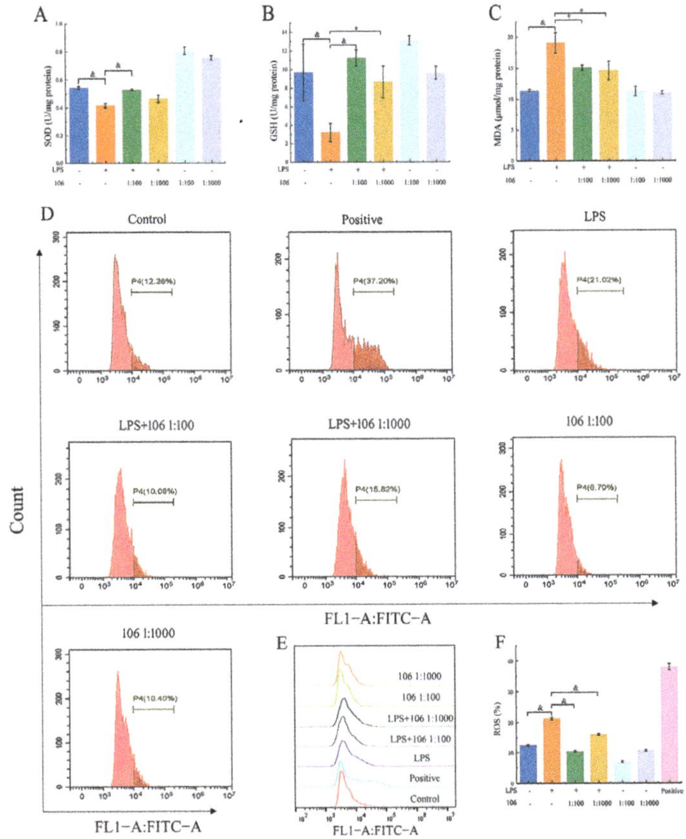

Figure 6. Effect of *L. plantarum* A106 treatment on the SOD (**A**), GSH (**B**), and MDA (**C**) activity and the cellular ROS levels of the LPS-induced oxidative stress in RAW264.7 macrophages; (**D**) ROS levels analyzed by flow cytometry; (**E**) overview of ROS levels in different groups; and (**F**) comparative histogram of the ROS levels in different groups. The data are expressed as the mean ± SEM; and the *p* value was determined by a one-way ANOVA followed by multiple comparisons. * denotes a significant difference at $p < 0.05$; & denotes a significant difference at $p < 0.01$; and 106 represents *L. plantarum* A106.

3.9. Effect of *L. plantarum* A106 on LPS-Induced ROS Levels

The results of the ROS levels are shown in Figure 6D. The ROS level was significantly elevated in the RAW264.7 macrophages stimulated by LPS compared to the control group, while the addition of *L. plantarum* A106 significantly reduced the ROS levels ($p < 0.01$), and the high dose group (1:100) was more effective than the low dose (1:1000). This result suggests that *L. plantarum* A106 had the potential to effectively reverse the increase in the ROS levels induced by LPS and alleviate the oxidative damage in LPS-induced RAW264.7 macrophages.

3.10. Effects of *L. plantarum* A106 on LPS-Induced Mitochondrial Membrane Potential (MMP)

The above analysis revealed that lipopolysaccharide (LPS) might influence the mitochondrial membrane potential (Figure 4D). Therefore, fluorescence microscopy (Olympus, 50X) was used to preliminarily observe the cells in each treatment group. As shown in Figure 7A, more cells were with green fluorescence in the LPS-induced RAW264.7 macrophages compared to the control group, indicating a loss of MMP; subsequently, the green fluorescence weakened visually after treatment with the *L. plantarum* A106, but it was difficult to calculate the exact ratio between the red and green fluorescence. Therefore, flow cytometry was employed to detect the precise fluorescence change. As shown in Figure 7B,C, the red/green fluorescence intensity ratio obviously decreased in the LPS-induced RAW264.7 macrophages ($p < 0.01$) but remarkably increased after the treatment with *L. plantarum* A106 ($p < 0.05$).

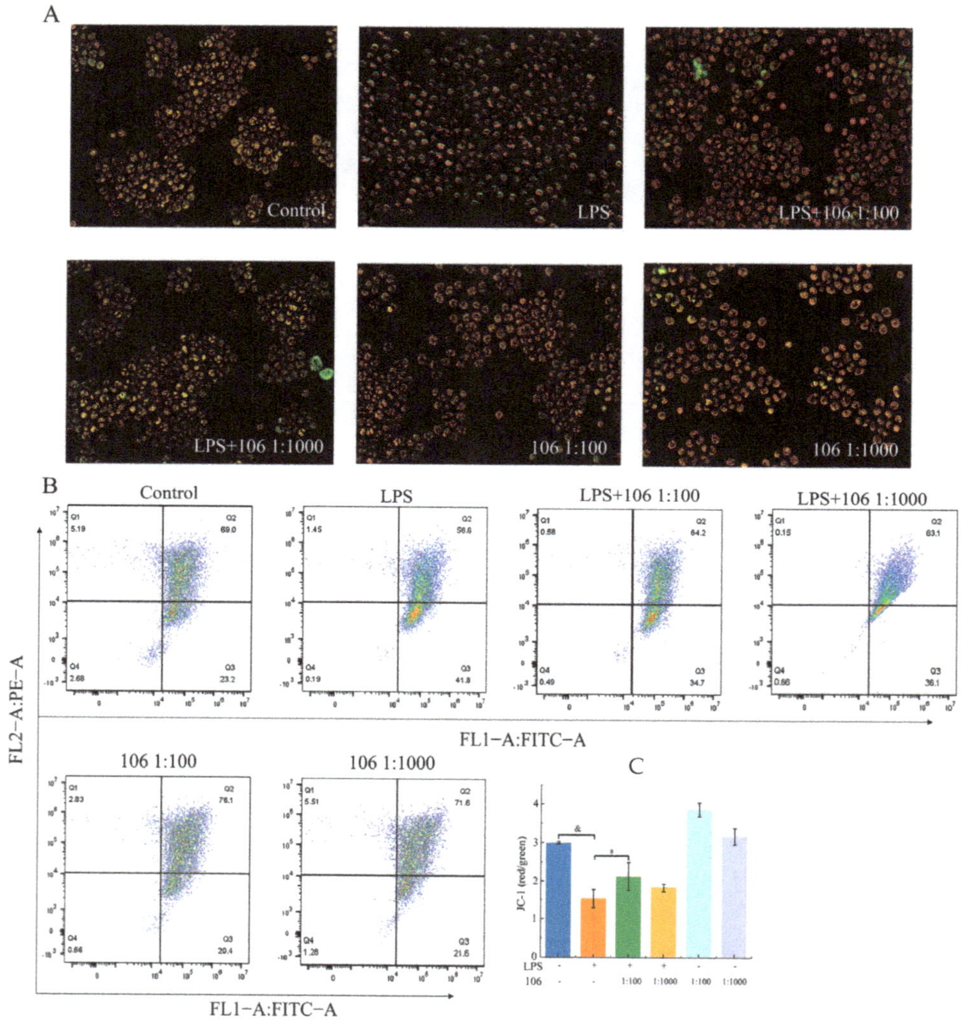

Figure 7. Effect of *L. plantarum* A106 treatment on the mitochondrial membrane potential of LPS-induced oxidative stress in RAW264.7 macrophages. (**A**) Changes of mitochondrial membrane potential in different groups under the fluorescence microscope; (**B**) the flow cytometric analysis of the mitochondrial membrane potential in different groups; and (**C**) a comparative histogram of the levels of mitochondrial membrane potential in different groups. The data are expressed as the mean ± SEM; and the p value was determined by a one-way ANOVA followed by multiple comparisons. * denotes a significant difference at $p < 0.05$; & denotes a significant difference at $p < 0.01$; and 106 represents *L. plantarum* A106.

3.11. Effects of *L. plantarum* A106 on LPS-Induced Apoptosis

The loss of mitochondrial membrane potential is often a precursor to apoptosis, leading to the activation of downstream caspases and resulting in cell apoptosis. In this study, flow cytometry was used to detect the apoptotic status of the cells in each treatment group. The results, as shown in Figure 8A,B, indicate that stimulation with LPS significantly increased the percentage of apoptotic cells compared to the control, and the addition of *L. plantarum* A106 significantly decreased the percentage of apoptotic cells ($p < 0.01$). Furthermore, there was no significant difference between the LPS-induced group with a low dose of *L. plantarum* A106 and with a high dose of *L. plantarum* A106.

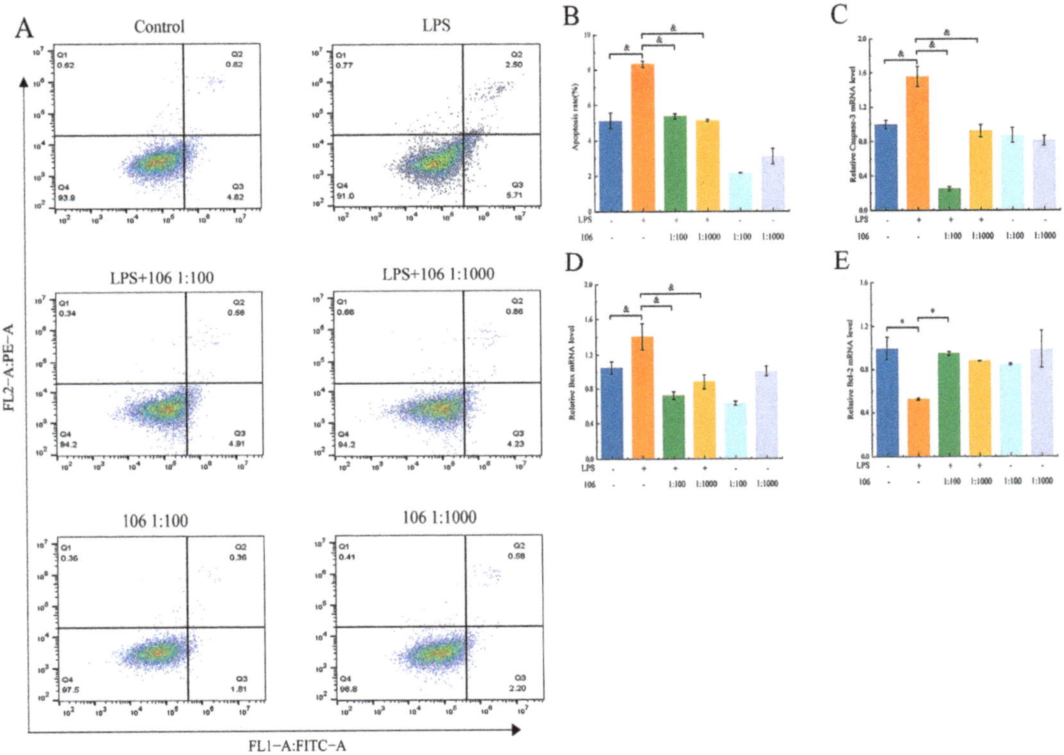

Figure 8. Effect of *L. plantarum* A106 treatment on cell apoptosis and apoptosis-related gene expression in LPS-induced oxidative stress in RAW264.7 macrophages. (**A**) Flow cytometry analysis of apoptosis in different groups; (**B**) combined analysis of the flow cytometry results among different groups; (**C**) Caspase–3 mRNA expression level; (**D**) Bax mRNA expression level; and (**E**) Bcl–2 mRNA expression level in LPS-induced oxidative stress in RAW264.7 macrophages. The data are expressed as the mean ± SEM; the p value was determined by a one-way ANOVA followed by multiple comparisons; and & denotes a significant difference at $p < 0.01$. * denotes a significant difference at $p < 0.05$.

Additionally, we assessed the impact of various treatments on the expression of apoptosis-related genes (Caspase–3, Bcl–2, and Bax) through a qPCR. As shown in Figure 8C,D, compared with the control group, the Bcl–2 gene expression was significantly reduced ($p < 0.01$), but the Caspase–3 and Bax gene expressions were significantly increased in the RAW264.7 macrophages stimulated by LPS ($p < 0.01$), while the addition of *L. plantarum* A106 significantly upregulated the Bcl–2 expression ($p < 0.01$) but significantly downregulated the Caspase-3 and Bax expressions ($p < 0.01$).

4. Discussion

More and more studies have provided reliable evidence for the biological activities of LAB, identifying their antioxidant, antitumor, and immunomodulatory properties. LAB have also been considered as a potential natural antioxidant [19,20]. However, there has been the lack of a comprehensive and precise research method to deeply explore the anti-inflammatory and antioxidative effects of LAB. Therefore, this study combined bioinformatics analysis and experiment validation to investigate the potential anti-inflammatory and antioxidative effects of *L. plantarum* A106 and its underlying mechanisms.

DPPH, $ABTS^+$, and hydroxyl radicals are electronic reduction products of a class of oxygen in organisms and common reactive oxygen species. Thus, the antioxidant activity of LAB can be measured with high ratios of DPPH, $ABTS^+$, and hydroxyl radical scavenging. Tang et al. (2017) assessed the antioxidant activity exhibited by the fermented product of *L. plantarum* MA2, which was isolated from the traditional kefir grains in Tibet, and demonstrated that the DPPH radical scavenging rates of both the MA2 fermentation supernatant and the intact cells were superior to those of a cell-free extract [21]. Interestingly, this results were similar to our results. Specifically, in this study, *L. plantarum* A106 supernatant or cell-free extracts also had a higher DPPH radical scavenging rate. Moreover, the scavenging rates of the $ABTS^+$ and hydroxyl radicals also presented similar results to the DPPH radical rate, indicating that *L. plantarum* A106 has a strong antioxidant activity.

It has also been reported that LAB have an immunological regulation effect [22,23]. Li et al. (2021) found that *L. plantarum* KSFY06 upregulated the anti-inflammatory factor IL–10 and downregulated the pro-inflammatory factors IL–6 and TNF–α to prevent the inflammatory response of an LPS-induced acute liver injury in mice [24]. Prolonged inflammatory processes increased the ROS production and caused oxidative stress; as a result, anti-inflammation might also mitigate the oxidative damage to cells [25]. In this study, the treatment with *L. plantarum* A106 upregulated the anti-inflammatory factor IL–4 and downregulated the gene expression of the pro-inflammatory factor TNF–α. However, further study was needed to explore the anti-inflammatory and antioxidative mechanisms of *L. plantarum* A106.

Subsequently, bioinformatics analysis was performed on the GSE186886 dataset, identifying 2465 differentially expressed genes (DEGs), including significantly upregulated inflammatory-related genes Csf3, Ccl5, and Ccl. Further KEGG enrichment analysis of the DEGs2 (genes from DEGs1 with $|\log2$ (fold change, FC)$| > 2$) revealed key inflammatory pathways including cytokine—cytokine receptor interaction, a TNF signaling pathway, and an IL–17 signaling pathway [26]. The GO BP and GO MF analyses additionally indicated enrichment in the innate immune response, inflammatory response, and cytokine activity [27]. These findings suggest the potential impact of LPS on the inflammatory signaling pathways.

The gene set enrichment analysis (GSEA) results underscored the significant roles of these genes in biological functions and pathways. Specifically, these DEGs were notably enriched in cytokine—cytokine interactions, the IL–17 signaling pathway, and cytokine signaling in the immune system through the KEGG and Reactome pathway analyses, highlighting their importance in inflammatory signal transduction. Moreover, the GO analysis revealed that these genes predominantly clustered in extracellular spaces and regions, as well as in the biological processes and molecular functions associated with the innate immune and inflammatory responses. These findings further unveiled the mechanism behind LPS's promotion of inflammatory responses in RAW264.7 macrophages. These insights provide a deeper understanding of the LPS-induced inflammatory mechanisms.

By integrating the analyses of the betweenness centrality and maximal clique centrality (MCC) through the use of the Cytoscape software (version 3.9.1), we identified several key hub genes: TNF, IL–6, Cxcl10, and IL–1β [28]. These genes were likely to play critical roles in the inflammatory signaling induced by LPS [29]. Furthermore, the enrichment analysis of the mitochondrial-related differentially expressed genes (DEGs) using Metscape suggested that LPS perturbed cellular respiration, mitochondrial membrane potential, the respiratory electron transport chain, and the process of apoptosis [30]. These findings reveal that while LPS promotes inflammatory responses, it also disrupts the cellular energy metabolism and redox reactions, affecting cellular homeostasis.

TNF and IL–1β are pro-inflammatory cytokines that play pivotal roles in the initiation and perpetuation of inflammation. They were involved in the upregulation of inflammatory pathways, enhancing the production of other cytokines and recruiting immune cells to sites of infection or injury, thereby amplifying the inflammatory response [31]. IL–6 was a multifunctional cytokine that not only facilitated inflammation but also possessed immunoregulatory functions [32]. *L. helveticus* SBT2171 significantly reduced the LPS-induced expression and secretion of IL–6 and IL–1β cytokines in peritoneal macrophages, leading to anti-inflammatory effects [33]. Cxcl10, a chemokine induced by IFN–γ, promoted the migration of immune cells to inflamed tissues during the LPS-induced inflammation, further exacerbating the inflammatory milieu [34]. Consistent with these findings, our bioinformatics analysis also identified these four molecules as hub genes in a macrophage model treated with LPS. Additionally, our qPCR results demonstrated an increase in the expression of these genes following the LPS stimulation. Compared to the LPS-treated group, the addition of *L. plantarum* A106 significantly downregulated the expression of these genes, suggesting that these cytokines may be potential targets for the therapeutic intervention of inflammatory diseases with *L. plantarum* A106.

Through a meta-analysis of mitochondrial-related DEGs, we discovered that LPS may significantly affect cellular respiration and the mitochondrial electron transport chain. Cellular respiration, a crucial process in cellular energy metabolism, is accompanied by the production of reactive oxygen species (ROS), particularly within the mitochondrial electron transport chain. ROS include free radicals such as superoxide anions (O2-) and non-radical forms like hydrogen peroxide (H_2O_2) [35]. Excessive ROS can lead to oxidative stress, damaging proteins, lipids, and nucleic acids, thereby triggering various diseases [36]. To counteract the potential harm caused by ROS, cells have developed a complex antioxidant system, comprising enzymes like superoxide dismutase (SOD), catalase, and glutathione peroxidase, as well as non-enzymatic antioxidants such as glutathione and vitamin C [37]. These antioxidants neutralize the ROS and maintain the redox balance within the cell [38]. In this study, we found that the plant-derived *L. plantarum* A106 could reduce the ROS and MDA levels in RAW264.7 macrophages and enhance the SOD and GSH activities, thus boosting the cellular antioxidant capacity and ameliorating the LPS-induced oxidative stress in RAW264.7 macrophages. This result was consistent with the result that *L. plantarum* DP189 significantly reversed the MPTP-induced decrease in SOD and GSH-Px activity, increased the MDA activity, and improved the antioxidant capacity in Parkinsonian mice [3].

The bioinformatics analysis suggested that LPS might interfere with the regulation of the mitochondrial membrane potential, a critical indicator of mitochondrial health and function and the driving force behind ATP synthesis via oxidative phosphorylation. Disruption of this potential was often a precursor to mitochondrial apoptosis, with changes triggering the release of pro-apoptotic factors such as cytochrome c, leading to the activation of downstream caspases and resulting in cell apoptosis [39,40]. In our study, we employed JC-1 staining to observe a significant decrease in the mitochondrial membrane potential (MMP) induced by LPS, suggesting an increase in the mitochondrial pathway apoptosis. Subsequently, through flow cytometry and qPCR assays, we detected an increase in the apoptosis rates and an elevated expression of the anti-apoptotic gene Bcl–2, alongside a reduced expression of the pro-apoptotic genes Caspase–3 and Bax following the LPS treatment of RAW264.7 macrophages. The treatment of RAW264.7 macrophages with *L. plantarum* A106 reversed the decline in the MMP and increased the apoptosis rates caused by LPS. This experimental result was consistent with Li et al. (2022b), who reported that dandelion extract reduced the apoptosis rate of RAW264.7 macrophages and attenuated the LPS-induced inflammatory response [41].

5. Conclusions

In summary, this study demonstrates that *L. plantarum* A106 possesses anti-inflammatory and antioxidant properties. The bioinformatics analysis and experimental results indicate that TNF, IL–1β, IL–6, and Cxcl10 may be potential targets for *L. plantarum* A106 in exerting its anti-inflammatory role. Additionally, *L. plantarum* A106 can improve the mitochondrial function in RAW264.7 macrophages, enhance the activities of SOD and GSH, and reduce the levels of MDA and ROS, thereby mitigating oxidative stress and reducing LPS-induced cell apoptosis.

Author Contributions: Data curation, L.M. and C.D.; funding acquisition, C.D.; investigation, M.S.; methodology, Y.X. and C.D.; resources, X.L.; supervision, D.L.; validation, F.M.; writing—original draft, M.Q.; and writing—review and editing, C.D. All authors have read and agreed to the published version of the manuscript.

Funding: This work was financially supported by the Scientific Research Project of the Education Department of Jilin Province (JJKH20230683KJ) and the Basic Research Project of Free Exploration, Science and Technology Department of Jilin Province (YDZJ202201ZYTS688).

Institutional Review Board Statement: Not applicable.

Informed Consent Statement: Not applicable.

Data Availability Statement: The original contributions presented in the study are included in the article, further inquiries can be directed to the corresponding author.

Conflicts of Interest: The authors declare that they have no known competing financial interests or personal relationships that could have appeared to influence the work reported in this paper.

References

1. Li, Y.; Yang, S.; Lun, J.; Gao, J.; Gao, X.; Gong, Z.; Wan, Y.; He, X.; Cao, H. Inhibitory Effects of the *Lactobacillus rhamnosus* GG effector protein HM0539 on inflammatory response through the TLR4/MyD88/NF-κB axis. *Front. Immunol.* **2020**, *11*, 551449. [CrossRef] [PubMed]
2. Sun, K.Y.; Xu, D.H.; Xie, C.; Plummer, S.; Tang, J.; Yang, X.F.; Ji, X.H. *Lactobacillus paracasei* modulates LPS-induced inflammatory cytokine release by monocyte-macrophages via the up-regulation of negative regulators of NF-kappaB signaling in a TLR2-dependent manner. *Cytokine* **2017**, *92*, 1–11.
3. Jang, A.Y.; Rod-In, W.; Monmai, C.; Sohn, M.; Kim, T.R.; Jeon, M.G.; Park, W.J. Anti-inflammatory potential of Lactobacillus reuteri LM1071 via eicosanoid regulation in LPS-stimulated RAW264.7 cells. *J. Appl. Microbiol.* **2022**, *133*, 67–75. [CrossRef]
4. Wang, L.; Zhao, Z.; Zhao, L.; Zhao, Y.; Yang, G.; Wang, C.; Gao, L.; Niu, C.; Li, S. *Lactobacillus plantarum* DP189 reduces α-SYN aggravation in MPTP-induced parkinson's disease mice via regulating oxidative damage, inflammation, and gut microbiota disorder. *J. Agric. Food Chem.* **2022**, *70*, 1163–1173. [CrossRef]
5. Lin, X.; Xia, Y.; Wang, G.; Xiong, Z.; Zhang, H.; Lai, F.; Ai, L. *Lactobacillus plantarum* AR501 alleviates the oxidative stress of D-Galactose-induced aging mice liver by upregulation of Nrf2-mediated antioxidant enzyme expression. *J. Food Sci.* **2018**, *83*, 1990–1998. [CrossRef] [PubMed]
6. Xu, C.; Bailly-Maitre, B.; Reed, J.C. Endoplasmic reticulum stress: Cell life and death decisions. *J. Clin. Investig.* **2005**, *115*, 2656–2664. [CrossRef]
7. Tian, Y.; Zhou, S.; Takeda, R.; Okazaki, K.; Sekita, M.; Sakamoto, K. Anti-inflammatory activities of amber extract in lipopolysaccharide-induced RAW264.7 macrophages. *Biomed. Pharmacother.* **2021**, *141*, 111854. [CrossRef] [PubMed]
8. Hwang, J.H.; Ma, J.N.; Park, J.H.; Jung, H.W.; Park, Y.K. Anti-inflammatory and antioxidant effects of MOK, a polyherbal extract, on lipopolysaccharide-stimulated RAW264.7 macrophages. *Int. J. Mol. Med.* **2019**, *43*, 26–36.
9. Cao, Y.; Chen, J.; Ren, G.; Zhang, Y.; Tan, X.; Yang, L. Punicalagin prevents inflammation in LPS-induced RAW264.7 macrophages by inhibiting FoxO3a/Autophagy signaling pathway. *Nutrients* **2019**, *11*, 2794. [CrossRef]
10. Viola, A.; Munari, F.; Sánchez-Rodríguez, R.; Scolaro, T.; Castegna, A. The metabolic signature of macrophage responses. *Front. Immunol.* **2019**, *10*, 1462. [CrossRef]
11. Locati, M.; Curtale, G.; Mantovani, A. Diversity, mechanisms, and significance of macrophage plasticity. *Annu. Rev. Pathol.* **2020**, *24*, 123–147. [CrossRef] [PubMed]
12. Tang, J.; Diao, P.; Shu, X.; Li, L.; Xiong, L. Quercetin and quercitrin attenuates the inflammatory response and oxidative stress in LPS-induced RAW264.7 cells: In vitro assessment and a theoretical model. *Biomed. Res. Int.* **2019**, *28*, 7039802. [CrossRef] [PubMed]
13. Yamazaki, T.; Yamada, S.; Ohshio, K.; Sugamata, M.; Morita, Y. *Lactobacillus paracasei* KW3110 prevents inflammatory-stress-induced mitochondrial dysfunction in mouse macrophages. *Int. J. Mol. Sci.* **2022**, *23*, 1443. [CrossRef]

14. Liu, J.Q.; Liu, J.; Tong, X.L.; Peng, W.J.; Wei, S.S.; Sun, T.L.; Wang, Y.K.; Zhang, B.K.; Li, W.Q. Network pharmacology prediction and molecular docking-based strategy to discover the potential pharmacological mechanism of Huai Hua San against ulcerative colitis. *Drug Des. Dev. Ther.* **2021**, *15*, 3255–3276. [CrossRef]
15. Xu, Y.; Cui, Y.; Wang, X.; Yue, F.; Shan, Y.; Liu, B.; Zhou, Y.; Yi, Y.; Lü, X. Purification, characterization and bioactivity of exopolysaccharides produced by *Lactobacillus plantarum* KX041. *Int J. Biol. Macromol.* **2019**, *128*, 480–492. [CrossRef]
16. Li, C.; Huang, Q.; Fu, X.; Yue, X.J.; Liu, R.H.; You, L.J. Characterization, antioxidant and immunomodulatory activities of polysaccharides from prunella vulgaris linn. *Int. J. Biol. Macromol.* **2015**, *75*, 298–305. [CrossRef] [PubMed]
17. Yang, M.; Jiang, R.; Liu, M.; Chen, S.J.; He, L.; Ao, X.L.; Zou, L.K.; Liu, S.L.; Zhou, K. Study of the probiotic properties of lactic acid bacteria isolated from chinese traditional fermented pickles. *J. Food Process Pres.* **2017**, *41*, e12954. [CrossRef]
18. Huang, J.C.; Yue, Z.P.; Yu, H.F.; Yang, Z.Q.; Wang, Y.S.; Guo, B. TAZ ameliorates the microglia-mediated inflammatory response via the Nrf2-ROS-NF-κB pathway. *Mol. Ther. Nucleic Acids* **2022**, *28*, 435–449. [CrossRef]
19. Li, W.; Gao, L.; Huang, W.; Ma, Y.; Muhammad, I.; Hanif, A.; Ding, Z.; Guo, X. Antioxidant properties of lactic acid bacteria isolated from traditional fermented yak milk and their probiotic effects on the oxidative senescence of caenorhabditis elegans. *Food Funct.* **2022**, *13*, 3690–3703. [CrossRef]
20. Méndez, U.V.E.; Pérez, V.D.; Perdigón, G.; de LeBlanc, A.d.M. Milk fermented by *Lactobacillus casei* CRL431 administered as an immune adjuvant in models of breast cancer and metastasis under chemotherapy. *Appl. Microbiol. Biotechnol.* **2021**, *105*, 327–340. [CrossRef]
21. Tang, W.; Xing, Z.; Li, C.; Wang, J.; Wang, Y. Molecular mechanisms and in vitro antioxidant effects of *Lactobacillus plantarum* MA2. *Food Chem.* **2017**, *221*, 1642–1649. [CrossRef] [PubMed]
22. Afridi, R.; Lee, W.H.; Suk, K. Microglia gone awry: Linking immunometabolism to neurodegeneration. *Front. Cell Neurosci.* **2020**, *14*, 246. [CrossRef] [PubMed]
23. Liu, Y.; Jiao, F.; Qiu, Y.; Li, W.; Lao, F.; Zhou, G.; Sun, B.; Xing, G.; Dong, J.; Zhao, Y.; et al. The effect of Gd@$C_{82}(OH)_{22}$ nanoparticles on the release of Th1/Th2 cytokines and induction of TNF-alpha mediated cellular immunity. *Biomaterials* **2009**, *30*, 3934–3945. [CrossRef]
24. Li, C.; Si, J.; Tan, F.; Park, K.Y.; Zhao, X. *Lactobacillus plantarum* KSFY06 prevents inflammatory response and oxidative stress in acute liver injury induced by D-Gal/LPS in mice. *Drug Des. Dev. Ther.* **2021**, *15*, 37–50. [CrossRef] [PubMed]
25. Agita, A.; Alsagaff, M.T. Inflammation, immunity, and hypertension. *Acta Med. Indones.* **2017**, *49*, 158–165. [PubMed]
26. Watanabe, T.; Juvet, S.C.; Berra, G.; Havlin, J.; Zhong, W.; Boonstra, K.; Daigneault, T.; Horie, M.; Konoeda, C.; Teskey, G.; et al. Donor IL-17 receptor A regulates LPS-potentiated acute and chronic murine lung allograft rejection. *JCI Insight* **2023**, *21*, e158002. [CrossRef] [PubMed]
27. Tomomatsu, N.; Aoki, K.; Alles, N.; Soysa, N.S.; Hussain, A.; Nakachi, H.; Kita, S.; Shimokawa, H.; Ohya, K.; Amagasa, T. LPS-induced inhibition of osteogenesis is TNF-alpha dependent in a murine tooth extraction model. *J. Bone Miner. Res.* **2009**, *24*, 1770–1781. [CrossRef]
28. Zhou, L.; Zhang, Y.F.; Yang, F.H.; Mao, H.Q.; Chen, Z.; Zhang, L. Mitochondrial DNA leakage induces odontoblast inflammation via the cGAS-STING pathway. *Cell Commun. Signal.* **2021**, *19*, 58. [CrossRef] [PubMed]
29. Rodriguez, A.E.; Ducker, G.S.; Billingham, L.K.; Martinez, C.A.; Mainolfi, N.; Suri, V.; Friedman, A.; Manfredi, M.G.; Weinberg, S.E.; Rabinowitz, J.D.; et al. Serine Metabolism Supports Macrophage IL-1β Production. *Cell Metab.* **2019**, *29*, 1003–1011.e4. [CrossRef]
30. Esteves, A.R.; Silva, D.F.; Banha, D.; Candeias, E.; Guedes, B.; Cardoso, S.M. LPS-induced mitochondrial dysfunction regulates innate immunity activation and α-synuclein oligomerization in Parkinson's disease. *Redox Biol.* **2023**, *63*, 102714. [CrossRef]
31. Sims, J.E.; Smith, D.E. The IL-1 family: Regulators of immunity. *Nat. Rev. Immunol.* **2010**, *10*, 89–102. [CrossRef] [PubMed]
32. Kishimoto, T. IL-6: From its discovery to clinical applications. *Int. Immunol.* **2010**, *22*, 347–352. [CrossRef] [PubMed]
33. Kawano, M.; Miyoshi, M.; Miyazaki, T. *Lactobacillus helveticus* SBT2171 induces A20 expression via Toll-Like receptor 2 signaling and inhibits the Lipopolysaccharide-induced activation of nuclear factor-kappa B and mitogen-activated protein kinases in peritoneal macrophages. *Front. Immunol.* **2019**, *10*, 845. [CrossRef] [PubMed]
34. Antonelli, A.; Ferrari, S.M.; Giuggioli, D.; Ferrannini, E.; Ferri, C.; Fallahi, P. Chemokine (C-X-C motif) ligand (CXCL)10 in autoimmune diseases. *Autoimmun. Rev.* **2014**, *13*, 272–280. [CrossRef] [PubMed]
35. Murphy, M.P. How mitochondria produce reactive oxygen species. *Biochem. J.* **2009**, *417*, 1–13. [CrossRef] [PubMed]
36. Liguori, I.; Russo, G.; Curcio, F.; Bulli, G.; Aran, L.; Della-Morte, D.; Gargiulo, G.; Testa, G.; Cacciatore, F.; Bonaduce, D.; et al. Oxidative stress, aging, and diseases. *Clin. Interv. Aging.* **2018**, *13*, 757–772. [CrossRef] [PubMed]
37. Couto, N.; Wood, J.; Barber, J. The role of glutathione reductase and related enzymes on cellular redox homoeostasis network. *Free Radic. Biol. Med.* **2016**, *95*, 27–42. [CrossRef]
38. Jomova, K.; Raptova, R.; Alomar, S.Y.; Alwasel, S.H.; Nepovimova, E.; Kuca, K.; Valko, M. Reactive oxygen species, toxicity, oxidative stress, and antioxidants: Chronic diseases and aging. *Arch. Toxicol.* **2023**, *97*, 2499–2574. [PubMed]
39. Green, D.R.; Galluzzi, L.; Kroemer, G. Mitochondria and the autophagy-inflammation-cell death axis in organismal aging. *Science* **2011**, *333*, 1109–1112. [CrossRef]

40. Hu, H.; Tian, M.; Ding, C.; Yu, S. The C/EBP homologous protein (CHOP) transcription factor functions in endoplasmic reticulum stress-induced apoptosis and microbial infection. *Front. Immunol.* **2019**, *9*, 3083. [CrossRef]
41. Li, W.; Luo, F.; Wu, X.; Fan, B.; Yang, M.; Zhong, W.; Guan, D.; Wang, F.; Wang, Q. Anti-inflammatory effects and mechanisms of dandelion in RAW264.7 macrophages and zebrafish larvae. *Front. Pharmacol.* **2022**, *13*, 906927. [CrossRef] [PubMed]

foods MDPI

Article

Microbiological, Functional, and Chemico-Physical Characterization of Artisanal Kombucha: An Interesting Reservoir of Microbial Diversity

Joel Armando Njieukam [1], Marianna Ciccone [1], Davide Gottardi [1,2], Arianna Ricci [1,2], Giuseppina Paola Parpinello [1,2], Lorenzo Siroli [1,2,*], Rosalba Lanciotti [1,2] and Francesca Patrignani [1,2]

1 Department of Agricultural and Food Sciences, Campus of Food Science, Alma Mater Studiorum, University of Bologna, Piazza Goidanich 60, 47521 Cesena, Italy; joelarmando.njieuka2@unibo.it (J.A.N.); marianna.ciccone2@unibo.it (M.C.); davide.gottardi2@unibo.it (D.G.); arianna.ricci4@unibo.it (A.R.); giusi.parpinello@unibo.it (G.P.P.); rosalba.lanciotti@unibo.it (R.L.); francesca.patrignani@unibo.it (F.P.)

2 Interdepartmental Centre for Agri-Food Industrial Research, Campus of Food Science, Alma Mater Studiorum, University of Bologna, Via Quinto Bucci 336, 47521 Cesena, Italy

* Correspondence: lorenzo.siroli2@unibo.it; Tel.: +39-0547-338145

Abstract: Kombucha is a trending tea fermented via a complex microflora of yeasts and acetic acid bacteria. It can be a valid low-calorie substitute for soft drinks due to its sour, naturally carbonated, and sweet taste. Despite increased interest, the microflora and functional properties of kombucha have not yet been fully understood. The aim of this work was to characterize, from a microbiological, chemico-physical, and functional point of view, three types of artisanal kombucha obtained by fermenting green tea containing sugar by means of different starter cultures. Metagenomic analysis revealed a predominance of yeasts compared to bacteria, regardless of the sample. In particular, *Brettanomyces* spp. was found to be the dominant yeast. Moreover, the different types of kombucha had different microbial patterns in terms of acetic acid bacteria and yeasts. Ethanol and acetic acid were the dominant volatile molecules of the kombucha volatilome; the samples differed from each other in terms of their content of alcohols, esters, and acids. All the samples showed a high antioxidant potential linked to the high content of phenols. This study confirmed the positive chemico-physical and functional properties of kombucha and indicated that the microflora responsible for the fermentation process can significantly affect the characteristics of the final product.

Keywords: kombucha; metagenomics; functionality; antioxidant activity; volatile molecule profile

1. Introduction

The consumer interest in functional fermented foods and drinks with beneficial health effects has significantly increased in recent years [1]. An example of a fermented beverage that is attracting considerable interest due to its functional properties is kombucha. In fact, it represents a valid substitute to soft drinks due to its acidic, naturally carbonated, and sweet taste, but with lower calories [2]. Kombucha is made starting from black or green tea infusions supplemented with sugars and a symbiotic culture of bacteria (specifically acetic acid and lactic acid bacteria) and yeasts (SCOBY) [3,4]. Kombucha is rich in bioactive compounds including natural antimicrobials, phenolic compounds (i.e., catechins and other flavonoids), amino acids (in particular, theanine), organic acids, and vitamins (E, K, B, and C), coupled with other micronutrients [3,5]. The potential health benefits of kombucha include antioxidant, antitumoral, antidiabetic, hepatoprotective, and antimicrobial properties [3]. However, most of the reported benefits should be proven via human clinical trials in order to determine whether kombucha consumption can be targeted to address specific health concerns [1].

The fermentation process lasts from 7 to 14 days at temperatures ranging between 20 and 30 °C [6,7]. Traditionally, kombucha is fermented in the presence of indigenous microorganisms, mainly yeasts (*Brettanomyces* spp., *Hanseniaspora* spp., *Saccharomyces* spp., *Pichia* spp., or *Zygosaccharomyces* spp.) and acetic acid bacteria (*Komagataeibacter* spp. *Acetobacter* spp., and *Gluconacetobacter* spp.), under aerobic conditions, and the resulting product is rich in organic acids (acetic acid, lactic acid, gluconic acid, and glucuronic acids) and CO_2 [8]. Several parameters, such as the fermentation time and temperature, sugar content, type of tea, and use of starter cultures, as well as environmental factors (geographic place) affect the microbial composition of kombucha [9]. In this system, yeasts and bacteria cooperate and compete inside kombucha, and their metabolisms strongly affect the chemical composition of the final product [10]. Consequently, kombucha microbiota and their dynamics during the fermentation process affect both the sensory and functional characteristics of the product [11], and high microbial variability depends on production conditions [12,13]. In the production of kombucha, it is customary to introduce both the SCOBY and a portion of the fermented liquid from the previous batch into the sweetened tea [14]. However, it is possible to produce kombucha without the addition of a SCOBY but only with the fermented liquid from the previous batch.

Therefore, the aim of this work was to characterize, from a microbiological, functional, and chemico-physical point of view, three different batches of kombucha obtained via the inoculum of the fermented liquid from a previous batch in order to evaluate the effect of the starter used on the final quality of kombucha. Specifically, the samples were analyzed during and at the end of the production process from a microbiological point of view, using both culture-dependent and independent (metagenomic) methods. In addition, physico-chemical and functional characteristics, as well as the total phenol content, antioxidant activity, and volatile molecule profiles, were analyzed.

2. Materials and Methods

2.1. Kombucha Fermentation

The kombucha samples were provided by a local artisanal producer (Frui-Lab, Bologna, Italy). Three different kombucha batches, P1, P2, and P3, were produced through different blending and aromatization techniques to obtain commercial products with different characteristics. The starter inoculation was performed in the absence of a SCOBY, utilizing only the fermented liquid derived from the prior batch.

The starter of batch P1 was obtained from spontaneously fermented sweetened sencha green tea to which was added ginger and beetroot, which was then periodically backslopped by the producer.

The starters of batches P2 and P3 were originally two different commercial starters periodically backslopped by the producer.

The different batches had the same ingredients, sencha green tea and sucrose, 10%, and they were obtained by adding the three different microbial starters used by the manufacturer (P1, P2, and P3). The kombucha batches were prepared by adding 5 g of dried sencha green tea into 1 L of deionized hot water for 10 min. The resulting hot tea, to which was added 10% sucrose, was used as a fermentation medium for kombucha production (sugared raw tea).

In all cases, the inoculum was performed with the liquid part of a previous batch (50% of the total) of the same type of kombucha.

All the fermentations were carried out in 150 L plastic fermenters, with a fermentation volume of 100 L. Fermentation was conducted at a controlled temperature of 25 °C for 5 days. Samples were collected immediately after the inoculum (T_0), at 48 h of fermentation (T_{48}), and at the end of fermentation (T_{end}).

2.2. Microbial Analysis Using Culture-Dependent Methods and Subsequent Identification of the Microbial Isolates

Ten mL of kombucha samples were diluted (range: 1:100–1:1,000,000) in a sterile saline solution (0.9% NaCl). The dilutions were plated on media selective for acetic acid bacteria (AAB), lactic acid bacteria (LAB), and yeasts. Specifically, for the detection of

AAB, Gluconobacter agar medium-GA (25 g/L of mannitol, 5 g/L of yeast extract, 3 g/L of peptone, and 15 of g/L agar), glucose yeast extract agar—GYA (50 g/L of glucose, 5 g/L of yeast extract, and 15 g/L agar), acetic acid medium agar—AA (10 g/L of glucose, 5 g/L of ethanol, 3 g/L of acetic acid, 15 g/L of peptone, 8 g/L of yeast extract, and 15 g/L of agar) were used, all supplemented with cycloheximide as a selective agent. Yeast extract peptone dextrose (YPD) (Oxoid, Basingstoke, UK) with chloramphenicol added was used for the enumeration of yeasts, while LAB were enumerated in de Man Rogosa and Sharpe (MRS) medium (Oxoid, Basingstoke, UK) to which cycloheximide was added. Plates for the detection of AAB and yeasts were incubated at 25 °C for 24–48 h, while MRS agar plates were incubated at 30 °C for 24 h. Microbiological analyses were performed immediately after the inoculum (T_0), at 48 h of fermentation (T_{48}), and at the end of fermentation (T_{end}).

The GA, GYA, AA, YPD, and MRS media allowed for the isolation of different colonies characterized by distinct morphologies. Afterwards, isolates were cultivated in their appropriate media and conditions for 24 h and then re-streaked onto appropriate agar media. Genomic DNA was extracted from each isolate using the InstaGene Matrix kit (Bio-Rad Laboratories, Milano, Italy). A total of 35 representative isolates were identified using RAPD-PCR (primer M13) and by sequencing the 16S and ITS1-ITS4 rRNA region for bacteria and yeasts, respectively, according to the protocol described by Siroli et al. [15].

2.3. *Metagenomic Analyses*

Metagenomic DNA from liquid kombucha samples was extracted at the beginning (T_0) and end of fermentation (T_{end}). gDNA was extracted with ZymoBIOMICS DNA Mini Kit (Zymo research, Irvine, CA, USA), and it was quantified with a spectrophotometer and a fluorimetric method using Qubit™ dsDNA HS Assay Kit (ThermoFisher Scientific, Waltham, MA, USA). Then, 20–200 ng/g of DNA was sheared in 10 mM of Tris-HCl with sonication to a fragment size ranging from 100 to 1000 bp. DNA libraries were generated following the standard protocol, including End Repair, dA-Tailing, Adaptor Ligation, and PCR Enrichment NEBNext® Ultra™ II DNA Library Prep Kit for Illumina® (New England Biolabs, Ipswich, MA, USA). The libraries were quantified with a Qubit™ dsDNA HS Assay Kit and analyzed with a High Sensitivity D1000 ScreenTape Assay kit (Agilent Technologies, Santa Clara, CA, USA) and the Agilent TapeStation 4200 instrument (Agilent Technologies, Santa Clara, CA, USA). The libraries were sequenced on an Illumina NovaSeq 6000 instrument (Illumina, San Diego, CA, USA) with paired-end reads 150 bp long. Fastq files were quality-tested using FastQC (https://www.bioinformatics.babraham.ac.uk/projects/fastqc/, accessed on 1 June 2024). A metagenomic analysis was performed using the CCMetagen pipeline, v. 1.3 [16].

2.4. *Physico-Chemical and Functional Analyses*

Samples were analyzed for ethanol (%), density (d), total and volatile acidity (g/L), organic acids (g/L, malic, lactic, and citric), sugars (g/L, glucose, and fructose), potassium (g/L), and catechins (g/L) using the multiparameter analyzer Bacchus 3 (Steroglass Srl, Perugia, Italy), consisting of a is5 FTIR Thermo Fisher Scientific spectrophotometer (Walthman, MS, USA) equipped with a hydraulic system for automatic sampling, a filter unit, and a "Bacchus Analysis" software (version 3) interface for analysis management. The pH (units) analysis was carried out using a pH meter (BASIC 20, Crison, Modena, Italy). The gluconic acid (mg/L) was determined with an enzymatic assay kit (K-GATE, Megazyme). Antioxidant activity (%) was evaluated through 2,2-diphenyl-1-picryl-hydrazyl (DPPH) by diluting samples in 80% methanol. For the calculation of the μmol of DPPH radical scavenged via the extracts, the absorbance value was measured at 517 nm after 10 min. A blank reagent was used to verify the stability of the DPPH radical dot during the test time. The kinetics of the antioxidant reaction were also determined over 30 min and compared with gallic acid as an antioxidant reference and expressed as mg/L of gallic acid equivalent.

The analysis of total phenolic compounds was carried out according to the Folin–Ciocalteu methodology, as reported by Slinkard and Singleton [17]. The mixture absorbance was read at 750 nm, and the concentration of total phenolic compounds was expressed as mg/L of gallic acid equivalents.

2.5. *Volatile Molecule Profiles*

Volatile molecule profile analyses were conducted on the kombucha samples using gaschromatography–mass spectrometry (GC-MS), coupled with a solid-phase microextraction (SPME) technique, according to Siroli et al. [18]. The samples (3 mL, placed in sterile vials), were added to 6 µL of standard 4-methyl-2-pentanol at 10,000 mg/kg and heated for 10 min at 45 °C. A Carboxen/polydimethylsiloxane (Carboxen/PDMS) fiber (85 µm, Stalleflex Supelco, Bellefonte, PA, USA) was exposed to the sample headspace for 30 min and then desorbed for 10 min in the injector of the GC, coupled with a quadrupolar mass-selective spectrometer (Agilent 7890 A GC and Agilent 5975 C MS, Agilent Technologies). Analytes were separated on a Chrompack CP-Wax 52 CB column (Chrompack, São Paulo, Brazil) using the conditions described by Siroli et al. [18]. The assignation of the volatile peaks was conducted by comparing the mass spectral data with those of the references present in the NIST library (NIST/EPA/NIH Mass spectral Library, Version 1.6, Gaithersburg, MD, USA) of 2011 and WILEY (sixth edition, New York, NY, USA) of 1995. Volatile compounds were quantified in mg/kg equivalents to the internal standard used (4-methyl-2-penthanol).

2.6. *Statistical Analysis*

The results are expressed as means of three independent measurements from three repeated experiments on different days. The data were statistically analyzed using a one-way ANOVA (statistica 8.0, Tulsa, OK, USA). The differences between mean values were detected via the HSD Tukey test, and evaluations were based on a significance level of $p \leq 0.05$. Principal component analysis (PCA) was performed using the Statistica software, version 8.0, to obtain a visual overview of the volatile molecule profile of the different kombucha samples.

3. Results

3.1. *Microbial Characterization of Kombucha Samples via Culture-Dependent Methods*

The three different batches of kombucha (P1, P2, and P3) were analyzed during fermentation for microbiological patterns using culture-dependent methods. Figure 1 shows the cell load of AAB, LAB, and yeasts in the three types of kombucha after the inoculum (T_0), 48 h later (T_{48}), and at the end of the fermentation process, 120 h later (T_{end}).

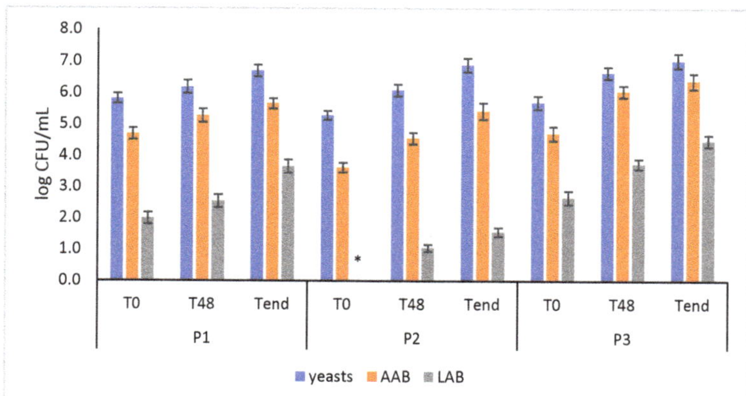

Figure 1. Cell load of yeasts, AAB, and LAB in the different kombucha samples (P1, P2, and P3) analyzed during the fermentation process (T_0, T_{48}, and T_{end}). *: below the limit of quantification (<1 log CFU/mL).

As shown in Figure 1, yeasts were found to be the dominant microflora independently of the type of kombucha and the time of sampling. They reached 6.7, 6.9, and 7.0 log CFU/mL in the P1, P2, and P3 samples, respectively, at the end of fermentation. The increase in the microbial load of the yeasts at the end of fermentation compared to the initial level ranged between 0.9 and 1.6 logarithmic cycles independently of the condition. Samples P1 and P3 showed an initial cell load of AAB (4.7 log CFU/mL) higher than P2 (3.6 log CFU/mL). All the samples increased the cell load of AAB during the fermentation process, with sample P3 reaching the highest cell load (6.4 log CFU/mL) of AAB at the end of fermentation. LABs were found on kombucha samples at lower levels than AAB and yeasts. Differences between the samples were found during and at the end of fermentation. The final LAB cell load was found to be 3.7, 1.6, and 4.5 log CFU/mL in samples P1, P2, and P3, respectively.

Colonies of different morphologies were isolated from the different culture media and identified. Among the 35 identified isolates, 15 were yeasts, 18 were acetic acid bacteria, and 2 were lactic acid bacteria (Table 1). The use of the traditional culture methods of isolation and identification resulted in the detection of limited biodiversity, especially in the case of yeasts that belonged only to the species *Starmerella davenportii* and *Brettanomyces bruxellensis*. *Starmerella* was detected in all samples and at all sampling times, while *Brettanomyces* was detected in all samples but only at the end of fermentation. Regarding LAB, only *Liquorilactobacillus nagelii* was detected in samples P1 and P3. AAB was detected in all samples, with *Acetobacter tropicalis* only at the beginning of fermentation and *Komagataeibacter* spp. at almost all sampling times, while *Gluconobacter oxydans* and *Komagataeibacter hansenii* were present only in samples P1 and P2, respectively.

Table 1. Isolated and identified strains from different kombucha samples at different sampling times.

Sample	Sampling Time	Strain Code	Identification
P1	T_0	1MB	*Acetobacter tropicalis*
		1DA	*Komagataeibacter* spp.
		1YB	*Starmerella davenportii*
		2MB	*Acetobacter tropicalis*
		1YC	*Starmerella davenportii*
	T_{48}	3DB	*Acetobacter tropicalis*
		3DA	*Gluconobacter oxydans*
		3YC	*Starmerella davenportii*
	T_{end}	4MA	*Liquorilactobacillus nagelii*
		4MF	*Komagataeibacter* spp.
		4YF	*Brettanomyces bruxellensis*
		4YB	*Starmerella davenportii*
P2	T_0	5YB	*Starmerella davenportii*
		6MA	*Acetobacter tropicalis*
		6YC	*Starmerella davenportii*
	T_{48}	7DB	*Komagataeibacter* spp.
		7DC	*Komagataeibacter* spp.
		7MA	*Komagataeibacter hansenii*
		7YB	*Starmerella davenportii*
	T_{end}	8DB	*Komagataeibacter* spp.
		8YE	*Brettanomyces bruxellensis*
		8YC	*Starmerella davenportii*

Table 1. *Cont.*

Sample	Sampling Time	Strain Code	Identification
P3	T_0	9DA	*Acetobacter tropicalis*
		9DB	*Acetobacter tropicalis*
		9YB	*Starmerella davenportii*
		10DA	*Komagataeibacter* spp.
		10MA	*Liquorilactobacillus nagelii*
		10YB	*Starmerella davenportii*
	T_{48}	11DA	*Komagataeibacter* spp.
		11DB	*Komagataeibacter* spp.
		11YB	*Starmerella davenportii*
	T_{end}	12YB	*Starmerella davenportii*
		12YF	*Brettanomyces bruxellensis*
		12DA	*Komagataeibacter* spp.
		12DB	*Komagataeibacter* spp.

3.2. Metagenomic Analyses

Also, culture-independent analyses (metagenomic) were performed on the kombucha samples (T_0 and T_{end}) since traditional culture methods for microbial population detection can have numerous limitations due to the presence of viable but not cultivable cells and fastidious microorganisms that can lead to an underestimation of real microbial populations. Table 2 shows, for each sample, the total number of readings and those attributable to the kingdom of bacteria and fungi, as well as their percentages of the total.

Table 2. Relative abundance of bacteria, fungi, and bacteria/fungi ratio in Kombucha samples according to metagenomic analyses.

Sample	Sampling Day	Fungi (%)	Bacteria (%)	Fungi/Bacteria Ratio
P1	T_0	98.83	1.17	84.5
	T_{end}	96.93	3.07	31.6
P2	T_0	99.27	0.73	135.9
	T_{end}	99.34	0.66	150.5
P3	T_0	89.84	10.16	8.8
	T_{end}	69.94	30.06	2.3

As can be seen from Table 2, the fungi kingdom was the dominant one in all the samples, with abundances at T_0 of 98.83, 99.27, and 89.84% for P1, P2, and P3, respectively. Negligible reductions in fungi's relative abundance were observed in the P1 and P2 samples at the end of fermentation, while sample P3 presented a drastic decrease in fungi's relative abundance (final value: 69.9%). Bacteria were found to be less present than fungi, which was also in accordance with what was observed from traditional culture-dependent methods, and the ratio of fungi to bacteria was found to range between 84.2 and 31.5 for sample P1, between 135.6 and 150.7 for sample P2, and between 2.3 and 8.8 for sample P3.

Taxonomic assignments were performed at the phylum, family, genus and species levels. An average of 99.9% of fungi readings could be assigned at the phylum level, while, in the case of bacteria, the average percentage of assigned readings was found to be 73.5%, 38.3%, and 95.3%, respectively, for samples P1, P2, and P3. Figure 2 shows the relative abundance of fungi and bacteria phyla, families genera and species.

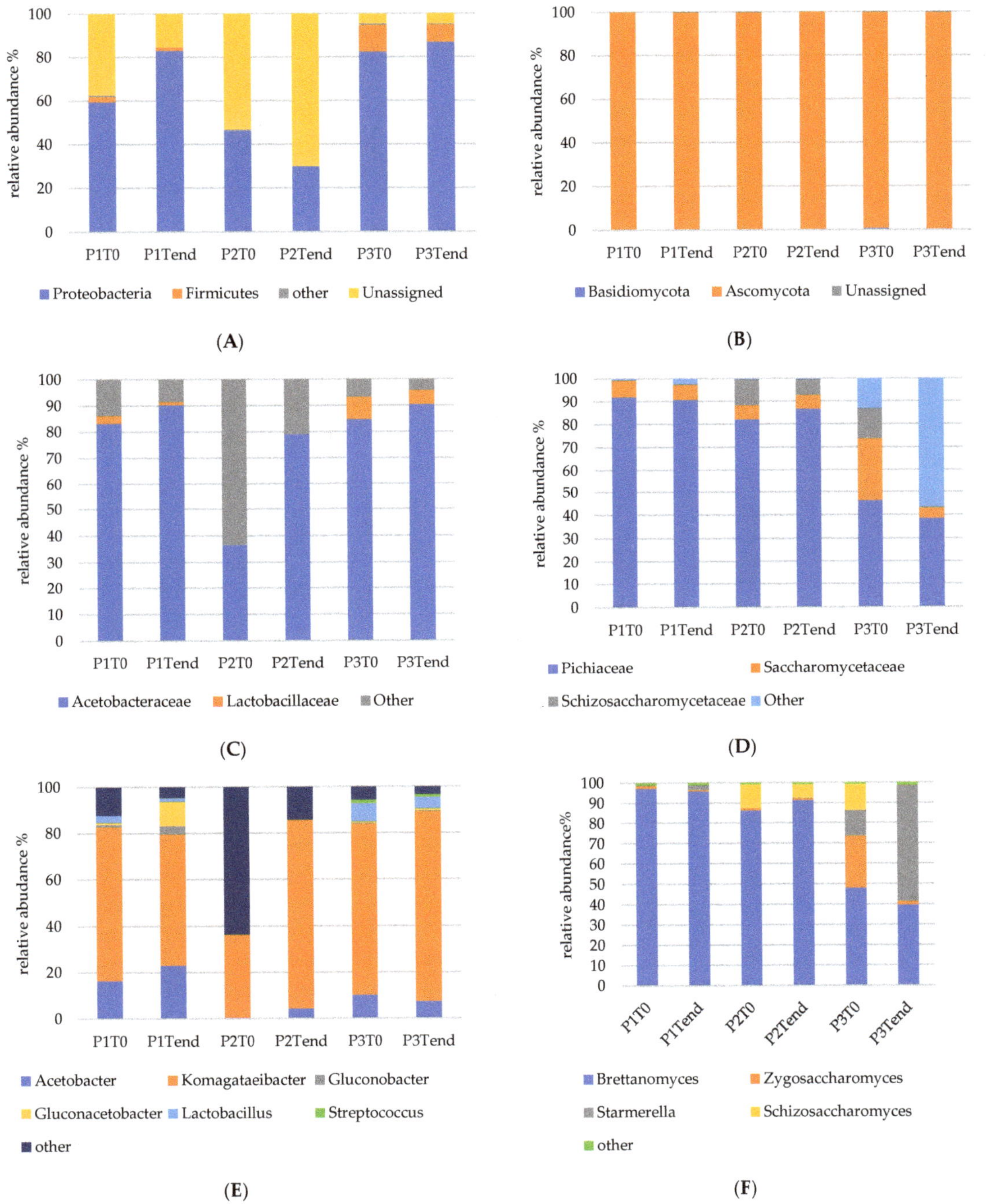

Figure 2. *Cont.*

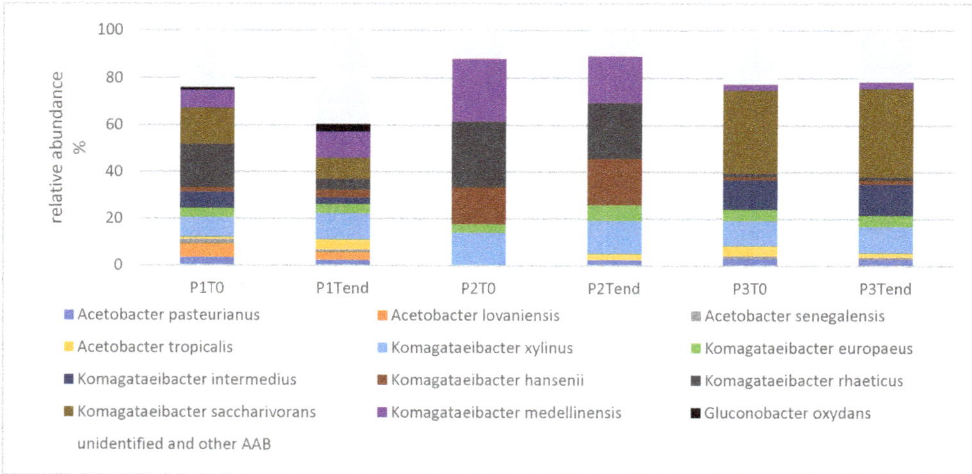

(G)

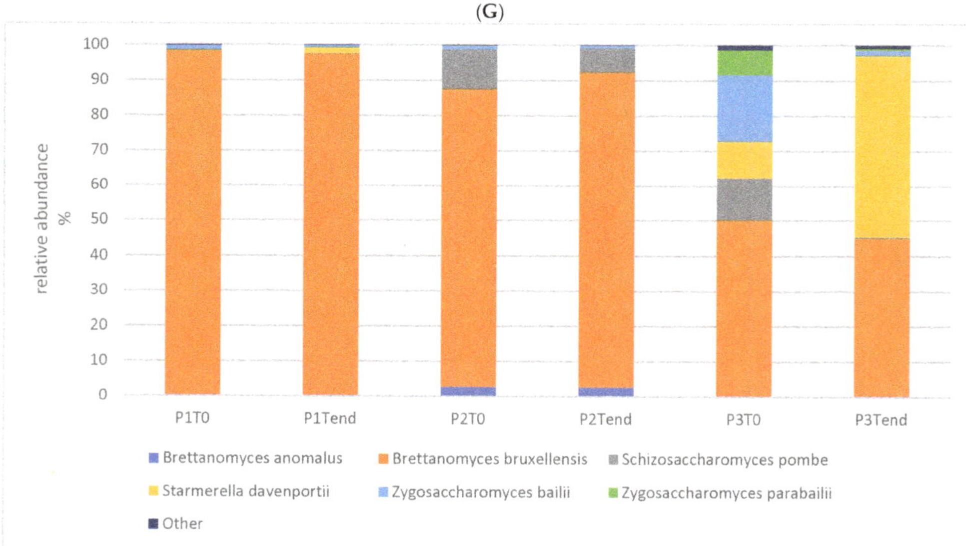

(H)

Figure 2. Relative abundance of bacterial (**A**) and fungal (**B**) phyla, bacterial (**C**) and fungal (**D**) families, bacterial (**E**) and fungal (**F**) genera, and bacterial (**G**) and fungal (**H**) species in different kombucha samples. Only phyla, families, and genera accounting for a minimum of 1% of reads in at least one sample were reported. For families and genera, unassigned bacteria and fungi were not considered.

As shown in Figure 2A,B, Ascomycota accounted for over 99% of the fungal community in all samples considered, while Proteobacteria was the dominant bacterial phylum, accounting for 60% and 83% of sample P1, 46% and 30% of sample P2, 82% and 87% of sample P3 at the beginning and end of fermentation, respectively. A significant presence of Firmicutes was detected in samples P3 (12% and 8%) and P1 (3% and 2%), while sample P1 was characterized by the highest level of unassigned bacteria (53% and 70%).

Concerning fungi, at the family level (Figure 2D), Pichiaceae were the most abundant in samples P1 (90–91%) and P2 (82–86%), while in sample P3, they accounted only for 46% at T_0 and 39% at the end of fermentation. Saccharomycetaceae were present at levels of about

6–7% in samples P1 and P2, regardless of the sampling time, whereas, in the P3 sample, they were present with an initial relative abundance of 28% and then decreased to 4% at the end of fermentation. Also, Schizosaccharomycetaceae were detected at percentages higher than 1%, but only in samples P2 and P3; in both cases, their relative abundance was higher at T_0, at 11% and 12%, respectively, compared to the end of fermentation (6.8% and 0.4%, respectively). In sample P3, particularly at the end of fermentation, other fungi families also showed a high abundance (57%). At the genus level (Figure 2F), *Brettanomyces* spp. were found to be the predominant yeast in all the samples, especially in sample P1, in which its relative abundance was 97% at T_0 and 95% at T_{end}, and sample P2 (86% at T_0 and 92% at T_{end}). The *Brettanomyces* abundance was lower in sample P3, in which the initial abundance of 48% decreased to 40% at the end of fermentation. In sample P1, the only other representative genera whose relative abundances were more than 1% were *Zygosaccharomyces* spp., present at 1.5% at the initial time and 0.8% at the end of fermentation, and *Starmerella* spp., found at 2% at the end of fermentation. In sample P2, *Schizosaccharomyces* spp. were detected at a relative abundance of 12% and 7%, respectively, at T_0 and T_{end}, while *Starmerella* was present at 0.02% and 0.07%, respectively, at T_0 and T_{end}. In sample P3, *Starmerella* was found to be the most representative yeast at T_{end}, showing a relative abundance of 57%. At the species level (Figure 2H), *B. bruxellensis* was predominant, accounting for over 97% of the P1 sample and over 85% of the P3 sample at both the beginning and end of fermentation. In the P2 samples, only *S. pombe* and *B. anomalus* were present at percentages higher than 2% at T_{end}. In the case of the P3 samples, *B. bruxellensis, S. pombe, S. davenportii*, and *Z. bailii* were detected at the beginning of fermentation at percentages higher than 10%. However, by the end of fermentation, *S. davenportii* and *B. bruxellensis* accounted for 97% of the yeast population.

Regarding bacteria (Figure 2C,E), Acetobacteraceae were the most abundant family, and they increased over time up to levels of 90%, 79%, and 91% in P1, P2, and P3, respectively. Lactobacillaceae were found only in sample P1 (1–2%) and to a major extent in P3 (5–8%). At the genus level, *Komagataeibacter* spp. was the most representative genus in all the kombucha samples since its relative abundance at T_{end} was higher than 80% in samples P2 and P3, and it was 56% in sample P1. The relative abundance of *Acetobacter* spp. increased in samples P1 and P2 during fermentation and was detected at T_{end} at an abundance of 23%, 4%, and 7% in samples P1, P2, and P3, respectively. *Gluconobacter* spp. and *Gluconacetobacter* spp. were present at a relative abundance higher than 1% only in sample P1. Finally, *Lactobacillus* spp. was detected only in samples P1 and P3, while *Streptococcus* spp. was detected only in the P3 sample. At the species level (Figure 2G), in the P1 sample, *K. saccharivorans, K. rhaeticus, K. xylinus*, and *K. medellinensis* were the predominant species, but none of them were present at a percentage higher than 11%. However, these samples also showed high percentages of unidentified AAB. In the P2 samples, *K. rhaeticus* and *K. medellinensis* were the predominant AAB at the beginning of fermentation, while at the end of fermentation, in addition to these species, *K. hansenii* and *K. xylinus* were also present at percentages close to 20%. Concerning the P3 samples, *K. saccharivorans* was the dominant AAB at both T_0 and T_{end} (>30%), but *K. xylinus* and *K. intermedius* also increased their presence at T_{end}.

3.3. Physico-Chemical and Functional Analyses

Table 3 shows the chemical composition of the different kombucha samples, including the unfermented sencha green tea (control). An increase in alcohol content compared to the sugared raw material was observed in all the samples (P1, P2, and P3). The alcohol content ranged between 0.90% and 1.54%, regardless of the time of fermentation. P1 and P2 showed a higher alcohol content compared to P3, while no significant differences were observed within each batch.

Table 3. Chemical profiles of the different kombucha samples (P1, P2, and P3) at different times of fermentation (T_0, T_{48}, and T_{end}) and sugared raw sencha tea.

	P1			P2			P3			
	T_0	T_{48}	T_{end}	T_0	T_{48}	T_{end}	T_0	T_{48}	T_{end}	**Sugared Raw Tea**
Alcohol (%)	1.28 ± 0.14^{cd}	1.32 ± 0.15^{cd}	1.46 ± 0.16^{d}	1.42 ± 0.16^{d}	1.54 ± 0.17^{d}	1.52 ± 0.17^{d}	0.90 ± 0.10^{b}	1.00 ± 0.11^{b}	1.08 ± 0.12^{bc}	0.45 ± 0.05^{a}
Density (d)	2.0 ± 0.1^{a}	2.0 ± 0.1^{a}	2.0 ± 0.2^{a}	2.0 ± 0.2^{a}	2.0 ± 0.1^{a}	2.0 ± 0.1^{a}	2.0 ± 0.1^{a}	2.0 ± 0.1^{a}	2.0 ± 0.1^{a}	2.0 ± 0.1^{a}
Total acidity (g/L)	4.7 ± 0.3^{c}	4.9 ± 0.3^{cd}	5.0 ± 0.3^{cd}	4.3 ± 0.5^{bc}	3.7 ± 0.5^{b}	4.5 ± 0.5^{bc}	5.3 ± 0.4^{cd}	5.5 ± 0.3^{d}	8.4 ± 0.6^{e}	0.30 ± 0.05^{a}
pH (U)	3.07 ± 0.10^{cd}	3.06 ± 0.08^{cd}	3.06 ± 0.09^{cd}	3.16 ± 0.10^{d}	3.02 ± 0.11^{bcd}	3.17 ± 0.08^{d}	2.93 ± 0.07^{bc}	2.88 ± 0.05^{b}	2.76 ± 0.05^{a}	7.46 ± 0.09^{e}
Volatile acidity (g/L)	5.1 ± 0.3^{d}	5.2 ± 0.2^{d}	4.8 ± 0.3^{d}	3.9 ± 0.2^{c}	3.0 ± 0.3^{b}	3.4 ± 0.3^{bc}	3.5 ± 0.3^{bc}	3.6 ± 0.3^{bc}	5.2 ± 0.3^{d}	0.11 ± 0013^{a}
Malic acid (g/L)	2.22 ± 0.12^{b}	2.24 ± 0.21^{b}	2.18 ± 0.18^{b}	2.08 ± 0.15^{b}	1.72 ± 0.12^{a}	2.28 ± 0.16^{b}	2.82 ± 0.22^{c}	2.92 ± 0.23^{c}	4.50 ± 0.33^{d}	*
Lactic acid (g/L)	0.44 ± 0.09^{abc}	0.26 ± 0.09^{a}	0.38 ± 0.08^{ab}	0.44 ± 0.10^{abc}	0.64 ± 0.08^{d}	0.62 ± 0.07^{d}	0.62 ± 0.10^{cd}	0.52 ± 0.08^{bcd}	0.26 ± 0.09^{a}	0.56 ± 0.10^{bcd}
Gluconic acid (g/L)	1.05 ± 0.12^{b}	1.20 ± 0.08^{bc}	1.27 ± 0.07^{c}	0.30 ± 0.05^{a}	0.31 ± 0.04^{a}	0.42 ± 0.10^{a}	1.21 ± 0.15^{bc}	1.08 ± 0.08^{b}	2.28 ± 0.19^{d}	*
Citric acid (g/L)	*	*	*	*	*	*	*	*	*	0.44 ± 0.10
Glucose (g/L)	22.6 ± 1.6^{bc}	22.2 ± 1.4^{bc}	24.1 ± 1.5^{c}	21.0 ± 1.1^{b}	20.9 ± 0.9^{b}	21.2 ± 1.0^{b}	29.1 ± 1.5^{d}	30.0 ± 1.6^{d}	31.1 ± 1.7^{d}	17.0 ± 0.7^{a}
Fructose (g/L)	15.4 ± 0.8^{cd}	16.1 ± 0.7^{d}	16.5 ± 1.0^{d}	13.6 ± 1.0^{bc}	12.6 ± 0.7^{ab}	11.6 ± 0.8^{a}	15.3 ± 1.0^{cd}	14.6 ± 0.9^{cd}	12.4 ± 1.1^{ab}	11.2 ± 0.8^{a}
Sucrose (g/L)	69.4 ± 3.4^{e}	37.26 ± 3.1^{d}	20.6 ± 1.6^{b}	66.7 ± 2.8^{e}	33.2 ± 2.7^{cd}	17.3 ± 1.8^{ab}	67.8 ± 2.4^{e}	31.6 ± 2.3^{c}	14.3 ± 1.1^{a}	92.1 ± 3.2^{f}
Potassium (g/L)	1.51 ± 0.10^{a}	1.48 ± 0.12^{a}	1.57 ± 0.14^{ab}	1.78 ± 0.12^{bc}	1.84 ± 0.14^{bc}	2.01 ± 0.13^{cd}	2.2 ± 0.13^{de}	2.19 ± 0.15^{de}	2.42 ± 0.12^{e}	1.70 ± 0.14^{abc}
Catechins (g/L)	0.23 ± 0.02^{c}	0.24 ± 0.02^{c}	0.23 ± 0.03^{c}	0.19 ± 0.02^{abc}	0.16 ± 0.03^{ab}	0.17 ± 0.02^{ab}	0.20 ± 0.02^{bc}	0.20 ± 0.02^{bc}	0.25 ± 0.02^{c}	0.16 ± 0.02^{a}
Total phenolics (g/L, gallic acid equivalent)	0.29 ± 0.04^{bc}	0.27 ± 0.04^{abc}	0.21 ± 0.03^{a}	0.47 ± 0.05^{d}	0.48 ± 0.06^{d}	0.33 ± 0.04^{c}	0.29 ± 0.03^{bc}	0.32 ± 0.04^{c}	0.22 ± 0.05^{ab}	0.35 ± 0.05^{c}
DPPH (g/L, gallic acid equivalent)	0.32 ± 0.03^{b}	0.32 ± 35^{b}	0.30 ± 0.04^{ab}	0.36 ± 0.05^{bc}	0.34 ± 0.04^{bc}	0.34 ± 0.03^{bc}	0.25 ± 0.04^{a}	0.29 ± 0.04^{ab}	0.25 ± 0.03^{a}	0.41 ± 0.05^{c}

*: below the detection limit (0.1 g/L). $^{a–f}$: For the same parameter, average values lacking a common letter (a,b...) significantly differ from each other.

At the end of fermentation, sample P3 was characterized by the lowest pH (2.76) and the highest total acidity (8.4 g/L). Samples P1 and P3 showed significantly higher volatile acidity (4.8 and 5.2 g/L, respectively) compared to sample P2 (3.4 g/L). Sample P3 at T_{end} showed a significant increase in malic acid, while in the other samples, the malic acid content did not deviate much from the T_0 values. The lactic acid content did not change significantly in sample P1, while it increased in P2 and decreased in P3 at T_{end}. Gluconic acid was not detected in the sencha green tea, while it increased at the end of fermentation in samples P1 and P3.

The glucose concentration was found to be higher in the P3 sample at both the beginning and end of fermentation (29–31 g/L), while it ranged between 21 and 24 g/L in samples P1 and P2. Fructose was found to be present in lower quantities than glucose (12–17 g/L), but in this case, a significant decrease was observed at T_{end} in samples P2 and P3. The glucose and fructose content was lower in the sugared sencha green tea (sugared raw tea). The highest concentration of sucrose was observed in the sugared raw tea. In the samples at T_0, the concentrations ranged from 67 to 69 g/L. A reduction in the concentration of sucrose was observed in all samples at both T_{48} and T_{end}. The residual sucrose at the end of fermentation was found to be higher in the P1 samples (20.6 g/L), followed by P2 (17 g/L) and P3 (14 g/L).

The potassium content was affected by the kombucha batch (P3 > P2 > P1). In particular, P3 samples at T_{end} showed a significant increase in potassium compared to the other samples. The catechin content did not change significantly during the fermentation process within each sample; however, samples P1 and P3 at the end of fermentation showed significantly higher catechin content compared to the sencha green tea and sample P2 at T_{end}. Regarding the total phenolic compounds and DPPH scavenging activity, samples P1 and P3 showed a significant decrease in these parameters at the end of fermentation compared to the starting sencha green tea.

3.4. Volatile Molecule Profiles

Specific volatile molecule profiles were detected based on the considered kombucha sample (Table 4). A total of 37, 32, 31, and 21 volatile molecules were identified in the aromatic profiles of P1, P2, P3, and the sugared raw tea samples, respectively. The molecules detected in the kombucha samples belonged to the chemical classes of alcohols, acids, ketones, aldehydes, and esters. The initial concentration of volatile molecules in the sencha green tea was lower (6 ppm) than in the fermented samples (83–133 ppm).

Table 4. Volatile compounds (mean value expressed as ppm equivalent) detected through GC-MS-SPME in the different kombucha samples (P1, P2, and P3) during fermentation (T_0, T_{48}, and T_{end}).

Compounds	ppm Equivalents									
	Samples									
	P1			P2			P3			Sugared Raw Tea
	T_0	T_{48}	T_{end}	T_0	T_{48}	T_{end}	T_0	T_{48}	T_{end}	
Pentanal	-	-	-	-	-	-	-	-	-	0.07
Acetaldehyde	0.02	0.06	0.05	0.14	0.08	0.10	0.95	0.88	1.32	-
Hexanal	-	-	-	-	-	-	-	-	-	0.45
Nonanal	0.06	0.11	0.12	0.03	0.05	-	0.11	0.08	0.15	0.18
Decanal	-	-	-	-	-	-	-	-	-	0.07
2,4 Heptadienal (E,E)	-	-	-	-	-	-	-	-	-	0.33
Furfural	0.29	0.40	0.35	0.11	-	-	-	-	-	0.36
Benzaldehyde	-	-	-	-	-	-	0.10	-	-	0.09
Phenylacetaldehyde	-	-	-	-	-	-	0.16	0.14	0.05	-
Total aldehydes	0.37	0.58	0.52	0.28	0.13	0.10	1.32	1.10	1.52	1.54

Table 4. *Cont.*

Compounds	ppm Equivalents									
	Samples									
	P1			P2			P3			Sugared Raw Tea
	T_0	T_{48}	T_{end}	T_0	T_{48}	T_{end}	T_0	T_{48}	T_{end}	
Acetic acid	37.62	31.19	28.67	37.79	34.31	44.50	31.20	49.96	56.22	-
2-Methyl-propionic acid	-	0.17	-	-	0.28	-	-	-	-	-
Butanoic acid	0.19	0.26	-	-	-	-	-	-	-	-
Isopentanoic acid	1.08	1.09	0.96	0.84	0.91	1.22	0.68	0.90	0.62	-
Hexanoic acid	0.94	1.16	1.15	0.92	0.95	0.18	0.66	0.60	0.44	-
Propanoic acid, 2-methyl-	0.17	0.25	0.21	-	-	-	-	-	-	-
Caprylic acid	4.95	6.35	5.99	2.18	3.20	0.07	1.02	1.73	2.62	-
Pelargonic acid	0.23	0.41	0.30	0.07	0.10	0.06	0.12	0.09	0.03	-
Decanoic acid	1.55	2.36	1.81	0.03	0.56	0.08	0.24	0.40	0.72	-
Total acids	46.72	43.24	39.09	41.83	40.31	46.10	33.91	53.67	60.64	-
Methyl isobutyl ketone	0.32	0.37	0.29	0.34	0.30	0.27	0.39	0.32	0.58	0.49
3-Hexanone, 5-methyl-	-	-	-	-	-	-	0.09	-	-	0.09
2-Hexanone, 4-methyl-	0.13	0.14	0.13	0.11	0.11	0.02	0.16	0.13	0.25	0.30
4 Heptanone 2,6 dimethyl	-	-	-	-	-	-	-	-	-	1.24
Diisobutyl ketone	0.89	1.26	0.89	0.88	-	-	1.50	0.92	0.07	-
2-Heptanone	-	0.01	0.05	-	-	-	0.07	-	-	-
2-Heptanone, 4,6-dimethyl-	0.02	0.05	-	-	-	-	0.05	-	-	0.06
3-Hydroxy-2-butanone	-	-	-	-	-	-	0.05	0.10	-	-
Hydroxyacetone	-	0.08	0.05	-	-	-	0.02	0.01	0.17	-
3,5-Octadien-2-one	0.02	0.03	-	0.04	-	-	0.78	-	-	0.20
1,3-Cyclopentanedione	0.05	0.05	-	-	-	-	-	-	-	0.05
Total ketones	1.43	1.99	1.40	1.37	0.41	0.28	3.11	1.47	1.07	2.43
Ethanol	25.42	21.40	31.77	43.57	51.84	72.27	17.74	37.41	42.30	-
Isoamyl alcohol	0.65	0.71	1.04	1.23	1.76	1.63	0.53	0.88	0.61	-
2 Penten 1 ol (Z)	-	-	-	-	-	-	-	-	-	0.27
1-Hexanol	-	-	-	0.08	-	-	-	-	-	0.29
2-Hexanol	0.06	0.08	0.08	0.03	0.06	-	0.06	-	-	0.30
2-Ethylhexanol	0.21	0.20	0.17	-	-	-	1.23	0.07	0.04	0.11
1,6-Octadien-3-ol, 3,7-dimethyl-	0.41	0.65	0.66	0.29	0.28	0.12	0.09	0.54	0.35	0.66
Alpha.-Terpineol	0.55	0.65	0.81	0.38	0.31	-	0.68	0.63	0.15	-
Phenylethyl alcohol	3.50	4.79	5.89	5.40	4.57	3.97	7.23	7.90	4.55	-
4-Ethylguaiacol	0.31	0.34	0.41	0.11	0.86	0.83	-	-	-	-
4-Ethylphenol	0.20	0.27	0.32	0.67	0.88	0.84	-	-	-	-
2,4-Di-tert-butylphenol	0.36	0.43	0.26	0.27	0.12	0.11	0.54	0.47	0.14	0.26
Farnesol	0.58	0.66	0.38	0.31	-	-	-	-	-	-
Total alcohols	32.25	30.18	41.80	52.34	60.69	79.77	28.11	47.89	48.14	1.89
Ethyl acetate	4.51	4.65	6.51	4.98	4.18	3.68	4.77	9.01	3.53	-
Ethyl hexanoate	-	-	-	-	0.11	-	-	-	-	-
Ethyl octanoate	0.17	0.28	0.48	2.90	2.83	1.30	0.10	0.06	0.07	-
Ethyl decanoate	0.10	0.16	0.29	1.39	0.71	0.53	-	-	-	-
Ethyl 9-hexadecenoate	-	0.05	0.09	0.52	0.77	0.29	-	-	-	-
2-Phenylethyl acetate	0.24	0.29	0.39	0.47	0.43	0.33	0.30	0.38	0.20	-
Butyl butanoate	0.41	0.52	0.46	0.40	0.27	0.28	0.17	0.16	0.08	0.11
Total esters	5.43	5.94	8.23	10.66	9.31	6.40	5.34	9.60	3.88	0.11
Other molecules	0.87	1.14	0.94	0.77	0.72	0.51	1.09	0.88	0.85	0.16
Total molecules	87.1	83.2	92.0	107.3	111.6	133.2	72.9	114.6	116.2	6.1

The coefficients of variability, expressed as the percentage ratios between the standard deviations and the mean values, ranged between 2% and 5%. "-": not detected.

In the kombucha samples, the predominant molecules were alcohols and acids, followed to a lesser extent by esters. In general, samples P2 and P3 showed a higher concentration of volatile molecules than P1. The total amount of volatile molecules did not increase over time in sample P1, while an increase in the amount of alcohols and acids was observed during fermentation in samples P2 and P3. The total amounts of alcohols, esters, and acids were already high at the beginning of fermentation. This is not surprising since the initial inoculum was performed with a 50% liquid phase of kombucha coming from a previous batch. In the kombucha samples, the characterizing molecules were acetic acid, ethanol, phenylethyl alcohol, ethyl acetate, and caprylic acid.

The GC-MS-SPME data were subjected to a principal component analysis (PCA) to better disclose the different kombucha samples. Figure 3 shows the projection of cases on the factorial plane (1 × 2) spanned by the first two factors (PC1 and PC2), where the main effect was the kombucha starter (P1, P2, and P3). The P1 samples were well separated from the P2 ones, mainly along PC1, which explains 33.74% of the variance, while they were well separated from those of P3, mainly along PC2, which explains 23.32% of the variance. The P2 samples, on the other hand, were well separated from the P3 samples both along the PC1 and, above all, along the PC2. In the case of the P3 sample, T_0 was separated along PC1 from samples T_{48} and T_{end}, which were found to be very similar. In contrast, samples P1 and P2 at T_0 and T_{48} grouped together and were separated from sample T_{end} along both PC1 and PC2.

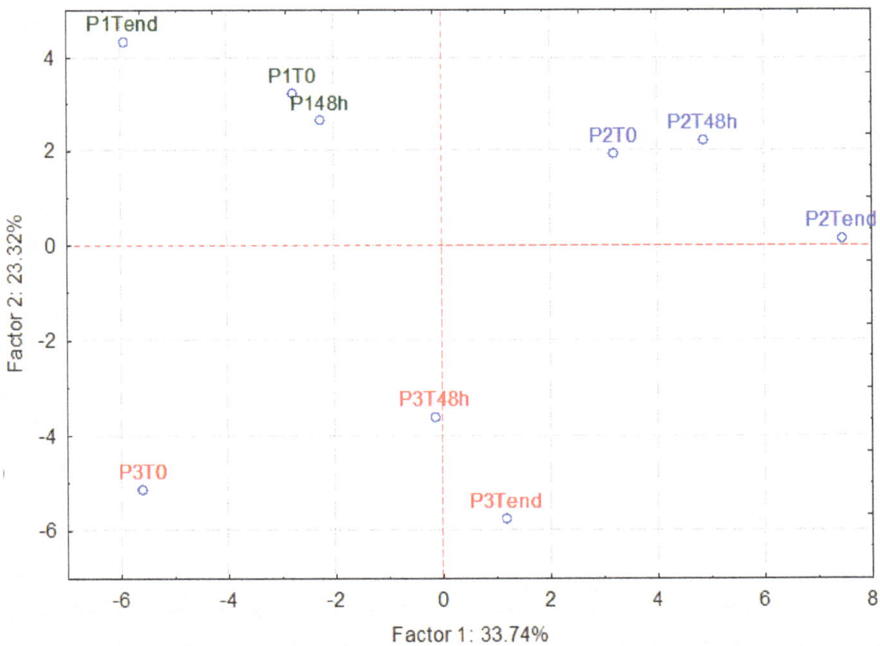

Figure 3. Projection of the different kombucha samples (P1, P2, and P3) on the factor plane (1−2) at different times of fermentation (T_0, T_{48}, and T_{end}) on the basis of the volatile profiles detected via solid-phase microextraction combined with a chromatography-mass spectrometry technique (GC/MS SPME).

Regarding the molecules responsible for the clustering of the samples, sample P2 was characterized by a higher amount of esters (mainly ethyl acetate and ethyl octanoate) and alcohols (ethanol, isoamyl alcohol, 4-ethylguaiacol, and 4-ethylphenol). Sample P3 was characterized by the highest amount of acids (acetic acid) and the lowest amount of esters.

Finally, the P1 sample showed the highest amount of short-chain fatty acids (caprylic acid, hexanoic acid, and decanoic acid).

4. Discussion

In this work, different samples of kombucha, obtained from an artisanal manufacturer (P1, P2, and P3) and at different steps/stages of fermentation (T_0, T_{48}, and T_{end}) were characterized from a microbiological and chemico-physical point of view in order to understand the microbial biodiversity and differences in the samples' compositions. Microbiological analyses were carried out using both traditional methods and a culture-independent metagenomic approach. Previous studies indicate that the microbial community in terms of yeasts and AAB abundance may vary between different kombucha fermentations across the globe, depending on the source and method of the inoculum used and the type of tea [19,20]. However, it is widely reported that AAB usually dominates the bacterial community of both SCOBY and the liquid phase of kombucha, driving the fermentative process [21]. Nevertheless, the microbiological data of the present work, deriving from culture-dependent and independent methods, have shown that yeasts were the predominant microflora in the liquid phase of the analyzed kombucha samples. On the other hand, the artisanal kombucha samples studied in this work were obtained from previous batches that were backslopped utilizing only the fermented liquid and not the SCOBY; this probably affected the microbial community, favoring a higher presence of yeasts than AAB. In fact, as reported by other authors, the larger amount of AAB would be entrapped in the pellicle on the surface of the kombucha [22,23]. Interestingly, the cell load of cultivable AAB differed based on the type of kombucha and ranged between 5.4 and 5.7 log CFU/mL in samples P1 and P2, while it increased considerably in sample P3 (6.5 log CFU/mL). The literature's data indicate AAB levels ranging between 6.5 and 7.5 log CFU/mL in kombucha liquor [21]. Lactic acid bacteria were only detected in samples P1 and P3, and this was also confirmed via molecular methods. On the other hand, it is reported that lactic acid bacteria may be present in kombucha even though they do not represent the dominant microflora [24]. The microbial colonies included limited genera and species of yeasts, mostly identified as *Starmerella davenportii* and *Brettanomyces bruxellensis*. Among bacteria, *Acetobacter tropicalis*, *Komagataeibacter saccharivorans*, *Gluconobacter oxydans*, *Komagataeibacter hansenii*, and *Liquorilactobacillus nagelii* were identified. Certainly, some differences in the microorganisms identified via culture-dependent methods and metagenomic analysis were observed. As expected, metagenomic analysis showed a higher biodiversity within both the fungi and bacteria kingdoms. This is not surprising since the traditional culture-dependent methods used for microbial population enumeration have several drawbacks. For instance, the presence of viable but not cultivable cells and fastidious microorganisms may lead to an underestimation of the microbial population. On the other hand, metagenomic analyses allow for the observation of greater diversity in the microbial population [24]. In addition, culture-dependent methods can promote the growth of certain microbial species over others, as observed in this study. In fact, the metagenomic approach showed that *Brettanomyces* spp. was the dominant genus in kombucha samples P1 and P2, while *Starmerella* spp. and *Brettanomyces* spp. dominated the yeast microflora of sample P3. However, the samples also contained *Schizosaccharomyces* spp. and *Zygosaccharomyces* spp., with relative abundances that depended on the sample considered. The data obtained are in agreement with the literature; in fact, several species of *Saccharomyces*, *Saccharomycodes*, *Schizosaccharomyces*, *Zygosaccharomyces*, *Brettanomyces*, *Candida*, *Torulaspora*, *Kloeckera*, *Pichia*, *Mycotorula*, and *Mycoderma* have been associated with kombucha [25,26]. On the other hand, the yeasts population of kombucha can be extremely variable, depending on the production area of the starter used and the fermentation conditions [5,25]. At the species level, *Brettanomyces bruxellensis* was the predominant yeast in samples P1 and P2, while *Starmerella davenportii* was predominant in sample P3. *Brettanomyces bruxellensis* is widely reported among the predominant yeast species in the kombucha population [13,27] due to its high resistance to osmotic and ethanol stress [13]. There is less literature available on the impact of *Starmerella* on kombucha fermentation. This is partly because *Starmerella* is a relatively new genus, resulting from the reclassification of several *Candida* species. Consequently, earlier kombucha studies that identified *Candida* at the genus level might have included species now classified as *Starmerella* [22]. A *Starmerella davenportii* strain was recently isolated from kombucha, and it has been shown to produce volatile aroma compounds and organic acids during black tea fermentation, showing its capability to be a potential starter for kombucha production [28].

Metagenomic analyses revealed *Komagataeibacter* spp. as the dominant genus regardless of the kombucha sample and the time of fermentation considered, as previously reported in other studies [19,22]. In fact, kombucha represents the most important reservoir of the genus *Komagataeibacter*, widely described as cellulose-producing bacteria [29,30]. Interest in bacterial cellulose has increased considerably in recent years due to excellent technological properties, including high purity, water retention capacity, high mechanical strength, and biocompatibility [30]. In fact, the isolation and selection of the highly bacterial cellulose producers *Komagataeibacter* spp. have become a matter of primary attention [30]. However, *Acetobacter* spp., *Gluconacetobacter* spp., and *Gluconobacter* spp. were also detected. In the kombucha samples analyzed, specific species of AAB demonstrated distinct patterns of predominance. For the P1 sample, *K. saccharivorans*, *K. rhaeticus*, *K. xylinus*, and *K. medellinensis* were the predominant species. In the P2 sample, the AAB population was dominated by *K. rhaeticus*, *K. medellinensis*, *K. hansenii*, and *K. xylinus*. Conversely, the P3 sample was primarily composed of *K. saccharivorans*, *K. xylinus*, and *K. intermedius*. These findings highlight the variability in AAB communities across different kombucha samples, potentially influencing the fermentation process and the final products' properties [13,19,31]. LAB were detected only in samples P1 and P3 with an abundance lower than 8%. According to the literature, their abundance in artisanal kombucha is negligible [5,19,21,31,32], while LAB can reach higher quantities in commercial products [33].

Sample P3 showed a higher load of AAB and the major biodiversity of microbial populations during fermentation. This resulted in a lower pH and higher volatile and total acidity for the P3 sample. However, all samples had pH values in line with the literature, generally between 2.5 and 3.8 [26,34]. A significant decrease in pH during fermentation was observed only in sample P3. This behavior can be attributed to the very high inoculum level (50% from a previous batch) used by the artisanal manufacturer, which lowered the initial pH of the kombucha. In all samples, a significant decrease in sucrose content over time was observed. Typically, sucrose is hydrolyzed by yeasts into fructose and glucose, which are then metabolized by yeast to produce ethanol and carbon dioxide. Additionally, due to the high osmotic pressure, glycerol is produced by yeasts [35]. Acetic acid bacteria can utilize glucose and fructose to primarily produce acids, including acetic acid, gluconic acid, and glucuronic acid. However, both glucose and fructose serve as substrates for the synthesis of the cellulose film by *Komagataeibacter* spp., which comprises the SCOBY [4,36]. Sample P3 showed an increase in the total and volatile acidity of fermented kombucha over time, while slight modifications were observed during fermentation in samples P1 and P2. The acidic content, particularly the content of volatile acids, such as acetic acid, is generally linked to fermentation conditions, particularly temperature and sugar concentrations, rather than to the type of microflora and, in particular, AAB [37]. At T_{end}, sample P3 showed a significantly higher content of malic acid compared to the other samples. The same trend was also observed by Nie et al. [38] in the production of traditional vinegar. Other authors showed an increase in malic acid during kombucha fermentation associated with the activity of AAB [23,39]. Lactic acid was found at low concentrations in all samples and did not change throughout the fermentation process; this is consistent with the literature's data reporting a very small increase in the concentration of lactic acid in kombucha fermentation using sucrose [40]. Conversely, the concentration of gluconic acid was dependent on the sample, with P3, which contained a higher amount of AAB, showing the highest level at the end of fermentation (T_{end}). On the other hand, gluconic acid production is correlated with the activity of AAB, and its content depends on strain interactions, inoculum ratios, and inoculum sizes [21,41]. Gluconic acid not only enhances the health benefit of kombucha but also positively impacts the sensory and flavor properties of kombucha [41]. The amounts of glucose and fructose were dependent on the sample but did not undergo any major changes during fermentation. On the other hand, during kombucha fermentation, sucrose is converted into glucose and fructose and, further, into ethanol, acetic acid, lactic acid, and a large number of other compounds [40]. As reported by other authors, the sugar content of commercial kombucha per 100 mL ranged from 0.2 to 6 g [42]. The total sugar content observed in the analyzed samples resulted in this range. To qualify as low-calorie, kombucha should ideally have no more than 5 g of sugar per 240 milliliters [43]. Even though the sugar content in kombucha may overcome the limit to classify the beverage as a low-calorie soft drink, it is important to consider that kombucha is often unpasteurized and unfiltered. This allows microbial activity to continue during storage, leading to further sugar consumption. Additionally, kombucha can be subjected to flavor infusion or dilution processes, which can reduce its sugar content and, consequently, its caloric content. All the samples exceeded 1% alcohol by volume (ABV) at T_{end}; this is in line with the literature's data since naturally fermented kombucha may contain 0 to 3% ABV, depending on the microflora and fermentation conditions [44]. The European Union generally defines non-alcoholic beverages as those containing less than 0.5% ABV; however, each country has specific regulations. For example, in Italy and France, non-alcoholic beverages must contain less than 1.2% ABV [45]. However, kombucha may not be marketed as-is but is subjected to flavor infusion and dilution processes that allow it to meet the standards for non-alcoholic beverages.

All the samples showed a high content of total phenolic compounds and catechins associated with high antioxidant activity. The beneficial impact of phenolic compounds on human health is widely reported since they improve lipid profiles, blood pressure, insulin resistance, and systemic inflammation [46]. However, although differences were observed based on the sample type and fermentation time, all samples showed an increase in catechin content and a decrease in total phenolic compounds compared to the sugared raw tea. On the other hand, during the initial stages of kombucha fermentation, the concentration of catechins may decrease due to enzymatic oxidation and microbial metabolism. However, as fermentation progresses, the concentration of some catechins, such as epigallocatechin gallate, may increase due to the metabolic activities of AAB and yeast [36]. In addition, the concentration of total phenolic compounds was higher in the P2 samples, while during fermentation, it remained constant in all samples at T3 before showing a decrease at the end of the fermentation process. Antioxidant activity, on the other hand, was higher in the P1 and P2 samples. The literature data indicate that the antioxidant activity and total phenolic compound content of kombucha is strongly linked to the type of tea used in preparation (green tea has a greater antioxidant power than black tea) and to the duration of fermentation [47,48]. In general, an initial increase in antioxidant activity is observed, but during excessively prolonged fermentations, a decrease in this parameter can be detected [7,32]. The kombucha microbial consortium can cause chemical and structural alterations in phenolic acids due to biotransformation and metabolic activities [36].

The three types of kombucha analyzed showed specific profiles in volatile molecules. The aromatic compounds in kombucha can be affected by the tea constituents and derived ingredients, the sugar substrates, the microbial population, and the phase and temperature of fermentation [49]. This study showed a prevalence of acids, followed by alcohols, in the P3 sample, which had the highest abundance of AAB. This is in accordance with the literature's data showing that carboxylic acids are also the major components of the fermentation process, followed by alcohols [49,50]. Acetic acid is normally produced by AAB via the oxidation of ethanol, and it contributes to the sour and distinctive flavor of kombucha [35]. Caprylic and hexanoic acid were found to be the most abundant carboxylic acids, and they were present in higher concentrations in samples P1 and P2, which were characterized by the highest presence of *Brettanomyces* spp. Other authors reported that fermentations with *Brettanomyces bruxellensis* produced the highest overall concentration of fatty acids, and their presence is related to yeast autolysis, which could be involved in the final yeasty aroma of kombucha, together with high alcohol [9,51]. Also, the presence of 4-ethylguaiacol, 4-ethylphenol was observed only in samples P1 and P2, and these molecules are generally produced by *Brettanomyces* spp., starting from coumaric acid [51]. High concentrations of these compounds are generally associated with the odor perception of horse sweat, barnyards, and medicine, and they are negatively perceived in wine and bottom-fermented pilsner beers [52,53]. However, their presence is essential for the overall flavor perception of many top-fermented blond and dark beers, as well as kefir, kombucha, and cider [54]. Together with ethanol, isoamyl alcohol and phenylethyl alcohol were the most abundant volatile alcohols, and their presence in kombucha is widely reported; they are produced from the respective amino acids via yeast activity [9]. Ethyl acetate was the most representative ester detected in accordance with the literature's data [49]. The high presence of esters in fermented beverages is very important since it is generally associated with pleasant floral and fruity aromas, enhancing the tea and white fruit odor perception in kombucha [55].

5. Conclusions

Microbiological analyses revealed yeasts as the predominant microflora of the analyzed artisanal kombucha samples, contrary to the common belief that AAB dominate the kombucha microflora. This deviation could be attributed to the artisanal production method, through which only the fermented liquid, not the SCOBY, was utilized. The presence of yeasts such as *Starmerella davenportii* and *Brettanomyces bruxellensis* was notable, alongside various bacterial genera including *Acetobacter*, *Komagataeibacter*, and *Gluconobacter*. Chemico-physical analyses revealed variations in acidity, organic acids, and volatile compounds among the samples, influenced by microbial composition and fermentation conditions. Notably, a higher AAB load in sample P3 was correlated with increased total acidity, as well as malic and gluconic acid content. All the samples exhibited high polyphenol and catechin content with substantial antioxidant activity. Acids predominated the volatile profile of the samples, followed by alcohols and esters, contributing to the characteristic aroma and flavor of kombucha. PCA analysis showed that the three analyzed kombucha samples had specific and distinct profiles in volatile molecules and appeared well separated from each other in the factorial plan. Based on the results obtained, the P3 starter appears to be the most interesting one due to the more balanced microflora in terms of yeasts and lactic acid bacteria, resulting in a kombucha with a higher content of gluconic acid and a lower ethanol content, as well as a more balanced aromatic profile in terms of acids and alcohols. Overall, this study provides valuable insights into the microbial dynamics and chemical composition of artisanal kombucha, emphasizing the role of fermentation parameters and microbial interactions in shaping its properties. The present work also allowed the isolation of strains, such as those belonging to the genera *Komagataeibacter*, gifted with very interesting technological features that are potentially exploitable in the production of bio-based materials.

Author Contributions: Conceptualization, L.S., F.P. and R.L.; methodology, D.G. and F.P.; software, J.A.N. and M.C.; validation, A.R. and G.P.P.; formal analysis, J.A.N., M.C. and A.R.; investigation, D.G.; resources, F.P. and R.L.; data curation, M.C. and L.S.; writing—original draft preparation, L.S., A.R. and J.A.N.; writing—review and editing, D.G., F.P., G.P.P. and R.L.; visualization, F.P.; supervision, L.S., D.G. and R.L.; project administration, F.P. and G.P.P.; funding acquisition, L.S., R.L. and F.P. All authors have read and agreed to the published version of the manuscript.

Funding: This research was funded by the Agritech National Research Center and received funding from the European Union Next-GenerationEU (PIANO NAZIONALE DI RIPRESA E RESILIENZA (PNRR)—MISSIONE 4 COMPONENTE 2, INVESTIMENTO 1.4—D.D. 1032 17/06/2022, CN00000022). This manuscript reflects only the authors' views and opinions; neither the European Union nor the European Commission can be considered responsible for them.

Institutional Review Board Statement: Not applicable.

Informed Consent Statement: Not applicable.

Data Availability Statement: The original contributions presented in the study are included in the article, further inquiries can be directed to the corresponding author.

Acknowledgments: The authors would like to thank Alessandro Oliviero from Frui-Lab (Bologna, Italy) for providing us with the kombucha samples.

Conflicts of Interest: The authors declare no conflicts of interest.

References

1. Diez-Ozaeta, I.; Astiazaran, O.J. Recent advances in Kombucha tea: Microbial consortium, chemical parameters, health implications and biocellulose production. *Int. J. Food Microbiol.* **2022**, *377*, 109783. [CrossRef] [PubMed]
2. Martínez Leal, J.; Valenzuela Suárez, L.; Jayabalan, R.; Huerta Oros, J.; Escalante-Aburto, A. A review on health benefits of kombucha nutritional compounds and metabolites. *CYTA-J. Food* **2018**, *16*, 390–399. [CrossRef]
3. Esatbeyoglu, T.; Sarikaya Aydin, S.; Gültekin Subasi, B.; Erskine, E.; Gök, R.; Ibrahim, S.A.; Yilmaz, B.; Özogul, F.; Capanoglu, E. Additional advances related to the health benefits associated with kombucha consumption. *Crit. Rev. Food Sci. Nutr.* **2022**, *20*, 1–18. [CrossRef] [PubMed]
4. de Miranda, J.F.; Ruiz, L.F.; Silva, C.B.; Uekane, T.M.; Silva, K.A.; Gonzalez, A.G.M.; Fernandes, F.F.; Lima, A.R. Kombucha: A review of substrates, regulations, composition, and biological properties. *J. Food Sci.* **2022**, *87*, 503–527. [CrossRef] [PubMed]
5. Bishop, P.; Pitts, E.R.; Budner, D.; Thompson-Witrick, K.A. Kombucha: Biochemical and microbiological impacts on the chemical and flavor profile. *Food Chem. Adv.* **2022**, *1*, 100025. [CrossRef]
6. Vitas, J.S.; Cvetanović, A.D.; Mašković, P.Z.; Švarc-Gajić, J.V.; Malbaša, R.V. Chemical composition and biological activity of novel types of kombucha beverages with yarrow. *J. Funct. Foods* **2018**, *44*, 95–102. [CrossRef]
7. Ahmed, R.F.; Hikal, M.S.; Abou-Taleb, K.A. Biological, chemical and antioxidant activities of different types Kombucha. *Ann. Agric. Sci.* **2020**, *65*, 35–41. [CrossRef]
8. Kaashyap, M.; Cohen, M.; Mantri, N.; Pelaez, M.; Li, L.; Muc-Wierzgon, M. Microbial Diversity and Characteristics of Kombucha as Revealed by Metagenomic and Physicochemical Analysis. *Nutrients* **2021**, *13*, 4446. [CrossRef] [PubMed]
9. Tran, T.; Roullier-Gall, C.; Verdier, F.; Martin, A.; Schmitt-Kopplin, P.; Alexandre, H.; Grandvalet, C.; Tourdot-Maréchal, R. Microbial Interactions in Kombucha through the Lens of Metabolomics. *Metabolites* **2022**, *12*, 235. [CrossRef]
10. May, A.; Narayanan, S.; Alcock, J.; Varsani, A.; Maley, C.; Aktipis, A. Kombucha: A novel model system for cooperation and conflict in a complex multi-species microbial ecosystem. *PeerJ* **2019**, *7*, e7565. [CrossRef]
11. Wang, X.; Wang, D.; Wang, H.; Jiao, S.; Wu, J.; Hou, Y.; Sun, J.; Yuan, J. Chemical Profile and Antioxidant Capacity of Kombucha Tea by the Pure Cultured Kombucha. *LWT* **2022**, *168*, 113931. [CrossRef]
12. Landis, E.A.; Fogarty, E.; Edwards, J.C.; Popa, O.; Eren, A.M.; Wolfe, B.E. Microbial Diversity and Interaction Specificity in Kombucha Tea Fermentations. *mSystems* **2022**, *7*, e0015722. [CrossRef]
13. Villarreal-Soto, S.A.; Beaufort, S.; Bouajila, J.; Souchard, J.P.; Taillandier, P. Understanding Kombucha Tea Fermentation: A Review. *J. Food Sci.* **2018**, *83*, 580–588. [CrossRef]
14. de Oliveira, Í.A.C.L.; Rolim, V.A.d.O.; Gaspar, R.P.L.; Rossini, D.Q.; de Souza, R.; Bogsan, C.S.B. The Technological Perspectives of Kombucha and Its Implications for Production. *Fermentation* **2022**, *8*, 185. [CrossRef]
15. Siroli, L.; Patrignani, F.; Serrazanetti, D.I.; Tabanelli, G.; Montanari, C.; Gardini, F.; Lanciotti, R. Lactic acid bacteria and natural antimicrobials to improve the safety and shelf-life of minimally processed sliced apples and lamb's lettuce. *Food Microbiol.* **2015**, *47*, 74–84. [CrossRef] [PubMed]
16. Marcelino, V.R.; Clausen, P.T.L.C.; Buchmann, J.P.; Wille, M.; Iredell, J.R.; Meyer, W.; Lund, O.; Sorrell, T.C.; Holmes, E.C. CCMetagen: Comprehensive and accurate identification of eukaryotes and prokaryotes in metagenomic data. *Genome Biol.* **2020**, *21*, 1–15. [CrossRef] [PubMed]
17. Slinkard, K.; Singleton, V.L. Total Phenol Analysis: Automation and Comparison with Manual Methods. *Am. J. Enol. Vitic.* **1977**, *28*, 49–55. [CrossRef]
18. Siroli, L.; Giordani, B.; Rossi, S.; Gottardi, D.; McMahon, H.; Augustyniak, A.; Menon, A.; Vannini, L.; Vitali, B.; Patrignani, F.; et al. Antioxidant and Functional Features of Pre-Fermented Ingredients Obtained by the Fermentation of Milling By-Products. *Fermentation* **2022**, *8*, 722. [CrossRef]
19. Chakravorty, S.; Bhattacharya, S.; Chatzinotas, A.; Chakraborty, W.; Bhattacharya, D.; Gachhui, R. Kombucha tea fermentation: Microbial and biochemical dynamics. *Int. J. Food Microbiol.* **2016**, *220*, 63–72. [CrossRef]
20. Malbaša, R.V.; Lončar, E.S.; Vitas, J.S.; Čanadanović-Brunet, J.M. Influence of starter cultures on the antioxidant activity of kombucha beverage. *Food Chem.* **2011**, *127*, 1727–1731. [CrossRef]
21. De Filippis, F.; Troise, A.D.; Vitaglione, P.; Ercolini, D. Different temperatures select distinctive acetic acid bacteria species and promotes organic acids production during Kombucha tea fermentation. *Food Microbiol.* **2018**, *73*, 11–16. [CrossRef] [PubMed]
22. Harrison, K.; Curtin, C. Microbial composition of scoby starter cultures used by commercial kombucha brewers in North America. *Microorganisms* **2021**, *9*, 1060. [CrossRef] [PubMed]

23. Tran, T.; Grandvalet, C.; Verdier, F.; Martin, A.; Alexandre, H.; Tourdot-Maréchal, R. Microbial Dynamics between Yeasts and Acetic Acid Bacteria in Kombucha: Impacts on the Chemical Composition of the Beverage. *Foods* **2020**, *9*, 963. [CrossRef] [PubMed]
24. Laavanya, D.; Shirkole, S.; Balasubramanian, P. Current challenges, applications and future perspectives of SCOBY cellulose of Kombucha fermentation. *J. Clean. Prod.* **2021**, *295*, 126454. [CrossRef]
25. Jayabalan, R.; Malbaša, R.V.; Lončar, E.S.; Vitas, J.S.; Sathishkumar, M. A Review on Kombucha Tea—Microbiology, Composition, Fermentation, Beneficial Effects, Toxicity, and Tea Fungus. *Compr. Rev. Food Sci. Food Saf.* **2014**, *13*, 538–550. [CrossRef] [PubMed]
26. Villarreal-Soto, S.A.; Bouajila, J.; Pace, M.; Leech, J.; Cotter, P.D.; Souchard, J.P.; Taillandier, P.; Beaufort, S. Metabolome-microbiome signatures in the fermented beverage, Kombucha. *Int. J. Food Microbiol.* **2020**, *333*, 108778. [CrossRef] [PubMed]
27. Mas, P.; Tran, T.; Verdier, F.; Martin, A.; Alexandre, H.; Grandvalet, C.; Tourdot-Maréchal, R. Evolution in Composition of Kombucha Consortia over Three Consecutive Years in Production Context. *Foods* **2022**, *11*, 614. [CrossRef] [PubMed]
28. Tu, C.; Hu, W.; Tang, S.; Meng, L.; Huang, Z.; Xu, X.; Xia, X.; Azi, F.; Dong, M. Isolation and identification of Starmerella davenportii strain Do18 and its application in black tea beverage fermentation. *Food Sci. Hum. Wellness* **2020**, *9*, 355–362. [CrossRef]
29. Gullo, M.; La China, S.; Falcone, P.M.; Giudici, P. Biotechnological production of cellulose by acetic acid bacteria: Current state and perspectives. *Appl. Microbiol. Biotechnol.* **2018**, *102*, 6885–6898. [CrossRef]
30. La China, S.; De Vero, L.; Anguluri, K.; Brugnoli, M.; Mamlouk, D.; Gullo, M. Kombucha Tea as a Reservoir of Cellulose Producing Bacteria: Assessing Diversity among Komagataeibacter Isolates. *Appl. Sci.* **2021**, *11*, 1595. [CrossRef]
31. Marsh, A.J.; O'Sullivan, O.; Hill, C.; Ross, R.P.; Cotter, P.D. Sequence-based analysis of the bacterial and fungal compositions of multiple kombucha (tea fungus) samples. *Food Microbiol.* **2014**, *38*, 171–178. [CrossRef] [PubMed]
32. Gaggìa, F.; Baffoni, L.; Galiano, M.; Nielsen, D.S.; Jakobsen, R.R.; Castro-Mejía, J.L.; Bosi, S.; Truzzi, F.; Musumeci, F.; Dinelli, G.; et al. Kombucha Beverage from Green, Black and Rooibos Teas: A Comparative Study Looking at Microbiology, Chemistry and Antioxidant Activity. *Nutrients* **2018**, *11*, 1. [CrossRef] [PubMed]
33. Coton, M.; Pawtowski, A.; Taminiau, B.; Burgaud, G.; Deniel, F.; Coulloumme-Labarthe, L.; Fall, A.; Daube, G.; Coton, E. Unraveling microbial ecology of industrial-scale Kombucha fermentations by metabarcoding and culture-based methods. *FEMS Microbiol. Ecol.* **2017**, *93*, fix048. [CrossRef] [PubMed]
34. Greenwalt, C.J.; Steinkraus, K.H.; Ledford, R.A. Kombucha, the fermented tea: Microbiology, composition, and claimed health effects. *J. Food Prot.* **2000**, *63*, 976–981. [CrossRef] [PubMed]
35. Wang, B.; Rutherfurd-Markwick, K.; Zhang, X.X.; Mutukumira, A.N. Kombucha: Production and Microbiological Research. *Foods* **2022**, *11*, 3456. [CrossRef]
36. Sanwal, N.; Gupta, A.; Bareen, M.A.; Sharma, N.; Sahu, J.K. Kombucha fermentation: Recent trends in process dynamics, functional bioactivities, toxicity management, and potential applications. *Food Chem. Adv.* **2023**, *3*, 100421. [CrossRef]
37. Antolak, H.; Piechota, D.; Kucharska, A. Kombucha Tea-A Double Power of Bioactive Compounds from Tea and Symbiotic Culture of Bacteria and Yeasts (SCOBY). *Antioxidants* **2021**, *10*, 1541. [CrossRef]
38. Nie, Z.; Zheng, Y.; Xie, S.; Zhang, X.; Song, J.; Xia, M.; Wang, M. Unraveling the correlation between microbiota succession and metabolite changes in traditional Shanxi aged vinegar. *Sci. Rep.* **2017**, *7*, 1–12.
39. Wang, S.; Li, C.; Wang, Y.; Wang, S.; Zou, Y.; Sun, Z.; Yuan, L. Changes on physiochemical properties and volatile compounds of Chinese kombucha during fermentation. *Food Biosci.* **2023**, *55*, 103029. [CrossRef]
40. Malbaša, R.; Lončar, E.; Djurić, M.; Došenović, I. Effect of sucrose concentration on the products of Kombucha fermentation on molasses. *Food Chem.* **2008**, *108*, 926–932. [CrossRef]
41. Li, R.; Xu, Y.; Chen, J.; Wang, F.; Zou, C.; Yin, J. Enhancing the proportion of gluconic acid with a microbial community reconstruction method to improve the taste quality of Kombucha. *LWT* **2022**, *155*, 112937. [CrossRef]
42. Kim, J.; Adhikari, K. Current Trends in Kombucha: Marketing Perspectives and the Need for Improved Sensory Research. *Beverages* **2020**, *6*, 15. [CrossRef]
43. FDA Food and Drug Administration. *Guidance for Industry: Food Labeling Guide*; FDA Food and Drug Administration: Washington, DC, USA, 2013.
44. Jang, S.S.I.K.; McIntyre, L.; Chan, M.; Brown, P.N.; Finley, J.; Chen, S.X. Ethanol Concentration of Kombucha Teas in British Columbia, Canada. *J. Food Prot.* **2021**, *84*, 1878–1883. [CrossRef] [PubMed]
45. Okaru, A.O.; Lachenmeier, D.W. Defining No and Low (NoLo) Alcohol Products. *Nutrients* **2022**, *14*, 3873. [CrossRef] [PubMed]
46. Rana, A.; Samtiya, M.; Dhewa, T.; Mishra, V.; Aluko, R.E. Health benefits of polyphenols: A concise review. *J. Food Biochem.* **2022**, *46*, e14264. [CrossRef] [PubMed]
47. Abaci, N.; Senol Deniz, F.S.; Orhan, I.E. Kombucha—An ancient fermented beverage with desired bioactivities: A narrowed review. *Food Chem. X* **2022**, *14*, 100302. [CrossRef] [PubMed]
48. Jakubczyk, K.; Kałduńska, J.; Kochman, J.; Janda, K. Chemical Profile and Antioxidant Activity of the Kombucha Beverage Derived from White, Green, Black and Red Tea. *Antioxidants* **2020**, *9*, 447. [CrossRef] [PubMed]
49. Suffys, S.; Richard, G.; Burgeon, C.; Werrie, P.Y.; Haubruge, E.; Fauconnier, M.L.; Goffin, D. Characterization of Aroma Active Compound Production during Kombucha Fermentation: Towards the Control of Sensory Profiles. *Foods* **2023**, *12*, 1657. [CrossRef] [PubMed]

50. Zhao, Z.J.; Sui, Y.C.; Wu, H.W.; Zhou, C.B.; Hu, X.C.; Zhang, J. Flavour chemical dynamics during fermentation of kombucha tea. *Emirates J. Food Agric.* **2018**, *30*, 732–741.
51. Ferremi Leali, N.; Binati, R.L.; Martelli, F.; Gatto, V.; Luzzini, G.; Salini, A.; Slaghenaufi, D.; Fusco, S.; Ugliano, M.; Torriani, S.; et al. Reconstruction of Simplified Microbial Consortia to Modulate Sensory Quality of Kombucha Tea. *Foods* **2022**, *11*, 3045. [CrossRef]
52. Suárez, R.; Suárez-Lepe, J.A.; Morata, A.; Calderón, F. The production of ethylphenols in wine by yeasts of the genera Brettanomyces and Dekkera: A review. *Food Chem.* **2007**, *102*, 10–21. [CrossRef]
53. Steensels, J.; Daenen, L.; Malcorps, P.; Derdelinckx, G.; Verachtert, H.; Verstrepen, K.J. Brettanomyces yeasts—From spoilage organisms to valuable contributors to industrial fermentations. *Int. J. Food Microbiol.* **2015**, *206*, 24–38. [CrossRef] [PubMed]
54. De Roos, J.; De Vuyst, L. Microbial acidification, alcoholization, and aroma production during spontaneous lambic beer production. *J. Sci. Food Agric.* **2019**, *99*, 25–38. [CrossRef] [PubMed]
55. Savary, O.; Mounier, J.; Thierry, A.; Poirier, E.; Jourdren, J.; Maillard, M.B.; Penland, M.; Decamps, C.; Coton, E.; Coton, M. Tailor-made microbial consortium for Kombucha fermentation: Microbiota-induced biochemical changes and biofilm formation. *Food Res. Int.* **2021**, *147*, 110549. [CrossRef] [PubMed]

foods

MDPI

Article

Impact of Lactic Acid Bacteria Fermentation Based on Biotransformation of Phenolic Compounds and Antioxidant Capacity of Mushrooms

Eda Nur Ayar-Sümer [1,2], Yannick Verheust [1], Beraat Özçelik [2] and Katleen Raes [1,*]

[1] Research Unit VEG-i-TEC, Department of Food Technology, Safety and Health, Faculty of Bioscience Engineering, Ghent University, St-Martem Latemlaan 2B, 8500 Kortrijk, Belgium; ayare15@itu.edu.tr (E.N.A.-S.); yannick.verheust@ugent.be (Y.V.)

[2] Department of Food Engineering, Faculty of Chemical and Metallurgical Engineering, Istanbul Technical University, Maslak, TR-34469 Istanbul, Turkey; ozcelik@itu.edu.tr

* Correspondence: katleen.raes@ugent.be

Abstract: Mushrooms contain phenolic compounds that possess health-promoting properties, including antioxidant effects. However, the low solubility and form of phenolic compounds affect their bioactivity and bioaccessibility. To overcome this limitation, our study investigates the fermentation of mushrooms to increase their free phenolic content and enhance their bioactivity. Our research focused on the impact of fermentation on both free and bound phenolic fractions (FPs and BPs, respectively) in *Lentinula edodes* and *Lactarius deliciosus*, which were successively fermented with *Lactiplantibacillus plantarum LMG 17673* for 72 h. We examined the total phenolic content (TPC), phenolic profile, and antioxidant activity of both FPs and BPs. Our results showed that the TPC of BPs was higher than that of FPs in both mushrooms, with strong antioxidant capabilities. Fermentation significantly increased the TPC of FPs in both mushrooms, particularly after 24 h of fermentation. The TPC of BPs in mushrooms decreased during fermentation, indicating their release from the matrix. Additionally, we identified 30 bioactive compounds using UPLC-Q-TOF-MS/MS. Our study demonstrates for the first time that lactic acid bacteria fermentation of mushrooms with high phenolic content leads to the liberation of bound phenolics, enhancing their bioactivity and bioaccessibility.

Keywords: fermented mushrooms; lactic fermentation; bound phenolics; free phenolics; bioactive compounds; antioxidant activities; UPLC-Q-TOF-MS/MS

1. Introduction

Consumption of edible mushrooms (cultivated and wild varieties) has been a global practice for centuries. *Lentinula edodes*, commonly referred to as shiitake mushrooms, is the second most cultivated edible mushroom worldwide [1]. In Europe, the commercialisation of 268 distinct edible wild mushroom species has been officially sanctioned, and among these, *Lactarius deliciosus* is one of the top five best-selling varieties. Recognised as saffron milkcap, *La. deliciosus* is a popular and extensively consumed wild edible mushroom [2,3].

The global mushroom cultivation market size is expected to reach approximately 24 million tons by 2028, which is 57% higher than that of 2021 [4]. However, this growth poses environmental challenges for the market because of the short shelf life of the mushrooms. In addition, a significant amount of by-products are produced during mushroom production, leading to high environmental impact and management costs for the industry. These by-products consist of caps, stipes, mushrooms that do not meet commercial standards for size, shape, or calibration, and spent mushroom substrates. Approximately 5 kg of spent mushroom substrate by-products are generated per kilogram of fresh mushroom produced [5], whereas up to 20% of fresh mushroom production volume is attributed to misshaped mushroom by-products [6]. The misshapen mushrooms have basically the same chemical composition as that of the normal mushroom [7]. Therefore, applying food processing to mushrooms may not only serve as an effective means of valorisation to by-product mushrooms but also contribute to extending the shelf life of all mushrooms.

Consumers are becoming interested in the potential health benefits of adding mushrooms to their diet. Mushrooms are recognised for their nutritional value as plant-based protein sources containing essential vitamins, minerals, fibre, and antioxidants. Additionally, mushrooms contain bioactive compounds with anti-inflammatory, antioxidant, antitumour, antiviral, and antimicrobial properties that actively promote health and reduce the risk of diseases in the human body [8]. These positive effects on health are due to the presence of bioactive compounds, such as phenolic compounds, terpenoids, steroids, lectins, nucleotides, glycoproteins, and polysaccharides [9], and much attention has been given to phenolic compounds as secondary metabolites in mushrooms [10]. Phenolic compounds are usually present in food matrices in both free and bound forms. As the bound phenolic compounds are linked to the cell wall matrices with covalent bonds, they cannot be absorbed in the small intestine, resulting in lower bioavailability compared to the free phenolic compounds. To increase the bioavailability of phenolic compounds, it is necessary to use additional processing techniques, such that bound phenolic compounds can be released and become available for absorption, similar to free phenolic compounds. This would also contribute to enhancing the health benefits of mushrooms [11].

Fermentation is a traditional non-thermal food processing method that involves the metabolic activities of microorganisms such as bacteria and fungi. Lactic acid bacteria (LAB) are gram-positive bacteria known for their ability to convert sugars to lactic acid during food fermentation. In addition to increasing the shelf life and improving the sensory qualities of food, LAB degrades macronutrients such as carbohydrates and proteins, resulting in alterations in the nutritional composition of the food product [12]. This process results in food biotransformation by converting macronutrients, releasing antioxidative peptides, and altering phenolic compounds [13]. In particular, *Lactiplantibacillus plantarum LMG 17673*, a versatile and widely utilised LAB species, is particularly effective in fermenting fruits, vegetables, and dairy products, with beneficial effects on human health [14].

Many studies have investigated the antioxidant activity properties of cultivated and wild edible mushrooms. However, to our knowledge, no research has been conducted on the impact of LAB fermentation, as a possible processing technique for mushrooms, on the free and bound phenolic compounds in mushroom bodies. Since phenolic compounds are diverse and have complex structures, advanced analytical techniques are needed for which the ultra-performance liquid chromatography–electrospray ionisation quadrupole time-of-flight mass spectrometry (UPLC-ESI-QTOF-MS/MS) method is commonly used to tentatively identify extracts. In addition, this study is the first to use lactic acid bacteria and fermentation to enhance the antioxidant activity of mushrooms by liberating the bound phenolic compounds.

2. Materials and Methods

2.1. Media and Chemicals

De Man-Rogosa-Sharpe (MRS) and plate count agar were purchased from Oxoid LTD (Basingstoke, Hampshire, England). Folin-Ciocalteu reagent, DPPH (2,2-diphenyl-1-picrylhydrazyl), ABTS (2,2′-azinobis-(3-ethylbenzothiazoline-6-sulphonic acid)), trolox (6-hydroxyl-2,5,7,8-tetramethyl chroman-2-carboxylic acid), gallic acid, vanillic acid, salicylic acid, L-rhamnose monohydrate, DL-arabinose, D(+)-xylose, D(+)-mannose, D(−)-ribose, D(+)-glucose, D(+)-maltose monohydrate, D(−)-mannitol, trehalose, galacturonic acid, and glucuronic acid were purchased from Sigma–Aldrich Fine Chemicals (St. Louis, MO, USA). D(−)-Fructose was purchased from Acros Organics (Geel, Belgium). HPLC-grade water and methanol were purchased from VWR Chemicals (VWR International S.A.S., Briare, France). Aluminium chloride, sodium nitrite, sodium hydroxide, sodium carbonate, hydrochloric acid, and methanol were purchased from Chem-Lab (ChemLab NV, Zedelgem, Belgium).

2.2. Sample Preparation and Lactic Acid Fermentation

In this study, although misshapen mushroom by-products were the preferred source for fermentation, our experimental outcomes remained unaffected by the use of either mushrooms or their misshapen by-products, as their nutritional compositions were identical. Due to practical considerations, fermentation was performed directly with mushrooms, which was more feasible than obtaining them from the mushroom by-product. Therefore, *L. edodes* and *La. deliciosus* were purchased from a local market in Turkey, known for their high standards and reliability, to guarantee the correct identification of the mushrooms. Furthermore, the mushrooms bought were controlled for their identification based on the mushroom identification guides as described by Davis et al. (2013) and Stamets (2011) and thoroughly examined for their key morphological features (cap shape, colour, size, and spore print) [15,16]. The mushrooms were cleaned before being cut into small pieces measuring 1 × 1 cm, frozen at −20 °C, and lyophilised (Christ Alphna 1-2 LDplus, Osterode am Harz, Germany). These lyophilised samples were then finely ground into a mushroom powder using a grinder (IKA A11, Staufen, Germany). The mushroom powder was mixed with physiological water to obtain a 2% (*w*/*v*) mushroom solution and sterilised.

The lactobacillus strain of *Lactiplantibacillus plantarum LMG 17673* was purchased from the BCCM/LMG Bacteria Collection, an integral component of the Belgian Co-ordinated Collections of Microorganisms situated within the Laboratory for Microbiology of the Faculty of Sciences of Ghent University, Belgium. The inoculum of the strain was prepared according to the method described by He et al. (2021) with slight modifications. The lactobacilli culture was aseptically activated by transferring 100 µL of glycerol stock culture into 10 mL of sterile MRS broth and was then incubated at 30 °C for 24 h [17]. *Lp. plantarum LMG 17673* was propagated in MRS broth at 30 °C for 24 h with shaking at 110 rpm before being used as a working culture to inoculate the mushroom solutions. Flasks containing 25 mL of sterilised mushroom solution were inoculated with 1% (*v*/*v*) working culture and then placed on a rotary shaker (110 rpm) at 30 °C. During fermentation, samples were collected at different time points (0, 24, 48, and 72 h), freeze-dried, and the powders were stored at −20 °C until further analysis.

2.3. Microbial Analysis

Microbial composition during fermentation was evaluated by measuring the pH and counting the number of LAB. The standard plate count method analysed LAB counts in mushroom samples at specific time points (0, 24, 48, and 72 h). Serial dilutions were prepared using physiological water. Using the spread plate method, 100 µL diluted aliquots were plated onto MRS agar plates and then cultured at 30 °C for 48 h. Finally, the microbial population was expressed as logarithmic colony-forming units per mL (log CFU/mL).

2.4. Fermentation Metabolites

The fermentation metabolites of mushrooms were quantified at specific time points (0, 24, 48, and 72 h) using pre-column high-performance liquid chromatography (HPLC) [18]. The metabolite analysis employed an LC system (Agilent LC 1260 Infinity II, Gent, Belgium) equipped with a column (Agilent Hi-Plex 300 × 7.7 mm, 8 µm particle, Gent, Belgium). The column was maintained at 60 °C, the diode array detector was set to 210 nm, and the refractive index detector was set to 55 °C. A fixed flow rate of 0.7 mL/min was combined with a 5 mmol/L sulfuric acid mobile phase, and 20 µL injection volumes were used. Samples were prepared by mixing 1 g of fermented mushrooms with 5 mL of 5 mmol/L sulfuric acid using an Ultraturrax (IKA-T18, Staufen, Germany) for 1 min at 10,000 rpm at 21 °C. Then, the probe was washed with 5 mL of 5 mmol/L sulfuric acid and centrifuged (Hermle Z 366 K, Wehingen, Germany) at 4000 rpm for 10 min at 21 °C. The resulting mixture was then filtered through a 0.45 µm disc filter, and the supernatant was collected and preserved at −40 °C until analysis. The quantification of free sugars and organic acids was based on calibration curves established for each compound by injecting known concentrations of external standards. Finally, the results were expressed as mg per g of dry fermented mushrooms. In this study, the total free sugar compound refers to the sum of determined glucose, ribose, trehalose, and mannitol, whereas the total organic acid refers to the sum of specified formic acid, malic acid, lactic acid, citric acid, fumaric acid, and succinic acid.

2.5. Extraction of Free and Bound Phenolic Fractions

The method by Gonzales et al. (2014) [19] was used to extract the free and bound phenolics. In summary, 2 g of fermented mushroom powder were blended with 15 mL of 100% methanol using Ultra-Turrax (IKA-T18D, Staufen, Germany) at 3000 rpm for 45 s. The tubes were promptly cooled in an ice bath for 15 s before being centrifuged at 13,000× *g* for 10 min at 4 °C using a centrifuge (Z 300 K, Hermle Labortechnik GmbH, Wehingen, Germany). The residue was subjected to a second extraction with 10 mL of 80% methanol, using the same procedure as before. The resulting supernatant was filtered, and the volume was adjusted to 25 mL using 80% methanol. The phenolic content of these extracts was referred to as the free phenolics (FPs). The residue was air-dried in a fume hood overnight to extract the bound phenolics. Briefly, 0.1 g of the air-dried residue was subjected to hydrolysis by adding 2 mL of 2M NaOH using ultrasound (UP 400S, Hielscher, GmbH, Chamerau, Germany) for 30 min at maximum amplitude (100%) and 60 °C. The hydrolysed samples were neutralised with 2M HCl, followed by extraction with 4 mL of methanol containing 0.1% formic acid. After vortexing and centrifugation at 10,000× *g* for 10 min at 4 °C, the residue underwent a second extraction with 4 mL of 0.1% formic acid in 100% methanol. Finally, the supernatants from both extractions were combined and adjusted to 20 mL with 80% methanol. The phenolic content of these extracts is referred to as the bound phenolics (BPs).

2.6. Total Phenolic Content

The total phenolic content (TPC) of the free and bound phenolic extracts was determined using the Folin–Ciocalteu method [20]. Each extract (1 mL) was mixed with 0.5 mL of 10-fold diluted Folin–Ciocalteu reagent and then neutralised with 1.5 mL of 20% sodium carbonate solution. The mixture was then incubated for 2 h at room temperature. After incubation, the blue colour that formed was measured using a 760 nm spectrophotometer (Model 4001/4, Thermo Spectronic, Waltham, MA, USA). The TPC was calculated using gallic acid calibration curves with R^2 values > 0.99 and expressed as mg gallic acid equivalent (GAE) per gram of mushroom dry matter (dm). The limit of detection (LOD) and the limit of quantitation (LOQ) were determined to be 0.16 and 1.44 μg GAE/mL, respectively.

2.7. Antioxidant Properties

2.7.1. Determination of DPPH Free-Radical-Scavenging Activity

To evaluate the antiradical activity against DPPH radicals, the method devised by Kumaran and Karunakaran [21] was applied. Briefly, a 0.1 mM DPPH solution (2 mL) was mixed with 100 μL of the extracts by vortexing for 10 s. Then, the mixture was incubated in the dark for 30 min at room temperature and the absorbance of the solution was measured at 517 nm using a spectrophotometer. The DPPH free-radical-scavenging capacity was calculated using the Trolox calibration curve with R^2 > 0.99 and expressed as mg Trolox equivalent (TE)/100 g mushroom dm. LOD and LOQ were found to be 2.6 and 7.01 μg TE/mL, respectively.

2.7.2. Determination of ABTS Free-Radical-Scavenging Activity

ABTS radical-scavenging capacity was assessed according to the method described by Re et al. [22]. ABTS was dissolved in distilled water to create a stock solution with a concentration of 7 mM. Then, the reaction between the ABTS stock solution and potassium persulfate (2.45 mM) was initiated to generate an ABTS radical cation, which was left in the dark at room temperature for 12–16 h before utilisation. The ABTS radical cation was diluted with 90% methanol to prepare a working solution of fresh ABTS radical cation, achieving an absorbance of 0.70 ± 0.02 at 734 nm and incubating at 30 °C. Subsequently,

2 mL of fresh ABTS radical cation solution was mixed with 20 μL of sample extract and vortexed. Afterwards, the mixture was incubated for 5 min in the dark at 25 °C and the absorbance of the solution at 734 nm was measured. The results were calculated using the Trolox calibration curve with R^2 values > 0.99. and expressed as mg Trolox equivalent (TE)/g mushroom dm. LOD and LOQ were found to be 2.74 and 8.23 μg TE/mL, respectively.

2.8. Phenolic Compound Identification and Characterisation

2.8.1. UPLC-Q-TOF-MS/MS-Based Metabolite Analysis

Fermentation metabolomic measurements were performed using an ultra-high-performance liquid chromatography system (UPLC Infinity 1290, Agilent Technologies, Santa Clara, CA, USA) coupled to a quadrupole time-of-flight mass spectrometer (Q-TOF 6546, Agilent Technologies) [23]. The system was equipped with an online degasser (Model 590, Alltech elite degassing system, Agilent Technologies, Santa Clara, CA, USA), a sampler (G 7129B), a quaternary pump (G 7104A), Column18 5u (4.6 × 150 mm; GRACE, Deerfield, IL, USA), Column Oven (G 7130A), and photodiode array detector (DAD) (G 7117A, Agilent Technologies). Separation was carried out using a gradient elution method with two mobile phases: ultrapure water (0.1%, *v*/*v*, formic acid; eluent A) and LC-MS-graded methanol (0.1%, *v*/*v*, formic acid; eluent B). The gradient profile was as follows: 0–6 min, isocratic 20% B; 6–12 min, isocratic 20% B; 12–13 min, 20–30% B; 13–23 min, 30–50%; 23–30 min, 50–90%; 30–35 min, isocratic 90%; 35–40 min, 90–10%; and 40–45 min, isocratic 10%. Each sample was run for a total of 45 min at a flow rate of 0.25 mL/min and an injection volume of 20 μL. The eluent was directed into a Dual Agilent Jet Stream (AJS) electrospray ionisation (ESI) source within the mass spectrometer. The parameters for the MS were set as follows: capillary voltage, 3.5 kV; nozzle voltage, 4 V; gas temperature, 300 °C; vaporiser temperature, 350 °C; gas flow rate, 8 L min^{-1}; and vaporiser flow, 11 L min^{-1}. A complete mass scan covering m/z 100–1700 was performed. For target analysis and fragmentation, MS/MS analysis was performed in the negative mode at different collision energies of 10, 20, and 40 V. Instrument control, data acquisition, and processing were performed using Agilent MassHunter Workstation Qualitative Analysis software (version 10.0, Agilent Technologies, Santa Clara, CA, USA).

2.8.2. Non-Target Screening on UPLC-Q-TOF-MS

All samples were analysed in triplicate in the negative mode. The acquisition procedure followed the described protocol, utilising Agilent MassHunter Workstation Profinder software (version 10.0) to extract non-target compounds via the Batch Recursive Feature Extraction workflow. Initially, the software identified molecular features in the first sample, where each feature represented a group of corresponding ions, including isotopes and adducts of the same compound. These groups formed a chromatographic peak at a specific retention time (RT). These detected features, accompanied by their precise monoisotopic mass and RT, were stored in a list, serving as a reference source for re-mining sample data using the Find-by-Ion method. Subsequently, the software used a binning algorithm to scan all the features across the samples. Detected features from all samples were compared, aligning exact masses and RT, and grouping corresponding features. This process generated a comprehensive list of features and their presence in respective samples. In a subsequent step, the average exact masses and RT of the consensus feature list were more sensitively scanned in each sample to detect any missed features in the initial round, facilitating recursive feature extraction. This approach enhances the feature identification accuracy by searching for low-level yet high-quality features that may have been overlooked initially, thus reducing false negatives and accelerating the binning process.

2.8.3. Non-Target Metabolomics Analysis

The acquired feature list was imported into Agilent Mass Profiler Professional (MPP, version 15.1, Agilent Technologies), which offers a complete array of advanced statistical tools and facilitates the powerful visualisation of the results. Due to the impracticality of identifying all features, statistical analyses were employed to focus on the relevant compounds. Initially, principal component analysis (PCA) was used to assess metabolic variations among replicate samples. Following this, Fold Change analysis was applied to features demonstrating abundance variations between different conditions. Paired analyses were conducted using the Benjamini–Hochberg method, and the findings were visually represented through volcano plots.

2.8.4. Tentative Identification of Phenolic Compounds

Qualitative analysis of phenolic compounds in fermented mushroom fractions was carried out using UPLC-Q-TOF-MS/MS in negative ionisation mode. The Agilent MassHunter Workstation Qualitative Analysis software (version 10.0) was used to process the data and extract additional details about the features of interest through a target MS/MS approach. The acquired MS/MS spectra are compared with the METLIN Agilent MassHunter Personal Compound Database and Library (PCDL) via the find-by-formula search feature. The software tentatively identifies target compounds by matching observed spectra to reference spectra. Only compounds with good matching (score > 75/100) and mass errors within ±5 ppm of the metabolite results were tentatively identified. The software evaluates the similarity or match quality between the observed MS/MS spectra and reference spectra from the library and assigns scores to these matches. Identification of compounds not found in the library was attempted again using MS/MS data from the ChemSpider database. The MS/MS data of the available authentic standards were also compared with the experimentally measured mass of the ions for verification.

2.9. Statistical Analysis

Correlations between TPC and antioxidant activity were evaluated using Pearson's correlation coefficient test. Statistical analyses were performed using the SPSS version 28.0 program (IBM SPSS Statistics, SPSS Inc., Chicago, IL, USA). To examine variations through mushroom fermentation, a one-way analysis of variance (ANOVA) was used. Subsequently, a multiple range test, Tukey's Honestly Significant Difference (HSD) test, was used to compare the means and identify significant differences among them. A significance level of $p < 0.05$ was employed to determine the presence of statistically significant differences. The reported values are presented as the mean ± standard deviation (SD) based on three independent samples.

3. Results and Discussion

3.1. The Bacterial Growth and pH Changes during Mushroom Fermentation

The changes in the viable cell count and pH values of the fermented mushrooms during the 72 h fermentation period are shown in Figure 1. As shown in Figure 1a, *Lp. plantarum LMG 17673* demonstrated robust growth in the mushroom samples, with no significant changes in viable cell count observed during the latter stages of fermentation (from 48 to 72 h) ($p \geq 0.05$). Initially, the viable bacteria counts of *Lp. plantarum LMG 17673* in *L. edodes* and *La. deliciosus* were 7.52 ± 0.07 and 7.67 ± 0.06 log CFU/mL, respectively. After 24 h of fermentation, both mushrooms exhibited a high growth rate, which remained stable throughout the fermentation process. In the later stages (from 48 to 72 h), the viable cell counts of *Lp. plantarum LMG 17673* remained above 8.5 log CFU/mL, indicating sustained metabolic activity. These findings confirm that the selected mushrooms are suitable substrates for the fermentation of the particular species.

During fermentation, LAB strains can metabolise sugar and produce organic acids, leading to changes in the pH value of mushrooms. The pH changes of mushroom samples during fermentation with *Lp. plantarum LMG 17673* are shown in Figure 1b. The most rapid decrease in pH occurred during the initial 24 h fermentation, dropping to 4.51 and 5.53 for *L. edodes* and *La. deliciosus*, respectively. These results are comparable to the pH of 4.61 observed in pickled *Agaricus bisporus* fermented with *Lb. plantarum* [24]. A lower pH inhibits the growth of pathogenic bacteria, potentially extending shelf life and maintaining anaerobic conditions for preservation [25]. Additionally, the pH reduction during fermentation is attributed to the accumulation of organic acids, as noted in recent studies [26].

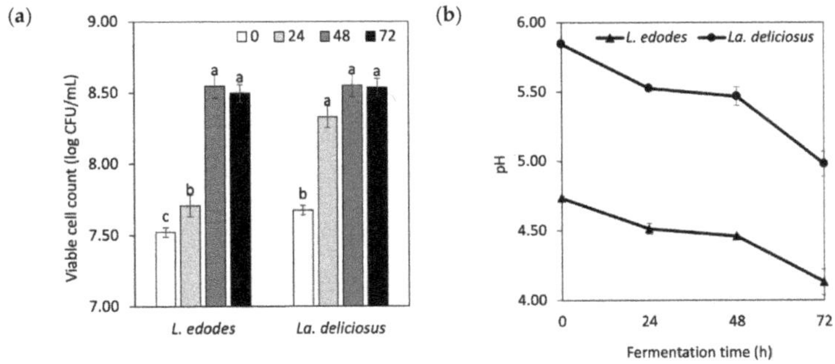

Figure 1. Microbial activity of *Lp. Plantarum LMG 17673*. (**a**) Viable cell count and (**b**) pH throughout the fermentation of *L. edodes* and *La. Deliciosus*. CFU: colony forming unit. Different lowercase letters within the same mushroom species indicate significant differences during fermentation ($p < 0.05$). Values are presented as mean ± SD ($n = 3$).

3.2. Changes in the Organic Acid and Sugar Levels during Mushroom Fermentation

The changes in organic acids and free sugars in fermented mushrooms are presented in Table S1. During mushroom fermentation, *Lp. Plantarum LMG 17673* can produce organic acids that lower the pH of the fermented mushroom matrixes. The organic acid contents of formic acid, malic acid, lactic acid, citric acid, fumaric acid, and succinic acid were assessed. During fermentation, the total organic acid content first increased, followed by a period of slight increase or stability (Figure 2a). The first 24 h of fermentation marked a crucial point for the alteration in organic acid content. Throughout the fermentation, there was an increase in lactic acid and formic acid levels and a corresponding decrease in malic acid, citric acid, fumaric acid, and succinic acid concentrations. These trends are consistent with a study by Yang et al. (2022), who fermented apple juice using LAB strains [27]. Malic acid can be transformed into lactic acid with the help of the malolactic enzyme produced by LAB, whereas the decomposition of citric acid may cause decreased citric acid content in various products. The fumaric and succinic acid contents in the fermented samples follow similar trends to those of malic and citric acids.

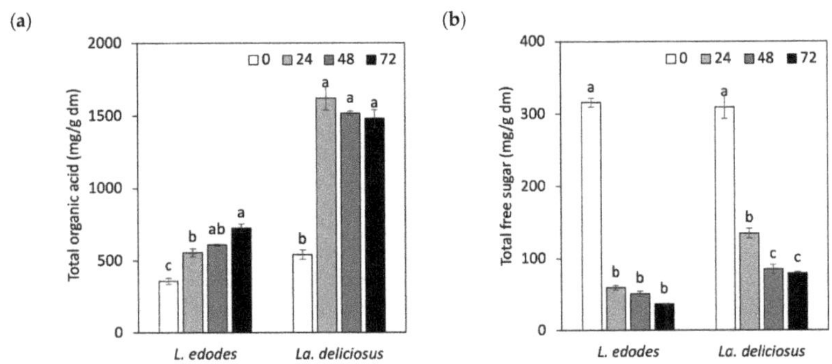

Figure 2. Effect of *Lp. plantarum LMG 17673* fermentation on mushrooms. (**a**) Total free sugar content and (**b**) total organic acid content. Different lowercase letters within the same mushroom species indicate significant differences during fermentation ($p < 0.05$). Values are presented as mean ± SD ($n = 3$).

The main mushroom monosaccharides, disaccharides, and sugar alcohols were glucose, ribose, trehalose, and mannitol, respectively, as has been reported for other edible mushrooms [28]. During the 72 h fermentation of mushrooms, the free sugar concentration in mushrooms decreased as shown in Figure 2b, primarily due to sugar consumption by LAB [29]. Previous studies have also reported decreased free sugar content in fermented foods such as cabbage during the initial fermentation stage [29]. Glucose and trehalose were rapidly consumed during the initial 24 h, while ribose was the only sugar found in the mushrooms after 72 h, similar to that reported for button mushrooms [24]. At the beginning of the fermentation process, mannitol, a sugar alcohol, was the most abundant sugar in the fermented mushrooms. After 72 h of fermentation, ribose was the highest level of sugar found in mushrooms.

3.3. Changes in Total Phenolic Content and Antioxidant Activity in Free and Bound Phenolics during Mushroom Fermentation

The TPC and antioxidant activity of the mushroom fractions are listed in Table 1. The TPC of *L. edodes* free phenolic fraction (LEFPs) and *La. deliciosus* free phenolic fraction (LDFPs), at the beginning of fermentation, were 1.86 $\pm$ 0.38 mg GAE/g dm and 1.68 $\pm$ 0.14 mg GAE/g dm, respectively. A significant increase in FPs during fermentation occurred within 24 h, with the highest TPC increase of 29% and 27.3% observed for LEFPs and LDFPs, respectively, compared to the unfermented samples (2.40 $\pm$ 0.06 mg GAE/g dm and 2.14 $\pm$ 0.39 mg GAE/g dm, respectively). Similarly, De Montijo-Prieto et al. (2023) also showed that 24 h fermented samples had the highest level of TPC while investigating the impact of *Lp. plantarum* fermentation on the TPC of ethanolic avocado leaf extracts during 96 h of fermentation [30]. In the present study, the TPC of LEFPs and LDFPs was steady between 24 h and 48 h of fermentation ($p > 0.05$), whereas from 48 h to 72 h, the TPC of the FPs slightly decreased ($p < 0.05$). The decrease in TPC during fermentation has been associated with the degradation of phenolic compounds by *Lp. plantarum* [31]. Comparatively, the bound fractions exhibited notably higher TPC levels than the free fractions did. After ultrasonic–alkaline hydrolysis, a greater amount of BPs were liberated from the cell wall components than FPs. Yeo et al. [11] also found that phenolic compounds were more present in the bound form rather than the free form in lentils. The highest TPC of *L. edodes* bound phenolic fraction (LEBPs) and *La. deliciosus* bound phenolic fraction (LDBPs) were measured at 0 h fermentation and amounted to 4.48 $\pm$ 0.26 mg GAE/g dm and 2.93 $\pm$ 0.53 mg GAE/g dm, respectively. Previous studies have indicated that the ratio between bound and free phenolic compounds can be altered depending on the processing technologies used, such as fermentation [32]. In the present study, the TPC of BPs in both mushrooms decreased during fermentation. This finding may be due to the availability of endogenous hydrolysing enzymes throughout fermentation and/or the enzymes produced by fermenting microorganisms. The endogenous enzymes benefit from LAB fermentation, which lowers the pH and creates a more optimal environment for their function. The enzymes can break down bonds and facilitate the mobilisation of phenolic compounds from their bound state to their free form [33,34]. LAB can also transform phenolic metabolites and release phenolic compounds from the food matrices during fermentation. This process helps to offset the degradation of the original parent phenolic compounds during fermentation [13].

In the study of Tu et al. (2021), the free phenolic content of *L. edodes* was reported as 7.16 mg GAE/g dry weight (dw), and the bond phenolic content of *L. edodes* was 7.47 $\pm$ 0.22 mg GAE/g dw, which corresponds with our results [35]. Similarly, the TPC of the methanolic extract of *L. edodes* was found to be around 2.7 mg GAE/g dw, which closely corresponded with our FPs results [36]. On the other hand, the high TPC of *L. edodes* fractions was opposite to the results reported by Wang et al. [37], who found the highest TPC of BPs of *L. edodes* as 26.67 mg GAE/100 g dm and FPs content as 20.62 mg/GAE 100 dm g. Our study showed that TPC values were 10-times higher than these values. Several factors can account for the observed differences, such as differences in the environment, mushroom variety, location, extraction conditions, substrates, and maturity [36,38]. In a study conducted by Yao et al. (2023), the TPC of ethanolic extract of *L. edodes* was reported as 127 mg GAE/g dw [39]; whereas, in the study by Pehlivan Karakas et al. (2023), the TPC of methanolic extract of *La. deliciosus* was measured as 193 $\pm$ 0.77 mg GAE/g dw [40]. The TPC values obtained in our study were lower than those reported in previous studies. It is clear that there are variations in extraction and measurement methods across different studies.

For instance, TPC in a previous study was measured using extracts obtained using 70% (*v*/*v*) ethanol and 80% (*v*/*v*) methanol. Despite these methodological differences, our study confirmed that *L. edodes* and *La. deliciosus* are rich in phenolic compounds.

Phenolic compounds can be used as reducing agents, free-radical scavengers, and singlet oxygen quenchers, and exhibit antioxidant activity primarily because of their ability to transfer hydrogen atoms or donate electrons to free radicals [30]. As shown in Table 1, although LAB fermentation did not increase ABTS radical-scavenging activity, it significantly increased DPPH radical-scavenging activity in the free fractions of both mushrooms, indicating that LAB fermentation may have the potential to enhance the antioxidant activity of mushrooms. An increase in TPC could lead to an enhancement in the antioxidant activity of fermented samples [26]. During the fermentation of both mushrooms, significant positive strong correlations ($p < 0.05$) were observed between DPPH and TPC for LEFPs and LDFPs ($R^2 = 0.95$ and $R^2 = 0.95$, respectively), and LEBPs and LDBPs ($R^2 = 0.98$ and $R^2 = 0.95$, respectively). Thus, the higher DPPH activity observed in fermented mushrooms in this study may be attributed to the elevated bioaccessible free total phenolic content modulated by *Lp. plantarum LMG 17673* during lactic acid fermentation. In this study, *Lp. plantarum LMG 17673* fermentation increased the FPs by releasing BFs from the food matrix, potentially enhancing the bioavailability of mushroom phenolics. In addition, the TPC and DPPH radical-scavenging activity of FPs increased, whereas the content in the BP fractions decreased. Similarly, fermentation with selenium-enriched *Lp. plantarum* markedly improves the total phenolic content and antioxidant properties of *Pleurotus eryngii* [32]. Interestingly, the ABTS activity of fermented mushrooms was not correlated with TPC and DPPH activities. The lack of correlation between ABTS levels and TPC and DPPH analysis is similar to the findings in pegaga, a medicinal herb [41], and galangal extracts [42]. The correlation between antioxidant capacities and antioxidant compounds is mostly complicated, as antioxidant capacity is determined by the number of antioxidants, as well as by environmental conditions, the interaction of antioxidants with one another and the matrix, and their physical location and structure [41]. These factors can also be attributed to the ambiguous correlation between ABTS activity and TPC. In addition, the quantified TPC values may include other components that can exert antioxidant activity. Furthermore, assessing antioxidant activity using varying methods based on different mechanisms may yield different results; ABTS relies on the hydrogen atom transfer mechanism, whereas DPPH functions through an electron transfer mechanism [12,34].

Table 1. Effect of fermentation on TPC and antioxidant activity of free and bound phenolic fractions of fermented mushrooms.

Fermentation Time	TPC (mg GAE/g)				DPPH (mg TE/g)				ABTS (mg TE/g)			
	L. edodes		*La. deliciosus*		*L. edodes*		*La. deliciosus*		*L. edodes*		*La. deliciosus*	
	Free	Bound	Free	Bound	Free	Bound	Free	Bound	Free	Bound	Free	Bound
0 h	1.86 ± 0.04 ^{c}	4.48 ± 0.08 ^{a}	1.68 ± 0.01 ^{c}	2.93 ± 0.05 ^{a}	2.55 ± 0.11 ^{c}	6.54 ± 0.26 ^{a}	2.26 ± 0.13 ^{b}	1.81 ± 0.14 ^{a}	6.29 ± 0.21 ^{b}	23.68 ± 1.51 ^{a}	5.69 ± 0.14 ^{a}	26.71 ± 1.43 ^{b}
24 h	2.40 ± 0.01 ^{a}	4.10 ± 0.06 ^{b}	2.14 ± 0.04 ^{a}	2.36 ± 0.04 ^{b}	3.77 ± 0.10 ^{a}	5.55 ± 0.16 ^{b}	2.88 ± 0.36 ^{a}	1.66 ± 0.15 ^{ab}	5.90 ± 0.45 ^{b}	24.04 ± 1.61 ^{a}	5.56 ± 0.31 ^{a}	28.05 ± 0.1 ^{a}
48 h	2.33 ± 0.04 ^{a}	3.79 ± 0.02 ^{c}	2.09 ± 0.07 ^{a}	2.35 ± 0.02 ^{b}	3.36 ± 0.13 ^{b}	5.21 ± 0.11 ^{b}	2.64 ± 0.03 ^{ab}	1.44 ± 0.12 ^{bc}	5.61 ± 0.71 ^{b}	23.40 ± 1.5 ^{a}	5.67 ± 1.23 ^{a}	25.45 ± 0.23 ^{c}
72 h	2.14 ± 0.04 ^{b}	3.50 ± 0.04 ^{d}	1.88 ± 0.03 ^{b}	2.21 ± 0.04 ^{b}	3.30 ± 0.14 ^{b}	4.24 ± 0.09 ^{c}	2.54 ± 0.06 ^{ab}	1.31 ± 0.11 ^{c}	6.10 ± 0.29 ^{a}	22.34 ± 2.21 ^{a}	5.47 ± 0.43 ^{a}	25.5 ± 0.14 ^{c}

TPC: total phenolic content; GAE: gallic acid equivalent antioxidant capacity; DPPH: 2,2-diphenyl-1-picrylhydrazyl; TE: trolox equivalent; ABTS: 2,2′azino-bis(3-ethylbenzthiazoline-6-sulphonic acid) diammonium salt. Different lowercase letters in the same column indicate significant differences between the samples in the same column ($p < 0.05$). Values are presented as mean ± SD ($n = 3$).

3.4. The Non-Targeted Analysis of UPLC-Q-TOF-MS Data from Mushroom Fermentation

An unsupervised PCA was applied to gain insights into the metabolic variations among different fermentation time points and phenolic fractions. A non-targeted metabolomics approach using UPLC-Q-TOF-MS datasets was employed for PCA, transforming the dataset into a series of values representing linearly uncorrelated variables. This widely utilised approach in metabolomic studies has allowed us to comprehensively overview metabolite variations among samples. Figure 3a shows the PCA score plot with PC1 and PC2, explaining 46.99% of the total variance (39.1% and 7.89%, respectively), which appeared to be between fermented *L. edodes* fractions. It could be observed that the BPs can be distinctly grouped from FPs. However, overlapping clusters at different fermentation time points in each fraction showed weakly explained variance. Figure 3b shows the PCA score plot with a higher explanation for *La. deliciosus*. PC1 and PC2 explained 57.2% of the total variance (48.97% and 8.23%, respectively), which also appeared to lie between those of fermented *La. deliciosus* fractions. The FPs of unfermented (0 h) *La. deliciosus* exhibited a clear difference from the fermented free fractions. PCA indicated that the form of phenolics was most affected by fermentation, leading to increased differentiation between the fermented and unfermented samples.

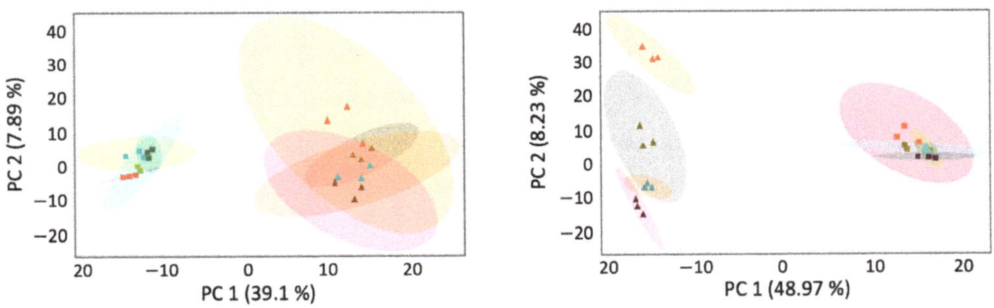

(a) (b)

Figure 3. Principal component analysis (PCA) plots with PC 1 and PC 2: (**a**) PCA plot of *L. edodes*; (**b**) PCA plot of *La. deliciosus*. The triangle represents bound phenolics, and the square represents free phenolics. Red colour represents 0 h fermented mushrooms, caramel colour represents 24 h fermented mushrooms, blue colour represents 48 h fermented mushrooms, and brown colour represents 72 h fermented mushrooms.

To further confirm the variance in metabolic profiles among fermented mushrooms, a volcano plot of fermented *L. edodes* (Figure 4a) and *La. deliciosus* (Figure 4b) showed the number of features which changed significantly during the fermentation. The horizontal green line represents the p-test limit, and the features positioned above this line were considered statistically significant ($p < 0.05$). The vertical green lines represent the fold-change limits. The volcano plot shows the passing features (coloured) and non-passing features (grey). Each square, coloured pale blue or orange, corresponds to a feature that has successfully passed the significance test yet failed to meet the fold-change cut-off criteria. During the fermentation of the mushrooms, pairwise group comparisons were conducted by comparing unfermented mushroom samples (0 h fermented) to samples at each further time point, including 24, 48, and 72 h of fermentation. Each red square represents upregulated features, each blue square represents downregulated features ($p < 0.05$ and FC ≥ 2.0), the upper right and left corner green squares represent the features ($p < 0.001$ and FC ≥ 2.0), and grey squares reveal metabolites that have no significant difference between different fermentation time point samples.

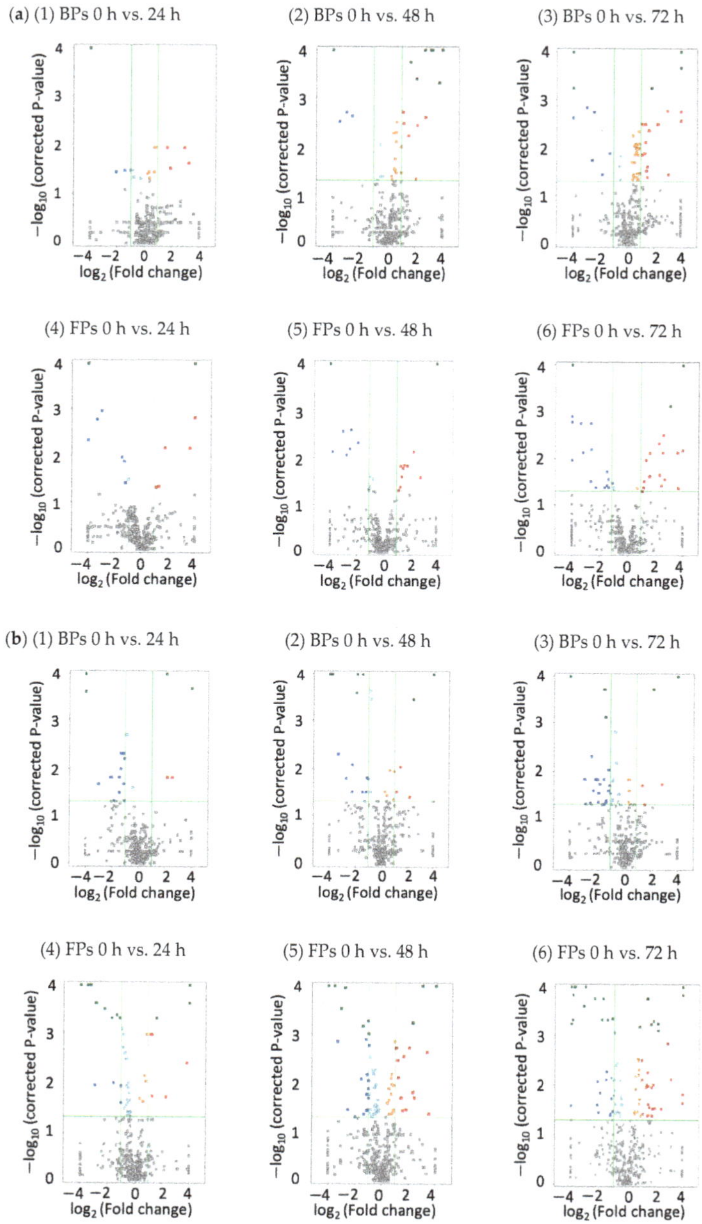

Figure 4. Volcanic plots of differential metabolites in pairwise comparisons between unfermented (0 h) and fermented samples: (**a**) *L. edodes* and (**b**) *La. deliciosus*. BPs; bound phenolics, FPs; free phenolic. The colours in the figure represent the following: grey squares reveal metabolites that show no significant difference between different fermentation time point samples; pale blue or orange squares indicate features that have successfully passed the significance test yet failed to meet the fold-change cut-off criteria. Red squares represent upregulated features, blue squares represent down-regulated features ($p < 0.05$ and FC ≥ 2.0), and green squares in the upper right and left corners represent features ($p < 0.001$ and FC ≥ 2.0).

A total of 468 features were evaluated in pairwise groups of fermented *L. edodes*. After the volcano plot analysis with $p < 0.05$ and FC $\geq$ 2.0, 216 out of 468 features were significantly different in *L. edodes* samples during fermentation. The volcano plots of fermented LEBPs showed that the number of features present at 0 h was significantly different from those present at 24, 48, and 72 h, with 17, 45, and 62 metabolites, respectively (as shown in Figures 1–3 and 4a) ($p < 0.05$). On the other hand, 0 h fermented LEFPs had 19 significantly different features compared to 24 h fermented LEFPs, which increased to 29 and 44 at 48 h and 72 h, respectively (Figures 4a, 5 and 6) ($p < 0.05$). In addition, 332 of the 469 potential features with significantly different amounts were screened within the pairwise groups for *La. deliciosus* during fermentation ($p < 0.05$ and FC $\geq$ 2.0). The volcano plot of fermented LDBPs revealed that at 0 h, the number of features was significantly different compared to 24 h, 48 h, and 72 h, with counts of 22, 32, and 49, respectively (Figures 1–3 and 4b) ($p < 0.05$). Similarly, LDFPs presenting the number of features at 0 h showed statistically significant differences compared to 24 h, 48 h, and 72 h, with results of 50, 85, and 94, respectively (Figures 4b, 5 and 6) ($p < 0.05$). As the fermentation duration increased, there was a corresponding increase in the number of features in the same fractions that exhibited statistically significant differences in their profiles. These findings suggest that LAB fermentation can alter both the composition and concentration of phenolic compounds, highlighting the fact that bound phenolics are predominantly affected by fermentation. This phenomenon may be linked to lower pH, which has the potential to influence the stability of phenolic compounds, leading to degradation or structural alterations [2]. Li et al. (2019) also observed that the phenolic composition of apple juice could undergo variations during fermentation with *Lp. plantarum* [43]. This alteration may result from the removal or hydrolysis of structural moieties in various phenolic compounds, leading to alterations in phenolic profiles.

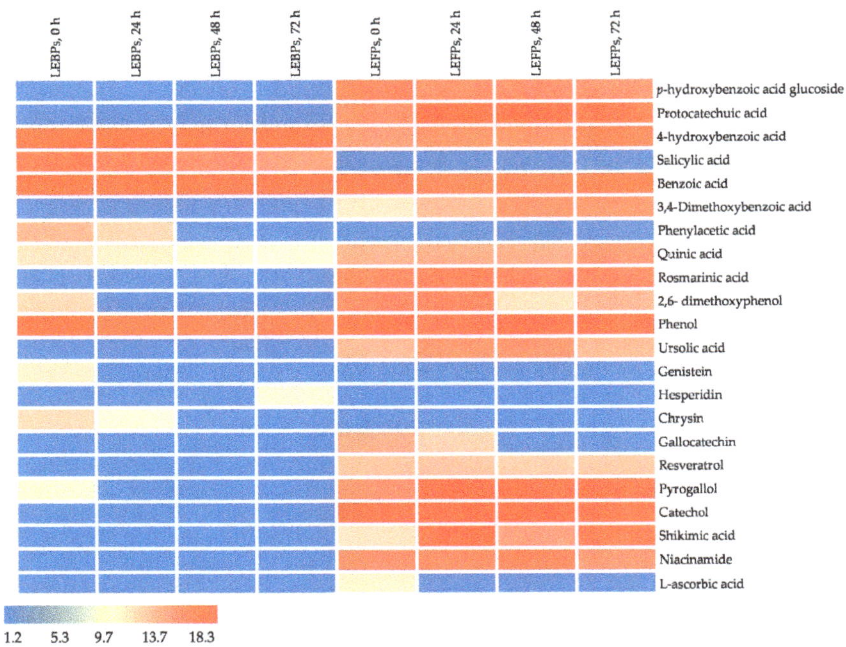

Figure 5. Level of phenolic compounds and vitamins in *L. edodes* fractions during fermentation. LEBPs: *Lentinula edodes* bound phenolics; LEFPs: *Lentinula edodes* free phenolics; LDBPs: *Lactarius delicious* bound phenolics; and LDFPs: *Lactarius delicious* free phenolics. 0 h, 24 h, 48 h, and 72 h represent fermentation time points.

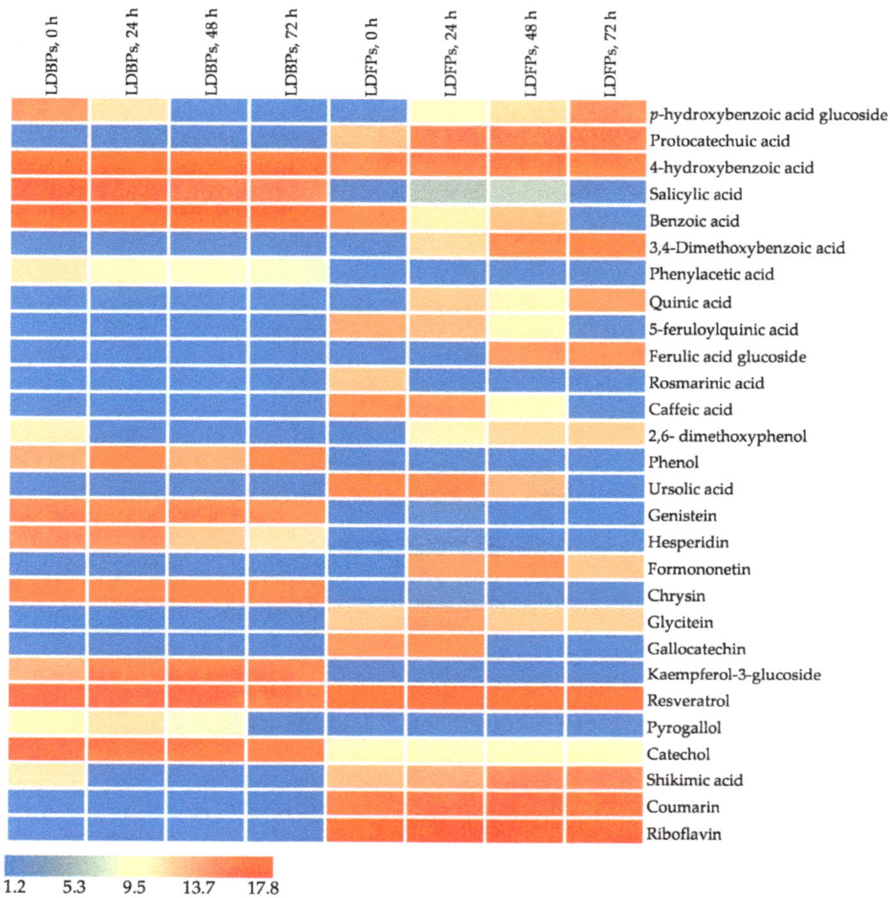

Figure 6. Level of phenolic compounds and vitamins in *La. deliciosus* fractions during fermentation. LEBPs: *Lentinula edodes* bound phenolics; LEFPs: *Lentinula edodes* free phenolics; LDBPs: *Lactarius delicious* bound phenolics; and LDFPs: *Lactarius delicious* free phenolics. 0 h, 24 h, 48 h, and 72 h represent fermentation time points.

3.5. Metabolite Profiling and Comparative Analysis of Mushrooms during Different Fermentation Durations via UPLC-Q-TOF-MS/MS

To prevent overestimation of TPC using the spectrophotometric method, we examined variations in individual phenolic compounds that occur during the fermentation process. Secondary metabolites, such as phenolic acids and flavonoids, were detected using Agilent UPLC-ESI-QTOF-MS/MS MassHunter Qualitative Software by considering the MS/MS spectral data. A total of 30 metabolites were tentatively identified in the fermented mushroom fractions (Table 2). Protocatechuic acid, 3,4-dimethoxybenzoic acid, 5-feruloylquinic acid, ferulic acid glucoside, rosmarinic acid, caffeic acid, ursolic acid, formononetin, glycitein, gallocatechin, coumarin, riboflavin, niacinamide, and L-ascorbic acid were only found in the free fractions (LEFPs and LDFPs). With the exception of protocatechuic acid, 3,4-dimethoxybenzoic acid, and phenylacetic acid, all hydroxybenzoic acids were present in the free and bound forms in mushrooms. The cinnamic acid derivatives were found only in their free forms, except for quinic acid. Consistent with our results, *L. edodes* has been reported to contain hydroxybenzoic acid [44], hesperidin [37],

3,4-dimethoxy benzoic acid, glycitein, l-ascorbic acid, riboflavin, niacinamide, salicylic acid, and shikimic acid [45], whereas *La. deliciosus* has been reported to contain hydroxybenzoic acid, pyrogallol, protocatechuic acid [9,46], resveratrol, benzoic acid, kaempferol, chrysin, and ferulic acid [47].

Table 2. Tentative identification of phenolic compounds, terpenic acids, and vitamins in the fermented *L. edodes* and *La. deliciosus* fractions.

No	Proposed Compound	Molecular Formula	RT (min)	Adduct	Molecular Weight	Theoretical (*m/z*)	Observed (*m/z*)	Mass Error (ppm)	MS/MS Product Ions Collision Energies			Mushroom Species	
									10 V	20 V	40 V	*L. edodes*	*La. deliciosus*
	Phenolic acids												
	i. Hydroxybenzoic acids												
1	*p*-hydroxybenzoic acid glucoside	$C_{13}H_{16}O_8$	13.42	(M−H)−	300.085	299.0779	299.0772	2.1	299	-	106	All-LEBPs All-LEFPs	LDBPs-0 h LDBPs-24 h All-LDFPs
2	Protocatechuic acid	$C_7H_6O_4$	4.30	(M−H)−	154.0264	153.0191	153.0193	−1.47	-	111	-	All-LEFPs	All-LDFPs
3	4-hydroxybenzoic acid *	$C_7H_6O_3$	7.91	(M−H)−	138.0318	137.0245	137.0244	0.76	109	111, 137	-	All-LEBPs All-LEFPs	All-LDBPs All-LDFPs
4	Salicylic acid *	$C_7H_6O_3$	19.83	(M−H)−	138.0317	137.0238	137.0244	−4.6	-	111	-	All-LEBPs	All-LDBPs LDFPs-24 h LDFPs-48 h
5	Benzoic acid	$C_7H_6O_2$	9.6	(M−H)−	122.0368	121.0295	121.0295	0.35	120	121	-	All-LEBPs All-LEFPs	All-LDBPs LDFPs-0 h LDFPs-24 h LDFPs-48 h
6	3,4-Dimethoxybenzoic acid	$C_9H_{10}O_4$	6.99	(M−H)−	182.0579	181.0506	181.0505	−0.16	101, 141	120	-	All-LEFPs	LDFPs-24 h LDFPs-48 h LDFPs-72 h
7	Phenylacetic acid *	$C_8H_8O_2$	11.41	(M−H)−	136.0516	135.0452	135.0452	0.14	135	117, 109	-	LEBPs-0 h LEBPs-24 h	All-LDBPs
	ii. Hydroxycinnamic acids												
8	Quinic acid	$C_7H_{12}O_6$	7.25	(M+HCOO)−	192.0638	237.057	237.062	1.95	156, 126	-	-	All-LEBPs All-LEFPs	LDFPs-24 h LDFPs-48 h LDFPs-72 h
9	5-feruloylquinic acid	$C_{17}H_{20}O_9$	4.35	(M+HCOO)−	368.1093	413.1227	413.0143	-3.8	312	-	-	-	LDFPs-0 h LDFPs-24 h LDFPs-48 h
11	Rosmarinic acid	$C_{18}H_{16}O_8$	11.18	(M+HCOO)−	360.087	405.0809	405.0827	−4.5	211	191, 227	139, 171	All-LEFPs	LDFPs-0 h
12	Caffeic acid	$C_9H_8O_4$	10.20	(M−H)−	180.0425	179.0352	179.0332	3.3	124	165	-	-	LDFPs-0 h LDFPs-24 h LDFPs-48 h
	Phenols												
13	2,6-dimethoxyphenol	$C_8H_{10}O_3$	8.77	(M+HCOO)−	154.0628	199.06	199.061	0.32	123	128	-	LEBPs-0 h All-LEFPs	LDBPs-0 h LDFPs-24 h LDFPs-48 h LDFPs-72 h
14	Phenol	C_6H_6O	30.87	(M+HCOO)−	94.0416	139.039	139.0398	−2.61	86	-	-	All-LEBPs All-LEFPs	All-LDBPs
	Terpenic acids												
15	Ursolic acid *	$C_{30}H_{48}O_3$	35.81	(M−H)− (M+HCOO)−	456.3603	455.3544 501.3584	455.3531 501.3585	2.82 −0.33	391 297, 337	255 152	163, 283 435	All-LEBPs	LDFPs-0 h LDFPs-24 h LDFPs-48 h
	Flavonoids												
16	Genistein	$C_{15}H_{10}O_5$	28.13	(M−H)−	270.053	269.0457	269.0455	0.42	151, 107	119	187	LEBPs-0 h	All-LDBPs
18	Formononetin	$C_{16}H_{12}O_4$	29.35	(M−H)−	268.0733	267.066	267.0663	−1	251	-	-	-	LDFPs-24 h LDFPs-48 h LDFPs-72 h
20	Glycitein	$C_{16}H_{12}O_5$	30.91	(M+HCOO)−	284.0745	329.0778	329.0774	−0.92	285	-	-	-	All-LDFPs
21	Gallocatechin	$C_{15}H_{14}O_7$	4.22	(M−H)−	306.0718	305.0702	305.0645	−2.8	156	-	-	LEFPs-0 h LEFPs-24 h	LDFPs-0 h LDFPs-24 h
22	Kaempferol-3-glucoside	$C_{21}H_{20}O_{11}$	22.99	(M−H)−	448.1001	447.0928	447.0933	−1.07	224, 115	-	-	-	All-LDBPs
	Stilbenes												
23	Resveratrol	$C_{14}H_{12}O_3$	29.02	(M−H)−	228.0785	227.0713	227.0714	−0.49	183, 199	155, 141	139, 167	All-LEFPs	All-LDBPs All-LDFPs
	Other polyphenols												
24	Pyrogallol *	$C_6H_6O_3$	3.57	(M−H)− (M+HCOO)−	126.0318	125.0244 171.03	125.0244 171.0299	0.21 0.59	124 103	108, 124 -	- -	LEBPs-0 h All-LEFPs	LDBPs-0 h LDBPs-24 h LDBPs-48 h
25	Catechol	$C_6H_6O_2$	5.17	(M−H)−	110.0366	109.032	109.0294	0.3	101	-	-	All-LEFPs	All-LDBPs All-LDFPs
26	Shikimic acid	$C_7H_{10}O_5$	2.9	(M−H)− (M+HCOO)−	174.0529	173.0457 219.0509	173.0455 219.051	0.82 −0.37	157, 109 113	- -	- -	All-LEFPs	LDBPs-0 h All-LDFPs
27	Coumarin	$C_9H_6O_2$	15.5	(M−H)−	146.0368	145.0295	145.0245	0.12	142	108	-	-	All-LDFPs
	Vitamins												
28	Riboflavin	$C_{17}H_{20}N_4O_6$	16.22	(M−H)−	376.1384	375.131	375,131	−0.08	255, 129	212, 165	198, 184	-	All-LDFPs

Table 2. *Cont.*

No	Proposed Compound	Molecular Formula	RT (min)	Adduct	Molecular Weight	Theoretical (*m*/*z*)	Observed (*m*/*z*)	Mass Error (ppm)	MS/MS Product Ions Collision Energies			Mushroom Species	
									10 V	20 V	40 V	*L. edodes*	*La. deliciosus*
29	Niacinamide	$C_6H_6N_2O$	15.62	(M+HCOO)−	122.048	167.0462	167.0462	−0.23	124, 128	108	-	All-LEFPs	-
30	L-ascorbic acid	$C_6H_8O_6$	1.66	(M−H)−	176.032	175.0247	175.0248	−0.92	117, 175	-	-	LEFPs-0 h	-

MS/MS: Tandem mass spectrometry; LEBPs: *Lentinula edodes* bound phenolics; LEFPs: *Lentinula edodes* free phenolics; LDBPs: *Lactarius delicious* bound phenolics; LDFPs: *Lactarius delicious* free phenolics. "All" represents all fermentation time points in the mentioned fraction. 0 h, 24 h, 48 h, and 72 h represent fermentation time points. "-" is not detected. "RT" stands for retention time. * The compounds were verified using authentic standards.

Heat map analysis was used to visualise the samples based on the concentration of bioactive compounds at distinct fermentation time points and phenolic fractions, where the colour scheme from red to blue shows the concentrations in decreasing order. Figures 3 and 4 show significant concentration differences ($p < 0.05$) in *L. edodes* and *La. deliciosus*, identifying 22 and 28 phenolic compounds, respectively, with three vitamins in each. The concentration differences during LAB fermentation can be explained by changes in the content and profile of bioactive compounds, modification of the parent phenolic compounds, and structural breakdown of the mushroom cell walls [48]. In our study, certain compounds that became undetectable or reduced within FPs during fermentation could be associated with the metabolic activities of *Lp. plantarum LMG 17673*. For instance, there was a decrease in p-hydroxybenzoic acid glucoside in both LEFPs and LDFPs, which corresponded to a higher accumulation of its reduced metabolite, 4-hydroxybenzoic acid. As a result of this concentration difference, the concentration of quinic acid increased in both LEFPs and LDFPs. In contrast, that of 5-feruloylquinic acid, a derivative of quinic acid, decreased during LAB fermentation. This decrease might have contributed to the increase in quinic acid content, suggesting a potential conversion. Similarly, the concentration of quinic acid was significantly higher in avocado seeds fermented with *Lp. plantarum*, indicating possible hydrolysis of hydroxycinnamic acids [49]. Additionally, hydroxycinnamic acids, such as caffeic acid, can undergo transformation into different compounds facilitated by reductase enzymes. This process may lead to the metabolism of dihydrocaffeic acids, which can further undergo decarboxylation to produce vinyl derivatives such as vinyl phenol and vinyl guaiacol [30,50]. Previous studies have reported that the *Lp. plantarum* may contribute to the degradation of phenolic compounds through processes such as depolymerisation, hydrolysis, decarboxylase-mediated metabolism, and reductase activity [51].

Fermentation increased the levels of certain components, such as protocatechuic acid, 4-hydroxybenzoic acid, 3,4-dimethoxybenzoic acid, quinic acid, phenol, pyrogallol, and shikimic acid in *L. edodes* FPs. Likewise, in the free fraction of *La. deliciosus*, the concentrations of protocatechuic acid, 4-hydroxybenzoic acid, salicylic acid, 3,4-dimethoxybenzoic acid, quinic acid, ferulic acid glucoside, 2,6-dimethoxyphenol, formononetin, glycitein, resveratrol, shikimic acid, and coumarin were significantly higher than those in the initial stages of fermentation. Furthermore, the biotransformation of ferulic acid to vanillin has been well-studied in various bacterial species, including LAB [52]. However, we found increased ferulic acid glucoside levels in LDFPs without identifying vanillin, possibly because of its low concentration below the threshold (score < 75). Moreover, during *L. edodes* fermentation, quinic acid, 2,6-dimethoxyphenol, phenol, and pyrogallol concentrations decreased in the bound fraction but increased in the free fraction. Similarly, during *La. deliciosus* fermentation, the levels of 4-hydroxybenzoic acid, salicylic acid, 2,6-dimethoxyphenol, and shikimic acid decreased in the bound fraction and increased in the free fraction. This phenomenon can be explained by the ability of LAB to break down large molecules in favour of phenolic compounds in mushrooms, which may include breaking down cell wall polysaccharides and proteins, wherein bound phenolics are linked [53]. Thus, LAB fermentation can potentially enhance biological activity by converting bound phenolics to a free state. Additionally, LAB fermentation can biotransform complex phenolic compounds and flavonoids into simple bioactive compounds, such as phenolic acids, thereby increasing their bioavailability [54].

4. Conclusions

This study demonstrated the efficiency of LAB fermentation in the biotransformation of phenolic compounds. *Lp. plantarum LMG 17673* was successfully fermented for 72 h on the substrate of *L. edodes* and *La. deliciosus*, resulting in a high viable cell count. *L. edodes* and *La. deliciosus* were rich in bound and free phenolics, with the former fraction being notably more abundant. LAB fermentation enhanced the releasing of bound phenolics and modified their parent phenolic compounds and structure. The phenolic compound composition and concentration changes were assessed using UPLC-MS-TOF-MS/MS. This study detected thirty phenolic compounds in the mushroom fractions during LAB fermentation. Notably, DPPH scanning activity analysis indicated that the release of bound phenolics significantly contributed to antioxidant activity through fermentation. Hence, this study demonstrated the effect of bound phenolics on the bioaccessibility of mushrooms. Subsequent investigations concerning the conversion mechanism of phenolic compounds within mushrooms through LAB fermentation should be pursued in future studies.

Supplementary Materials: The following supporting information can be downloaded at: https://www.mdpi.com/article/10.3390/foods13111616/s1, Table S1: Changes in free sugars and organic acids in fermented mushrooms during fermentation.

Author Contributions: Conceptualization, E.N.A.-S. and K.R.; Methodology, E.N.A.-S. and Y.V.; Validation, E.N.A.-S., Y.V. and K.R.; Formal Analysis, E.N.A.-S.; Investigation, E.N.A.-S.; Resources, K.R.; Data Curation, E.N.A.-S. and Y.V.; Writing—Original Draft Preparation, E.N.A.-S.; Writing—Review and Editing, K.R., E.N.A.-S. and Y.V.; Visualization, E.N.A.-S.; Supervision, K.R. and B.Ö.; Project Administration, E.N.A.-S.; Funding Acquisition, K.R. and E.N.A.-S. All authors have read and agreed to the published version of the manuscript.

Funding: This research received no external funding.

Institutional Review Board Statement: Not applicable.

Informed Consent Statement: Not applicable.

Data Availability Statement: The original contributions presented in the study are included in the article and Supplementary Materials, further inquiries can be directed to the corresponding author.

Acknowledgments: Eda Nur Ayar-Sümer is grateful to The Scientific and Technical Research Council of Turkey-TUBITAK-BIDEB 2214/A for supporting her research through the International Doctoral Research Fellowship Program (1059B142100530). Furthermore, the authors want to thank EFROFlanders, Hermes Fund, TUA WEST, Province West-Flanders, and Ghent University for the financial contribution to the OIP Voeding 1173 project.

Conflicts of Interest: The authors declare no conflicts of interest.

References

1. Lu, X.; Hou, H.; Fang, D.; Hu, Q.; Chen, J.; Zhao, L. Identification and Characterization of Volatile Compounds in *Lentinula edodes* during Vacuum Freeze-Drying. *J. Food Biochem.* **2022**, *46*, e13814. [CrossRef] [PubMed]
2. Wang, X.H.; Cheng, X.D.; Wang, D.; Wu, Z.; Chen, Y.; Wu, Q.X. Antioxidant and Anti-Aging Effects of Polysaccharide LDP-1 from Wild *Lactarius deliciosus* on Caenorhabditis Elegans. *Food Nutr. Res.* **2022**, *66*. [CrossRef]
3. Dogan, A.; Uyar, A.; Hasar, S.; Keles, O.F. The Protective Effects of the *Lactarius deliciosus* and *Agrocybe cylindracea* Mushrooms on Histopathology of Carbon Tetrachloride Induced Oxidative Stress in Rats. *Biotech. Histochem.* **2022**, *97*, 143–151. [CrossRef] [PubMed]
4. Dedousi, M.; Melanouri, E.M.; Karayannis, D.; Kaminarides, E.I.; Diamantopoulou, P. Utilization of Spent Substrates and Waste Products of Mushroom Cultivation to Produce New Crops of *Pleurotus ostreatus*, *Pleurotus eryngii* and *Agaricus bisporus*. *Carbon Resour. Convers.* **2024**, *7*, 100196. [CrossRef]
5. Leong, Y.K.; Ma, T.W.; Chang, J.S.; Yang, F.C. Recent Advances and Future Directions on the Valorization of Spent Mushroom Substrate (SMS): A Review. *Bioresour. Technol.* **2022**, *344*, 126157. [CrossRef] [PubMed]
6. Umaña, M.; Eim, V.; Garau, C.; Rosselló, C.; Simal, S. Ultrasound-Assisted Extraction of Ergosterol and Antioxidant Components from Mushroom by-Products and the Attainment of a β-Glucan Rich Residue. *Food Chem.* **2020**, *332*, 127390. [CrossRef]
7. Antunes, F.; Marçal, S.; Taofiq, O.; Morais, A.M.M.B.; Freitas, A.C.; Ferreira, I.C.F.R.; Pintado, M. Valorization of Mushroom By-Products as a Source of Value-Added Compounds and Potential Applications. *Molecules* **2020**, *25*, 2672. [CrossRef]

8. Cateni, F.; Gargano, M.L.; Procida, G.; Venturella, G.; Cirlincione, F.; Ferraro, V. Mycochemicals in Wild and Cultivated Mushrooms: Nutrition and Health. *Phytochem. Rev.* **2022**, *21*, 339–383. [CrossRef]
9. Çayan, F.; Deveci, E.; Tel-Çayan, G.; Duru, M.E. Identification and Quantification of Phenolic Acid Compounds of Twenty-Six Mushrooms by HPLC–DAD. *J. Food Meas. Charact.* **2020**, *14*, 1690–1698. [CrossRef]
10. Zhou, Y.; Chu, M.; Ahmadi, F.; Agar, O.T.; Barrow, C.J.; Dunshea, F.R.; Suleria, H.A.R. A Comprehensive Review on Phytochemical Profiling in Mushrooms: Occurrence, Biological Activities, Applications and Future Prospective. *Food Rev. Int.* **2024**, *40*, 924–951. [CrossRef]
11. Yeo, J.D.; Tsao, R.; Sun, Y.; Shahidi, F. Liberation of Insoluble-Bound Phenolics from Lentil Hull Matrices as Affected by *Rhizopus oryzae* Fermentation: Alteration in Phenolic Profiles and Their Inhibitory Capacities against Low-Density Lipoprotein (LDL) and DNA Oxidation. *Food Chem.* **2021**, *363*, 130275. [CrossRef] [PubMed]
12. Tao, R.; Chen, Q.; Li, Y.; Guo, L.; Zhou, Z. Physicochemical, Nutritional, and Phytochemical Profile Changes of Fermented Citrus Puree from Enzymatically Hydrolyzed Whole Fruit under Cold Storage. *LWT* **2022**, *169*, 114009. [CrossRef]
13. Leonard, W.; Zhang, P.; Ying, D.; Adhikari, B.; Fang, Z. Fermentation Transforms the Phenolic Profiles and Bioactivities of Plant-Based Foods. *Biotechnol. Adv.* **2021**, *49*, 107763. [CrossRef] [PubMed]
14. Yi, C.; Li, Y.; Zhu, H.; Liu, Y.; Quan, K. Effect of *Lactobacillus plantarum* Fermentation on the Volatile Flavors of Mung Beans. *LWT* **2021**, *146*, 111434. [CrossRef]
15. Davis, R.M.; Sommer, R.; Menge, J.A. *Field Guide to Mushrooms of Western North America*; University of California Press: Berkeley, CA, USA, 2012; ISBN 9780520271074.
16. Stamets, P. *Growing Gourmet and Medicinal Mushrooms*; Ten Speed Press: Berkeley, CA, USA, 2011.
17. He, C.; Sampers, I.; Raes, K. Dietary Fiber Concentrates Recovered from Agro-Industrial by-Products: Functional Properties and Application as Physical Carriers for Probiotics. *Food Hydrocoll.* **2021**, *111*, 106175. [CrossRef]
18. Wen, D.; Li, C.; Di, H.; Liao, Y.; Liu, H. A Universal HPLC Method for the Determination of Phenolic Acids in Compound Herbal Medicines. *J. Agric. Food Chem.* **2005**, *53*, 6624–6629. [CrossRef] [PubMed]
19. Gonzales, G.B.; Smagghe, G.; Raes, K.; Van Camp, J. Combined Alkaline Hydrolysis and Ultrasound-Assisted Extraction for the Release of Nonextractable Phenolics from Cauliflower (*Brassica oleracea var. botrytis*) Waste. *J. Agric. Food Chem.* **2014**, *62*, 3371–3376. [CrossRef] [PubMed]
20. Singleton, V.L.; Orthofer, R.; Lamuela-Raventós, R.M. [14] Analysis of total phenols and other oxidation substrates and antioxidants by means of folin-ciocalteu reagent. *Methods Enzymol.* **1999**, *299*, 152–178.
21. Kumaran, A.; Joel Karunakaran, R. In Vitro Antioxidant Activities of Methanol extracts of Five Phyllanthus Species from India. *LWT* **2007**, *40*, 344–352. [CrossRef]
22. Re, R.; Pellegrini, N.; Proteggente, A.; Pannala, A.; Yang, M.; Rice-Evans, C. Antioxidant activity applying an improved ABTS radical cation decolorization assay. *Free Radic. Biol. Med.* **1999**, *26*, 1231–1237. [CrossRef]
23. Gonzales, G.B.; Raes, K.; Coelus, S.; Struijs, K.; Smagghe, G.; Van Camp, J. Ultra(High)-Pressure Liquid Chromatography-Electrospray Ionization-Time-of-Flight-Ion Mobility-High Definition Mass Spectrometry for the Rapid Identification and Structural Characterization of Flavonoid Glycosides from Cauliflower Waste. *J. Chromatogr. A* **2014**, *1323*, 39–48. [CrossRef] [PubMed]
24. Jabłońska-Ryś, E.; Sławińska, A.; Skrzypczak, K.; Goral, K. Dynamics of Changes in PH and the Contents of Free Sugars, Organic Acids and LAB in Button Mushrooms during Controlled Lactic Fermentation. *Foods* **2022**, *11*, 1553. [CrossRef] [PubMed]
25. Liu, Y.; Xie, X.X.; Ibrahim, S.A.; Khaskheli, S.G.; Yang, H.; Wang, Y.F.; Huang, W. Characterization of *Lactobacillus pentosus* as a Starter Culture for the Fermentation of Edible Oyster Mushrooms (*Pleurotus* spp.). *LWT* **2016**, *68*, 21–26. [CrossRef]
26. Yang, J.; Sun, Y.; Chen, J.; Cheng, Y.; Zhang, H.; Gao, T.; Xu, F.; Pan, S.; Tao, Y.; Lu, J. Fermentation of Ginkgo Biloba Kernel Juice Using *Lactobacillus plantarum Y2* from the Ginkgo Peel: Fermentation Characteristics and Evolution of Phenolic Profiles, Antioxidant Activities in Vitro, and Volatile Flavor Compounds. *Front. Nutr.* **2022**, *9*, 1025080. [CrossRef] [PubMed]
27. Yang, J.; Sun, Y.; Gao, T.; Wu, Y.; Sun, H.; Zhu, Q.; Liu, C.; Zhou, C.; Han, Y.; Tao, Y. Fermentation and Storage Characteristics of "Fuji" Apple Juice Using *Lactobacillus acidophilus*, *Lactobacillus casei* and *Lactobacillus plantarum*: Microbial Growth, Metabolism of Bioactives and in Vitro Bioactivities. *Front. Nutr.* **2022**, *9*, 833906. [CrossRef] [PubMed]
28. Jabłońska-Ryś, E.; Skrzypczak, K.; Sławińska, A.; Radzki, W.; Gustaw, W. Lactic Acid Fermentation of Edible Mushrooms: Tradition, Technology, Current State of Research: A Review. *Compr. Rev. Food Sci. Food Saf.* **2019**, *18*, 655–669. [CrossRef] [PubMed]
29. Lee, M.A.; Choi, Y.J.; Lee, H.; Hwang, S.; Lee, H.J.; Park, S.J.; Chung, Y.B.; Yun, Y.R.; Park, S.H.; Min, S.; et al. Influence of Salinity on the Microbial Community Composition and Metabolite Profile in Kimchi. *Fermentation* **2021**, *7*, 308. [CrossRef]
30. De Montijo-Prieto, S.; del C. Razola-Díaz, M.; Barbieri, F.; Tabanelli, G.; Gardini, F.; Jiménez-Valera, M.; Ruiz-Bravo, A.; Verardo, V.; Gómez-Caravaca, A.M. Impact of Lactic Acid Bacteria Fermentation on Phenolic Compounds and Antioxidant Activity of Avocado Leaf Extracts. *Antioxidants* **2023**, *12*, 298. [CrossRef]
31. Carvalho, D.O.; Guido, L.F. A Review on the Fate of Phenolic Compounds during Malting and Brewing: Technological Strategies and Beer Styles. *Food Chem.* **2022**, *372*, 131093. [CrossRef]
32. Wang, B.; Zhao, N.; Li, J.; Xu, R.; Wang, T.; Guo, L.; Ma, M.; Fan, M.; Wei, X. Selenium-Enriched *Lactobacillus plantarum* Improves the Antioxidant Activity and Flavor Properties of Fermented *Pleurotus eryngii*. *Food Chem.* **2021**, *345*, 128770. [CrossRef]
33. Salar, R.K.; Purewal, S.S.; Bhatti, M.S. Optimization of Extraction Conditions and Enhancement of Phenolic Content and Antioxidant Activity of Pearl Millet Fermented with *Aspergillus awamori MTCC-548*. *Resour.-Effic. Technol.* **2016**, *2*, 148–157. [CrossRef]

34. Adebo, O.A.; Medina-Meza, I.G. Impact of Fermentation on the Phenolic Compounds and Antioxidant Activity of Whole Cereal Grains: A Mini Review. *Molecules* **2020**, *25*, 927. [CrossRef] [PubMed]
35. Tu, J.; Brennan, M.A.; Wu, G.; Bai, W.; Cheng, P.; Tian, B.; Brennan, C.S. Delivery of Phenolic Compounds, Peptides and β-Glucan to the Gastrointestinal Tract by Incorporating Dietary Fibre-Rich Mushrooms into Sorghum Biscuits. *Foods* **2021**, *10*, 1812. [CrossRef] [PubMed]
36. Xiaokang, W.; Brunton, N.P.; Lyng, J.G.; Harrison, S.M.; Carpes, S.T.; Papoutsis, K. Volatile and Non-Volatile Compounds of Shiitake Mushrooms Treated with Pulsed Light after Twenty-Four Hour Storage at Different Conditions. *Food Biosci.* **2020**, *36*, 100619. [CrossRef]
37. Wang, X.; He, X.; Sun, C.; Peng, M.; Zhang, Q.; Brennan, C.S.; Guan, W.; Wang, F.; Zhang, N. Extraction of Bound Phenolics from Shiitake Mushrooms (*Lentinus edodes*) by Combined Acid and Base Hydrolysis Procedures and Analysis of Phenolic Profiles and Antioxidant Capacities. *Int. J. Food Sci. Technol.* **2022**, *57*, 6613–6620. [CrossRef]
38. Chu, M.; Khan, R.D.; Zhou, Y.; Agar, O.T.; Barrow, C.J.; Dunshea, F.R.; Suleria, H.A.R. LC-ESI-QTOF-MS/MS Characterization of Phenolic Compounds in Common Commercial Mushrooms and Their Potential Antioxidant Activities. *Processes* **2023**, *11*, 1711. [CrossRef]
39. Yao, F.; Gao, H.; Yin, C.M.; Shi, D.F.; Fan, X.Z. Effect of Different Cooking Methods on the Bioactive Components, Color, Texture, Microstructure, and Volatiles of Shiitake Mushrooms. *Foods* **2023**, *12*, 2573. [CrossRef] [PubMed]
40. Pehlivan Karakas, F.; Ucar Turker, A.; Gokce Bozat, B.; Abant, B. Phenolic Content, Antibacterial and Antioxidant Potential of Several Edible Agaricomycetes Mushrooms Sold in Public Bazaar in Bolu, Turkey. *Int. J. Med. Mushrooms* **2023**, *25*, 45–56. [CrossRef]
41. Tan, P.W.; Tan, C.P.; Ho, C.W. Antioxidant properties: Effects of solid-to-solvent ratio on antioxidant compounds and capacities of Pegaga (*Centella asiatica*). *Int. Food Res. J.* **2011**, *18*, 2.
42. Ozdemir, M.; Gungor, V.; Melikoglu, M.; Aydiner, C. Solvent Selection and Effect of Extraction Conditions on Ultrasound-Assisted Extraction of Phenolic Compounds from Galangal (*Alpinia officinarum*). *J. Appl. Res. Med. Aromat. Plants* **2024**, *38*, 100525. [CrossRef]
43. Li, Z.; Teng, J.; Lyu, Y.; Hu, X.; Zhao, Y.; Wang, M. Enhanced Antioxidant Activity for Apple Juice Fermented with *Lactobacillus plantarum ATCC14917*. *Molecules* **2019**, *24*, 51. [CrossRef] [PubMed]
44. Reis, F.S.; Martins, A.; Barros, L.; Ferreira, I.C.F.R. Antioxidant Properties and Phenolic Profile of the Most Widely Appreciated Cultivated Mushrooms: A Comparative Study between in Vivo and in Vitro Samples. *Food Chem. Toxicol.* **2012**, *50*, 1201–1207. [CrossRef] [PubMed]
45. Nam, M.; Choi, J.Y.; Kim, M.S. Metabolic Profiles, Bioactive Compounds, and Antioxidant Capacity in *Lentinula edodes* Cultivated on Log versus Sawdust Substrates. *Biomolecules* **2021**, *11*, 1654. [CrossRef] [PubMed]
46. Palacios, I.; Lozano, M.; Moro, C.; D'Arrigo, M.; Rostagno, M.A.; Martínez, J.A.; García-Lafuente, A.; Guillamón, E.; Villares, A. Antioxidant Properties of Phenolic Compounds Occurring in Edible Mushrooms. *Food Chem.* **2011**, *128*, 674–678. [CrossRef]
47. Kalogeropoulos, N.; Yanni, A.E.; Koutrotsios, G.; Aloupi, M. Bioactive Microconstituents and Antioxidant Properties of Wild Edible Mushrooms from the Island of Lesvos, Greece. *Food Chem. Toxicol.* **2013**, *55*, 378–385. [CrossRef] [PubMed]
48. Rocchetti, G.; Gregorio, R.P.; Lorenzo, J.M.; Barba, F.J.; Oliveira, P.G.; Prieto, M.A.; Simal-Gandara, J.; Mosele, J.I.; Motilva, M.J.; Tomas, M.; et al. Functional Implications of Bound Phenolic Compounds and Phenolics–Food Interaction: A Review. *Compr. Rev. Food Sci. Food Saf.* **2022**, *21*, 811–842. [CrossRef] [PubMed]
49. del C. Razola-Díaz, M.; De Montijo-Prieto, S.; Aznar-Ramos, M.J.; Jiménez-Valera, M.; Ruiz-Bravo, A.; Verardo, V.; Gómez-Caravaca, A.M. Effect of Lactic Acid Bacteria Fermentation on the Polar Compounds Content with Antioxidant and Antidiabetic Activity of Avocado Seed Extracts. *Fermentation* **2023**, *9*, 420. [CrossRef]
50. Degrain, A.; Manhivi, V.; Remize, F.; Garcia, C.; Sivakumar, D. Effect of Lactic Acid Fermentation on Color, Phenolic Compounds and Antioxidant Activity in African Nightshade. *Microorganisms* **2020**, *8*, 1324. [CrossRef] [PubMed]
51. Wang, Y.; Wu, J.; Lv, M.; Shao, Z.; Hungwe, M.; Wang, J.; Bai, X.; Xie, J.; Wang, Y.; Geng, W. Metabolism Characteristics of Lactic Acid Bacteria and the Expanding Applications in Food Industry. *Front. Bioeng. Biotechnol.* **2021**, *9*, 612285. [CrossRef]
52. Lubbers, R.J.M.; Dilokpimol, A.; Visser, J.; Mäkelä, M.R.; Hildén, K.S.; de Vries, R.P. A Comparison between the Homocyclic Aromatic Metabolic Pathways from Plant-Derived Compounds by Bacteria and Fungi. *Biotechnol. Adv.* **2019**, *37*, 107396. [CrossRef]
53. Zhang, D.; Ye, Y.; Tan, B. Comparative Study of Solid-State Fermentation with Different Microbial Strains on the Bioactive Compounds and Microstructure of Brown Rice. *Food Chem.* **2022**, *397*, 133735. [CrossRef] [PubMed]
54. Yang, J.; Gao, T.; Ge, F.; Sun, H.; Cui, Z.; Wei, Z.; Wang, S.; Show, P.L.; Tao, Y.; Wang, W. *Porphyra yezoensis* Sauces Fermented with Lactic Acid Bacteria: Fermentation Properties, Flavor Profile, and Evaluation of Antioxidant Capacity in Vitro. *Front. Nutr.* **2022**, *8*, 810460. [CrossRef] [PubMed]

foods

MDPI

Article

Inhibition of LPS-Induced Skin Inflammatory Response and Barrier Damage via MAPK/NF-κB Signaling Pathway by *Houttuynia cordata* Thunb Fermentation Broth

Zixin Song [1,†], Jiaxuan Fang [1,†], Dongdong Wang [1,*], Yuncai Tian [2], Yuhua Xu [2], Ziwen Wang [1], Jiman Geng [1], Changtao Wang [1] and Meng Li [1]

1 School of Light Industry Science and Engineering, Beijing Technology & Business University, Beijing 100048, China; 18801173697@163.com (Z.S.); 18515626356@163.com (J.F.); zwwang727@163.com (Z.W.); gengjiman@163.com (J.G.); wangct@th.btbu.edu.cn (C.W.); limeng@btbu.edu.cn (M.L.)

2 Shanghai AZ Science & Technology Co., Ltd., Shanghai 201100, China; tianyuncai@shzchzp.wecom.work (Y.T.); huahua2125@126.com (Y.X.)

* Correspondence: wdd@btbu.edu.cn; Tel.: +86-15811357756

† These authors contributed equally to this work.

Abstract: *Houttuynia cordata* Thunb is rich in active substances and has excellent antioxidant and anti-inflammatory activity. Scanning electron microscopy and gel permeation chromatography were used to analyze the molecular characteristics of the fermentation broth of *Houttuynia cordata* Thunb obtained through fermentation with *Clavispora lusitaniae* (HCT-f). The molecular weight of HCT-f was 2.64265×10^5 Da, and the polydispersity coefficient was 183.10, which were higher than that of unfermented broth of *Houttuynia cordata* Thunb (HCT). By investigating the active substance content and in vitro antioxidant activity of HCT-f and HCT, the results indicated that HCT-f had a higher active substance content and exhibited a superior scavenging effect on 2,2-diphenyl-1-picrylhydrazyl radicals and hydroxyl radicals, with IC50 values of 11.85% and 9.01%, respectively. Our results showed that HCT-f could effectively alleviate the increase in the secretion of inflammatory factors and apoptotic factors caused by lipopolysaccharide (LPS) stimulation, and had a certain effect on repairing skin barrier damage. HCT-f could exert an anti-inflammatory effect by down-regulating signaling in the MAPK/NF-κB pathway. The results of erythrocyte hemolysis and chicken embryo experiments showed that HCT-f had a high safety profile. Therefore, this study provides a theoretical basis for the application of HCT-f as an effective ingredient in food and cosmetics.

Keywords: *Houttuynia cordata* Thunb; LPS; skin inflammation; skin barrier damage; MAPK/NF-κB pathway

1. Introduction

Research on natural plants and foods with physiologically and pharmacologically active effects has been a hot topic in recent years. In addition to serving as food to meet people's daily needs, traditional natural plants can also be utilized as raw materials for medicines and cosmetics [1]. Finding plant components with exceptional antioxidant, anti-inflammatory, and other efficacious effects is becoming more and more crucial as the market for safe and environmentally friendly cosmetic ingredients grows.

Houttuynia cordata Thunb is a perennial herb that was first cultivated in the mountainous regions of China, Japan, Korea, and Southeast Asia [2]. As a medicinal and food plant, it has a long history of application with a high degree of safety [3]. *Houttuynia cordata* Thunb is rich in a variety of bioactive components such as polyphenols, vitamins, and trace elements. It exhibits various bioactive effects, including anti-inflammatory, antibacterial, anti-cancer and antiviral properties [4–7]. It has been demonstrated that the ethanol extract of *Houttuynia cordata* Thunb can successfully reduce the oxidative damage to the liver and atherosclerosis brought on by a high-fat diet in rats [8]. Fang demonstrated that *Houttuynia cordata* Thunb extract could decrease the initiation of neuronal inflammation and apoptosis induced by propofol by enhancing neuronal activity [9]. *Houttuynia cordata* Thunb has shown good bacteriostatic activity against bacteria such as Staphylococcus aureus and Casecoccus [10].

Clavispora lusitaniae is an ester-producing yeast with a wide range of applications in enhancing the flavor and quality of white wines [11]. It can tolerate simulated gastric juices at pH 2.5 for 2 h and can produce extracellular polysaccharides during its growth [12]. The food products fermented by *Clavispora lusitaniae* are safer and richer in nutrients [13]. Recent studies have demonstrated that *Clavispora lusitaniae* JARR-1 has the capability to produce erythritol and ethanol [14].

Microbial fermentation technology refers to the utilization of microorganisms, under suitable conditions, to convert raw materials through specific metabolic pathways into materials required by humans. It mainly includes two categories: aerobic fermentation and anaerobic fermentation. This technology has a wide range of applications in the chemical industry, pharmaceutical industry, and other sectors [15–17]. The active ingredients of plants are primarily distributed in the cell, but the cell wall can hinder the release of these active ingredients. The microbial fermentation process generates enzymes such as cellulase and pectinase, which aid in breaking down the cell wall, facilitating the better release of active ingredients [18–20].

The skin is the body's natural protective barrier. When exposed to foreign substances, it can not only effectively resist the invasion of pathogens but also maintain the body's balance and prevent the loss of useful substances [21]. When the skin is stimulated by external harmful factors, its own barrier function will be compromised. This, in turn, triggers a series of immune responses that help maintain the skin's structure and function by modifying intercellular molecular signaling processes and facilitating matrix remodeling [22]. Low doses of nitric oxide (NO) have many functions, such as promoting wound healing and exhibiting antimicrobial effects against bacterial pathogens. The expression of the inducible isoform of NOS (iNOS) predicts inflammation. The expression of iNOS promotes the production of large amounts of NO, which in turn promotes various acute and chronic inflammatory conditions [23]. Lipopolysaccharides (LPS), complexes of lipids and polysaccharides, consist of three main components that can trigger a cascade of immune-stimulatory responses and toxic pathophysiological activities in the body. Studies have shown that LPS induce oxidative stress by activating vascular endothelial cells, leading to the synthesis and release of NO and ROS, and promoting the secretion of inflammation-associated factors. This process can result in inflammatory damage to endothelial cells [24]. Wang and his team found that aloe extracts inhibited the activation of ERK and JNK caused by LPS-induced injury, which is one of the reasons for its anti-inflammatory activity [25].

Several studies have demonstrated that numerous plants and their extracts possess inhibitory effects on the activation of the nuclear factor-κB (NF-κB) signaling pathway [26]. This pathway is one of the mechanisms through which these plants exert their anti-inflammatory effects. When IκB, which binds to NF-κB, is phosphorylated or degraded, NF-κB is activated, leading to its translocation into the nucleus and promoting the release of inflammatory factors. This, in turn, triggers a series of inflammatory responses [27]. The mitogen-activated protein kinase (MAPK) signaling pathway can also regulate inflammatory responses by up-regulating cytokine expression [28]. It has been shown that *Houttuynia cordata* Thunb extract exhibits anti-inflammatory activity. The mechanism behind this effect may involve the inhibition of the activation of the MAPK/NF-κB pathway, which is closely associated with the inflammatory response [29].

The anti-inflammatory effects of *Houttuynia cordata* Thunb extracts have been studied by many scholars. However, the role of *Houttuynia cordata* Thunb fermentation obtained via *Clavispora lusitaniae* (HCT-f) in an LPS-induced HaCaT cell inflammation model has not been reported. In this study, HCT-f was prepared, and the active substance content, molecular weight size, apparent morphological characteristics, in vitro antioxidant activity,

ROS regulation ability, effect on apoptosis, and modulation of inflammatory factors were determined (Figure 1). This provided a theoretical basis and data support for the application of HCT-f.

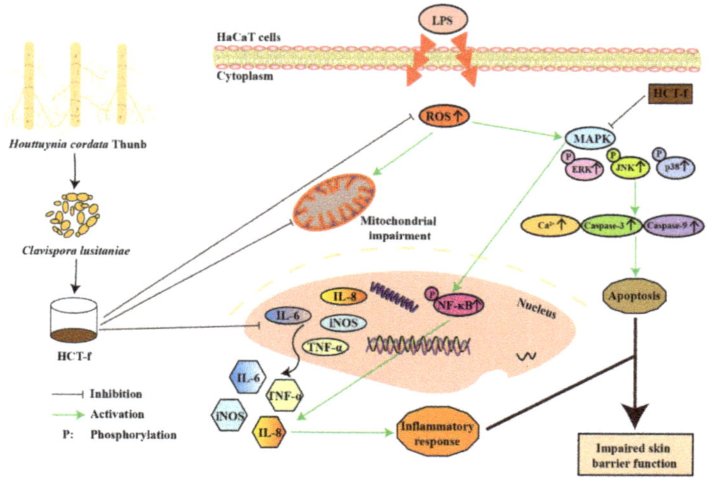

Figure 1. The role of HCT-f in LPS-induced injury in HaCaT cells.

2. Materials and Methods

2.1. Materials and Chemicals

Houttuynia cordata Thunb (Hubei Enshi, China); *Clavispora lusitaniae* (strain number: GDMCC64135, Guangdong Microbial Culture Collection Center); human immortalized keratinocytes (HaCaT) cells (National Experimental Cell Resources Sharing Service Platform, 4201HUM-CCTCC00106); hydrogen peroxide (H_2O_2), salicylic acid, ferrous sulfate ($FeSO_4$), and 2,2-diphenyl-1-picrylhydrazyl (Sinopharm Chemical Reagent Co., Ltd., Shanghai, China); ABTS and FRAP total antioxidant capacity assay kits (Beyotime Biotechnology, Shanghai, China); BCA protein assay kit (Biorigin Inc., Beijing, China); ROS (reactive oxygen species) assay kit (Beyotime Biotechnology, Shanghai, China); cell-counting kit-8 (Biorigin Inc., Beijing); DMEM (Dulbecco's modified Eagle medium) and PBS (phosphate-buffered saline) (Gibco Inc., Billings, MT, USA); 96-well plate (Corning, NY, USA); ELISA kits (Nanjing Jiancheng Bioengineering Research Institute, Nanjing, China).

2.2. Preparation of HCT and HCT-f

Houttuynia cordata Thunb was dried and crushed, passed through a 100-mesh sieve, added to distilled water (1:100), mixed well, sterilized at 121 °C for 20 min in an autoclave (BXM-30R Vertical Pressure Steam Sterilisation Kettle, Shanghai Boxun Industrial Co., Ltd., Shanghai, China), and then cooled to room temperature. *Clavispora lusitaniae* was inoculated and placed on a shaker (DQHZ-2001 A Full Temperature Shock Incubator, Taicang Huamei Biochemical Instrument Factory, Taicang, China) at 180 r/min, fermented at 28 °C for 48 h. It was then transferred into a high-pressure sterilization pot (Sigma, Steinheim, Germany) for sterilization (121 °C, 20 min) and centrifuged at 4800 r/min for 30 min, and the resulting supernatant was identified as the fermentation broth of *Houttuynia cordata* Thunb (HCT-f).

The preparation process for the unfermented broth of *Houttuynia cordata* Thunb (HCT) and HCT-f was essentially the same, except that it was not inoculated with *Clavispora lusitaniae*.

2.3. GPC (Gel Permeation Chromatography) and SEM (Scanning Electron Microscope)

The changes in molecular weight size of both HCT and HCT-f samples were determined using gel permeation chromatography (GPC). The samples were coated with gold on the sample surface and observed using the FEI Nova Nano SEM 450 instrument (Thermo Fisher Scientific, Waltham, MA, USA).

2.4. Determination of HCT-f Content

2.4.1. Total Sugars

The total sugar content of HCT-f was determined by using a total sugar content assay kit (Suzhou Grace Biotechnology Co., Suzhou, China). A master standard solution of glucose at a concentration of 1 mg/mL was prepared and then diluted to create six different concentration gradients: 0, 0.2, 0.4, 0.6, 0.8, and 1 mg/mL. These solutions were utilized to construct the standard curve. The specific experiments were carried out following the instructions provided in the kit.

2.4.2. Polysaccharides

The content of reducing sugar in HCT-f was determined by using a reducing sugar assay kit (Suzhou Grace Biotechnology Co., Suzhou, China). A master standard solution of glucose at a concentration of 1 mg/mL was prepared and then diluted to create six different concentration gradients: 0, 0.2, 0.4, 0.6, 0.8, and 1 mg/mL. These solutions were utilized to construct the standard curve as follows: Add 100 μL of different concentrations of samples and standards sequentially into the test tube (add an equal amount of distilled water to the blank tube). Then, add 100 μL of reagent I, mix uniformly, and heat it in a water bath (Beijing Jingke Huarui Instrument Co., Beijing, China) at 95 °C for 10 min. Remove it and immediately cool it down to room temperature with cold water. Add 1 mL of distilled water, mix it uniformly, and measure the absorbance value at 500 nm. Finally, use the standard curve to determine the amount of reducing sugar in the samples.

$$\text{Polysaccharide content} = \text{total sugar content} - \text{reducing sugar content} \quad (1)$$

2.4.3. Total Proteins

The total protein content in HCT-f was determined using a BCA protein content assay kit (Biorigin Inc., BN27109, Beijing, China), and the specific experimental procedures were performed according to the instructions.

2.4.4. Total Phenols

The total phenol content of the samples was determined using the forintol method [30]. Different concentrations of gallic acid standard solution and Na_2CO_3 (sodium carbonate, Sinopharm Chemical Reagent Co., Ltd., Shanghai, China) solution with a volume fraction of 26.7% were prepared. A total of 200 μL of gallic acid standard solution of different concentrations and the sample to be tested were added to a 2 mL EP tube. Next, 200 μL of deionized water, 100 μL of doubly diluted forintol reagent and 300 μL of 26.7% Na_2CO_3 solution were added sequentially, deionized water was added to reach 2 mL, and the reaction was carried out for 2 h. OD760 was measured by using an enzyme calibrator (Tecan Trading Co., Ltd., Männedorf, Switzerland), and the absorbance of the samples was brought into the standard curve for calculation.

2.5. Determination of HCT-f Antioxidant Capacity

2.5.1. DPPH Scavenging Assay

The determination of the scavenging capacity of the samples for DPPH (2,2-diphenyl-1-picrylhydrazyl) radicals was carried out with reference to the method of Su and her team [31]. The DPPH solution was prepared by dissolving 8 mg of DPPH in 100 mL of anhydrous ethanol. Six concentration gradients were set for each sample to be tested. The sample tube (A_1) was filled with 1 mL of various concentrations of the sample and 1 mL of DPPH. The blank tube (A_3) was filled with 1 mL of different concentrations of the sample and 1 mL of deionized water, while the control tube (A_2) contained 1 mL of deionized water and 1 mL of the DPPH solution (Table 1). After 30 min of reaction protected from light, OD517 was determined.

$$\text{DPPH free radical scavenging rate } \% = [(A_2 - A_1 + A_3)/A_2] \times 100\% \quad (2)$$

Table 1. DPPH spiking sequence.

Reagent	Volume (mL)		
	A_1	A_2	A_3
DPPH solution	1.0	1.0	0.0
Sample	1.0	0.0	1.0
Deionized water	0.0	1.0	1.0

2.5.2. Hydroxyl Radical Scavenging Assay

The scavenging capacity of the samples for hydroxyl radicals was determined using the salicylic acid method [31]. An 8 mmol/L $FeSO_4$ (ferrous sulfate) solution, a 3 mmol/L salicylic acid solution, and a 0.02 mol/L H_2O_2 (hydrogen peroxide) solution were prepared. To the blank tube (A_0), 0.3 mL of $FeSO_4$ solution, 1 mL of salicylic acid, 1.45 mL of deionized water, and 0.25 mL of H_2O_2 were added. The sample tube (A_1) was filled with 0.3 mL of $FeSO_4$ solution, 1 mL of salicylic acid, 1 mL of various concentrations of the sample, 0.45 mL of deionized water, and 0.25 mL of H_2O_2. The sample background tube (A_2) was filled with 0.3 mL of $FeSO_4$ solution, 1 mL of various concentrations of the sample, 1.45 mL of deionized water, and 0.25 mL of H_2O_2 (Table 2). The reaction was carried out at 37 °C for 1 h. The speed of the centrifuge was set to 4000 r/min and centrifuged for 10 min, and the OD510 was determined.

$$\text{Hydroxyl free radical scavenging rate } \% = [(A_0 - A_1 + A_2)/A_0] \times 100\% \quad (3)$$

Table 2. Hydroxyl radical spiking sequence.

Reagent	Volume (mL)		
	A_0	A_1	A_2
$FeSO_4$ (8 mmol/L)	0.3	0.3	0.3
Salicylic acid (3 mmol/L)	1.0	1.0	0.0
Sample	0.0	1.0	1.0
Deionized water	1.45	0.45	1.45
H_2O_2 (0.02 mol/L)	0.25	0.25	0.25

2.5.3. Determination of Total Antioxidant Capacity

The total antioxidant capacity of HCT-f was determined using the ABTS test kit (Beyotime Biotechnology, S0119, Shanghai, China) and FRAP test kit (Beyotime Biotechnology, S0116, Shanghai, China). Refer to the instructions for specific experimental steps.

2.6. Cell Culture and Viability Experiment

HaCaT cells were rapidly lysed in a water bath at 37 °C, and cultured in a T25 culture flask with 5 mL of medium (10% fetal bovine serum and 1% penicillin–streptomycin). The culture flask was incubated in a incubator (37 °C and 5% CO_2), and when the amount of cell apposition reached 80–90%, the cells were digested and centrifuged at 1500 r/min for 5 min for the next round of passaging culture.

HaCaT cells at a good growth state were selected and 100 µL of cell suspension was inoculated into each well of a 96-well plate (1×10^4 cells). The 96-well plates inoculated with cells were incubated for 12 h (no cells were added to the cell-free group), and 100 µL of serum-free medium was added to the control group (group C) and cell-free group. Then, 100 µL of a 200 µg/mL lipopolysaccharide (LPS) solution was added to each of the remaining wells. The plate was incubated for 24 h; next, 100 μL of serum-free medium was added to the control group (group C) and cell-free group, 100 μL of 200 μg/mL lipopolysaccharide (LPS) solution was added to each of the remaining wells, and the plate was incubated for 24 h. Additionally, 100 μL of samples at varying concentrations were added to the sample group. The plates were then incubated for 24 h in the incubator. The effect of samples on cell viability was assessed using the CCK8 method (Biorigin Inc., BN15201, Beijing, China). The plates were washed twice with Phosphate buffered saline (PBS). Then, 100 μL of serum-free medium and 10 μL of CCK8 reagent were added to each well. The plates were incubated in the incubator at 37 °C for 2–4 h. Absorbance values at 450 nm were measured, and cell viability was calculated using the formula below.

Cell viability = (absorbance value of the sample group − absorbance value of the cell free group)/(absorbance value of the group C − absorbance value of the cell free group) × 100% (4)

2.7. ROS Detection

In a 6-well plate, each well was inoculated with 2 mL of cell suspension (5×10^5 cells) and placed in incubator for 24 h. Next, 2 mL of serum-free medium was added to group C, and 2 mL out of 200 μg/mL of LPS was added to both group M and the sample group. The plate was incubated in an incubator for 24 h, and 2 mL of different samples was added to the sample group, and 2 mL of serum-free medium was added to both groups C and M. The plate was incubated in an incubator for 24 h. The effect of different samples on ROS levels was assayed by using a reactive oxygen detection kit (Beyotime Biotechnology, S0033S, Shanghai, China). The fluorescence intensity of ROS in the cells was observed using a fluorescence-inverted microscope, photographed, and recorded.

2.8. The Effect of HCT-f on Apoptotic Cell Death

2.8.1. Ca^{2+} Concentration Measurement

In a 6-well plate, each well was inoculated with 2 mL of cell suspension (5×10^5 cells) and placed in an incubator for 24 h. Changes in intracellular Ca^{2+} concentration levels were observed by the specific binding of a Fluo-4, AM ester Ca^{2+} fluorescent probe to Ca^{2+}. Firstly, a 4 μM Fluo-4, AM ester working solution was prepared, the 6-well plate was washed with PBS for 3 times, and 1 mL of Fluo-4, AM ester working solution was added to each well. It was then incubated at 37 °C for 20 min in an incubator maintained at constant temperature and humidity. The Fluo-4, AM ester working solution was discarded, the plate was washed 3 times, 1 mL PBS buffer was added, and incubated at 37 °C for 10 min. The changes in intracellular Ca^{2+} concentration were observed using a fluorescence-inverted microscope, and the experimental results were recorded.

2.8.2. Mitochondrial Membrane Potential Measurement

In a 6-well plate, each well was inoculated with 2 mL of cell suspension (5×10^5 cells) and placed in an incubator for 24 h. The mitochondrial membrane potential of HaCaT cells was measured using the Beyotime enhanced mitochondrial membrane potential detection kit as follows: Take 50 μL of JC-1 (200X) and add 8 mL of ultrapure water. Mix well to fully dissolve, then add 2 mL of JC-1 staining buffer (5X) and mix well to prepare the staining working solution. The specific operation was conducted following the instruction manual, and the experimental results were observed and recorded using a fluorescence microscope (CKX53, Olympus LS, Tokyo, Japan).

2.9. Enzyme-Linked Immunosorbent Assay (ELISA)

A 6-well plate was lined with 50×10^4 cells per well and incubated. The cell supernatant was collected in 2 mL centrifuge tubes, and the cells at the bottom of the 6-well plate were lysed to collect the cell lysate. Both cell supernatant and lysate were centrifuged at 10,000 r for 5 min, and the precipitate was discarded and stored at −20 °C for subsequent experiments. The experiments were performed according to the corresponding instructions of the IL-6, IL-8, iNOS, TNF-α, AQP3, FLG, LOR, Caspase-14, Caspase-3, and Caspase-9 assay kits.

2.10. qRT-PCR

Primer sequences were designed using PrimerExpress software (version 3.0). Total RNA extraction and cDNA synthesis were performed using Trizol reagent and a reverse transcription kit (Biorigin, BN12028, Beijing, China). Refer to the Fast Super EvaGreen® qPCR Master Mix kit (Biorigin, BN12008, Beijing, China) instructions for subsequent experiments. The reverse transcription system is shown in Table 3. Primer sequences are shown in Table 4. The PCR reaction system is shown in Table 5.

Table 3. Reverse transcription system.

Component Name	Volume (μL)
Anchored Oligo (dt) 18 Primer	1.0
2 × ES reaction Mix	10.0
Total RNA	2.0
EasyScript RT/RI Enzyme Mix	1.0
Gdna Remover	1.0
Rnase-free Water	5.0

Table 4. Primer sequences.

Gene	Direction	Primer (5′-3′)
IL-6	F	TTCTCCACAAGCGCCTTC
	R	AGAGGTGAGTGGCTGTCTGT
IL-8	F	GGAGAAGTTTTTGAAGAGGGCTG
	R	ACAGACCCACACAATACATGAAG
TNF-α	F	TCTCCTTCCTGATCGTGGCA
	R	CAGCTTGAGGGTTTGCTACAAC
iNOS	F	CGTGGAGACGGGAAAGAAGT
	R	GACCCCAGGCAAGATTTGGA
AQP3	F	CTTCTTTGACCAGGACCGGC
	R	GGGCCAGCTTCACATTCTCT
FLG	F	TGAGGCATACCCAGAGGACT
	R	CTGTATCGCGGTGAGAGGAT
LOR	F	CTTCCTGGTGCTTTGGGCTC
	R	CTGGGGGATCTATTTGGACGG
Caspase-14	F	ATTCCACGGTAGAGGGATACA
	R	TCAGGGTTCGTTTTCCTTGCT
Caspase-3	F	ACTCCACAGCACCTGGTTAT
	R	TCTGTTGCCACCTTTCGGTT
Caspase-9	F	GGTGGACATTGGTTCTGGCAGAG
	R	ACGTTGTTGATGATGAGGCAGTGG
JNK	F	ATTGTTTGTGCTGCGTTCG
	R	CTTTTTGCGTGGGGTTGATT
ERK	F	CATCGCGACCTCAAACCTTC
	R	TCCGGATCTGCAACACGAG
NF-κB	F	CCAGGTCCTCCAGGCAATCCAAAAG
	R	AGTGCAGGGCTGTCAAACCATCGTAG
P38	F	ATGTTCCCTGTCCCTGAGC
	R	GCCCTGAGTGTAGCCTGTGT
Beta-actin	F	CTGAAGCCCCACTCAATCCA
	R	GCCAAGTCAAGACGGAGGAT

Table 5. PCR reaction system.

Reagent	Volume (μL)
Template	1.5
Forward Primer (10 μM)	0.4
Reverse Primer (10 μM)	0.4
2 × TransStart® Top Green qPCR SuperMix	10.0
Passive Reference Dye (50×)	0.4
Nuclease-free Water	7.3

2.11. Safety Determination

2.11.1. Rabbit Erythrocyte Hemolysis Assay (RBC)

Fresh rabbit blood was centrifuged at 1500× *g* for 10 min, and the precipitate and PBS were taken and diluted in a 1:49 ratio. Blood washing was completed when 0.5 mL of cell suspension was added with 4.5 mL of deionized water and the OD541 was 0.5 ± 5%. The washed blood and the samples to be tested were mixed well in the ratio of 1:3, incubated for 60 min on a shaking bed, and the incubation was terminated by centrifugation for 10 min at 10,000× *g*. The absorbance value at 560 nm was measured. The above experimental procedure was repeated with PBS and deionized water instead of the sample as the negative control (NC) and with the positive control (PC) of the sample.

$$\text{Hemolysis} = (\text{sample OD}_{560} - \text{NC OD}_{560})/(\text{PC OD}_{560} - \text{NC OD}_{560}) \times 100\% \quad (5)$$

2.11.2. Chicken Embryo Chorionic Allantoic Membrane Eye Irritation Test (HET-CAM)

Fertilized eggs were incubated in an incubator for 10 days (38 °C, 60–70% humidity), and the position of the air chambers was observed under a light source. The eggshell in the upper part of the air chamber was removed with tweezers. The eggshell membrane was moistened with 0.9% NaCl solution and torn off with forceps to expose the blood vessels. Next, 300 μL of 0.9% NaCl solution, 0.1 mol/L NaOH solution, and the samples were added directly to the exposed blood vessels, and the blood vessel damage was observed for 3 min, which served as the negative control group, the positive control group, and the sample group, respectively, and the time of bleeding, coagulation, and dissolution of the blood vessels were recorded. The samples were evaluated for eye irritation using the endpoint evaluation method (Tables 6 and 7).

Table 6. Observation scoring record table.

Category	Score	Level	Performance
hemorrhage	0	None	-
	1	Mildly	Bleeding from small blood vessels and small amounts of hemorrhage only
	2	Moderate	Bleeding from small and large blood vessels with a noticeable amount of blood coming out
	3	Severe	Bleeding from almost all the blood vessels, a lot of blood coming out
coagulation	0	None	-
	1	Mildly	Mild intra/extravascular coagulation, mild clouding of CAM membranes
	2	Moderate	Moderate intra/extravascular coagulation with mild clouding of CAM membranes
	3	Severe	Heavy intra/extravascular coagulation with mild clouding of CAM membranes
angiolysis	0	None	-
	1	Mildly	Small vessel lysis only
	2	Moderate	Small and large vessel lysis
	3	Severe	Lysis of large vessels and the entire vascular tree

Table 7. ES scoring standards.

Score Range	Irritation Classification
<6	Non-irritating
$6 \leq S \leq 12$	Mildly irritating
$12 \leq S \leq 15$	Moderately irritating
≥ 15	Severe irritation

2.12. Data Analysis

All experiments were repeated three times. Data were analyzed and plotted using GraphPad Prism 8 (GraphPad Software, Inc., San Diego, CA, USA), the significance of the data was analyzed using conventional one-way analysis of variance (ANOVA), and the data were processed and IC50 calculated using SPSS software 23. Differences were considered significant when $p < 0.05$.

3. Results

3.1. Characterization of the Microscopic Morphology of HCT-f

The molecular weights (MWs) of HCT and HCT-f are shown in Table 8. The average molecular weight of HCT was 1.02758×10^5 Da, and that of HCT-f was 2.64265×10^5 Da. The molecular weight was increased after fermentation, which might be due to the fact that microorganisms produce enzymes to promote the rupture of the plant cell wall in the process of fermentation, which promotes the release of the macromolecular active substances. The polydispersity coefficient (Mw/Mn) of HCT-f is 183.10, which is higher than that of HCT, indicating a more complex molecular composition. The morphological characteristics of *Houttuynia cordata* Thunb under different levels of precision are shown in Figure 2. It can be seen that HCT-f has better structural homogeneity than HCT, and under the same precision, the surface morphological structure of HCT-f presents more pores and is rougher, indicating that HCT-f has better absorbability and adhesion.

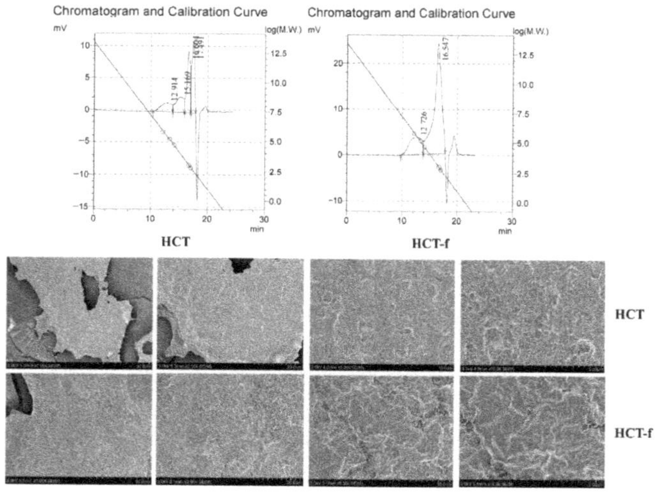

Figure 2. Changes in morphological and structural characteristics of *Houttuynia cordata* Thunb before and after fermentation. The magnifications of the HCT and HCT-f electron microscope images were 1000×, 2000×, 5000×, and 10,000× in that order. The diagonal lines in the figure represent standard curves. These curves were generated using a narrowly distributed set of prasinomannan-labeled samples with molecular weights of 5.8×10^5, 1.46×10^5, 4.42×10^4, 1000 and 600 for the GPC experiments. The small arrow on the graph is the point where the peak taking starts and ends. And the circles in the graph represent the efflux times of the narrowly distributed Pullulan polysaccharide labeled groups with molecular weights of 5.8×10^5, 1.46×10^5, 4.42×10^4, 1000 and 600.

Table 8. Determination of OB and OBF molecular weights.

	HCT	HCT-f
Mn (Da)	848	1443
Mw (Da)	1.02758×10^5	2.64265×10^5
Mz (Da)	1.807968×10^6	4.12595×10^6
Mw/Mn	121.12684	183.09849
Mz/Mw	17.59448	15.61292

Mn: Number-average molecular weight; Mw: weight-average molecular weight; Mz: Z-average molecular weight.

3.2. Active Substance Content and In Vitro Antioxidant Activity of *Houttuynia cordata* Thunb

The various bioactive effects of *Houttuynia cordata* Thunb, such as anti-inflammatory, antioxidant, and antimicrobial, are closely related to its rich content of bioactive substances. We determined the changes in the bioactive substance content and in vitro antioxidant activity of *Houttuynia cordata* Thunb before and after fermentation (Figure 3).

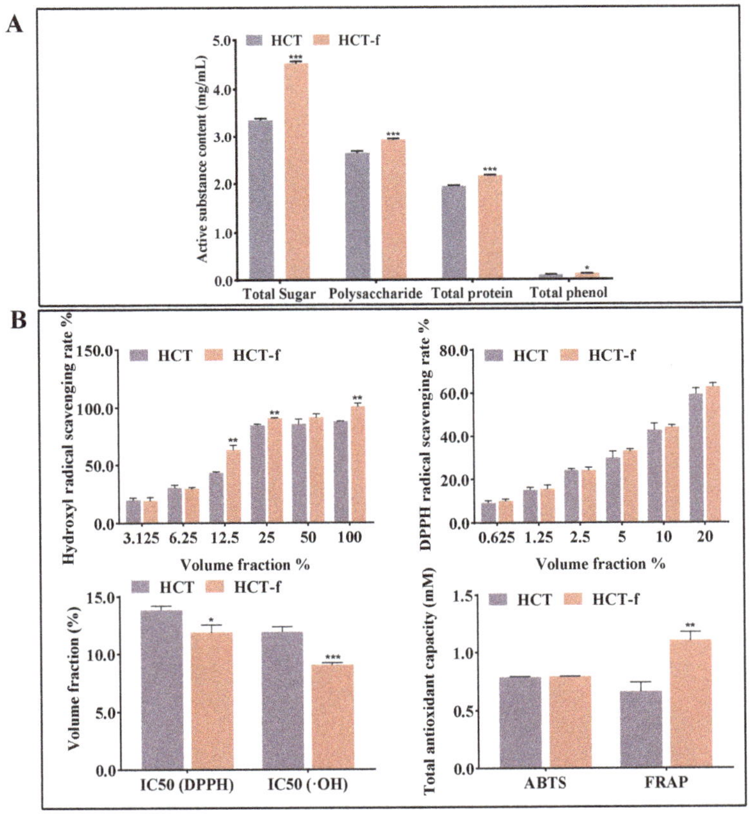

Figure 3. Determination of the content of bioactive substances in HCT and HCT-f and in vitro antioxidant capacity. (**A**) Determination of total sugars, polysaccharides, total proteins, and total phenols in HCT and HCT-f. (**B**) Determination of DPPH, hydroxyl radical, and ABTS radical scavenging capacity of HCT and HCT-f, and Fe^{2+} reduction capacity. One-way ANOVA was performed to determine statistical significance. * $p < 0.05$, ** $p < 0.01$, *** $p < 0.001$ vs. the HCT group.

As can be seen in Figure 3A, HCT and HCT-f were rich in total sugars, polysaccharides, and total proteins, and showed reduced polyphenolic compounds. After fermentation, a significant increase in the content of all four bioactive substances occurred. The total sugar content and polysaccharide content of HCT-f were both significantly higher compared to HCT, at 4.52 and 2.92 mg/mL, respectively.

Figure 3B showed that the in vitro antioxidant activity of HCT-f was superior to that of HCT. The scavenging ability of HCT-f with a volume fraction of 25% on hydroxyl radicals could reach more than 90%, which exceeded that of HCT with a volume fraction of 100%, and there was little difference in the scavenging effect of different concentrations of HCT and HCT-f on DPPH radicals. The smaller IC50 represents the better scavenging ability of the sample on free radicals, and it can be seen in the figure that the HCT-f with a volume fraction of 11.85% could scavenge 50% of DPPH radicals. The experimental results of the total antioxidant capacity assay showed that the scavenging ability of HCT-f and HCT on ABTS radicals did not differ much, and the reducing ability of HCT-f on Fe^{2+} was 1.67 times that of HCT, which could reach 1.11 mM Trolox equivalent.

In conclusion, we can see that after fermentation, the content of bioactive substances and the in vitro antioxidant activity of *Houttuynia cordata* Thunb were increased, which suggests that compared with HCT, HCT-f may have a higher application value.

3.3. Effect of HCT-f on Cell Viability

The CCK8 assay was utilized to assess whether HCT-f (0.625–20%) inhibited cell viability in a volume fraction-dependent manner. Figure 4A demonstrates that, compared to the blank control group C, varying volume fractions of HCT-f had a specific impact on HaCaT cells in promoting cell proliferation, with HCT-f showing a superior promotion of cell proliferation. To further investigate the impact of HCT-f treatment on the proliferation of HaCaT cells following LPS-induced injury, the LPS-injured HaCaT cells were exposed to varying concentrations of HCT-f. The results showed (Figure 4B) a significant decrease in cell viability in cells treated with LPS alone (group M) compared to group C without LPS-induced injury ($p < 0.001$). Compared with group M, HCT-f with a volume fraction of 1.25% to 10% restored cell viability in a dose-dependent manner, suggesting that treating cells with HCT-f attenuates LPS-induced damage in HaCaT cells. For the comprehensive assessment of cell viability, status, and the potential toxic effects of a high volume fraction (20%) of HCT-f on cells, we opted to utilize HCT-f with a volume fraction of 10% for the upcoming cell experiments.

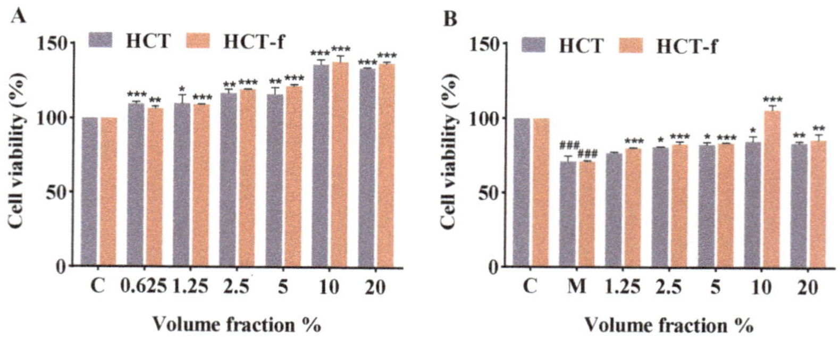

Figure 4. Effect of HCT and HCT-f on the viability of HaCaT cells. (**A**) Effect of 0.625–20% volume fraction of HCT and HCT-f on HaCaT cell viability. (**B**) Effect of 1.25–20% volume fraction of HCT and HCT-f on the viability of HaCaT cells with LPS-induced injury. (C: not subjected to LPS-induced injury; M: subjected to LPS-induced injury only, no sample added.) One-way ANOVA was performed to determine statistical significance. * $p < 0.05$, ** $p < 0.01$, *** $p < 0.001$ vs. the control group; ### $p < 0.001$ vs. the control group, * $p < 0.05$, ** $p < 0.01$, *** $p < 0.001$ vs. the model group.

3.4. Effect of HCT-f on Reactive Oxygen Species (ROS) Content in Cells

Under normal circumstances, ROS, as natural byproducts of oxygen metabolism, are present at low levels in the organism. They participate in intracellular signaling and regulation as redox messengers, playing a crucial role in the cell cycle and gene expression.

When the body is stimulated by exogenous stimuli such as LPS, the content of ROS will increase rapidly, surpassing the body's natural capacity to eliminate them. This disruption will then disturb the dynamic balance between ROS production and removal in the body, causing oxidative stress. Oxidative stress can result in DNA damage, changes in protein structure and function, and harm to cell membranes.

From Figure 5B, it can be seen that the intensity of green fluorescence was significantly enhanced in group M after LPS induction compared to group C without LPS induction. The intensity of green fluorescence was weakened after treatment with both HCT and HCT-f at a volume fraction of 10%. Additionally, the effect of HCT-f was superior to that of HCT. Figure 5A shows that after LPS injury, the relative expression of ROS in group M was significantly increased. Treatment with both HCT and HCT-f reduced the generation of ROS. The inhibitory effect of HCT-f on the generation of ROS was superior to that of HCT.

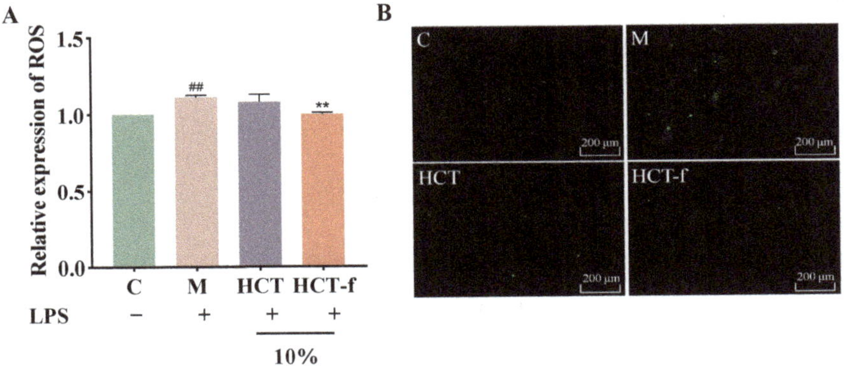

Figure 5. Effects of HCT and HCT-f on ROS content in LPS-induced damaged HaCaT cells. (**A**) Effects of HCT and HCT-f on the relative expression of ROS. (**B**) ROS generation (green fluorescence) was detected by fluorescence microscopy. One-way ANOVA was performed to determine statistical significance. ## $p < 0.01$ vs. the control group, ** $p < 0.01$ vs. the model group.

3.5. Effects of HCT-f on Apoptotic Cell Death

Mitochondrial membrane potential can be used to assess the proper functioning of mitochondria. It is widely recognized that ROS are the primary cause of mitochondrial dysfunction [32]. An increase in ROS content can disrupt the mitochondrial membrane, altering its permeability and ultimately resulting in a decrease in membrane potential. The concentration of Ca^{2+} in vivo also plays a crucial role in maintaining the permeability of the mitochondrial membrane. An increase in the concentration of Ca^{2+} in the mitochondrial leads to elevated levels of ROS, which subsequently activate the Caspase-dependent apoptotic pathway [33]. Under normal conditions, the inner mitochondrial membrane potential is high and remains at a negative potential, while the outer membrane potential is low and remains at a positive potential. The decrease in mitochondrial membrane potential is associated with autophagy, apoptosis, and necrosis of cells [34].

When the level of Ca^{2+} increases, it promotes ATP production, leading to more free electrons being leaked to form superoxide. This results in the generation of a large amount of ROS, which in turn leads to the oxidation of unsaturated fatty acids, amino acids, and DNA damage, ultimately resulting in apoptosis [35]. The detection of changes in intracellular Ca^{2+} concentration can be used to initially assess whether the cell is about to undergo apoptosis. The Figure 6C experimental results show that after LPS-stimulated damage, the intensity of green fluorescence significantly increased, and the relative expression of Ca^{2+} underwent a significant increase compared to group C. Following treatment with HCT and HCT-f, the intensity of green fluorescence decreased to varying degrees, and the relative expression of Ca^{2+} was reduced compared to group M. The inhibitory effect of HCT-f on Ca^{2+} expression was superior to that of HCT.

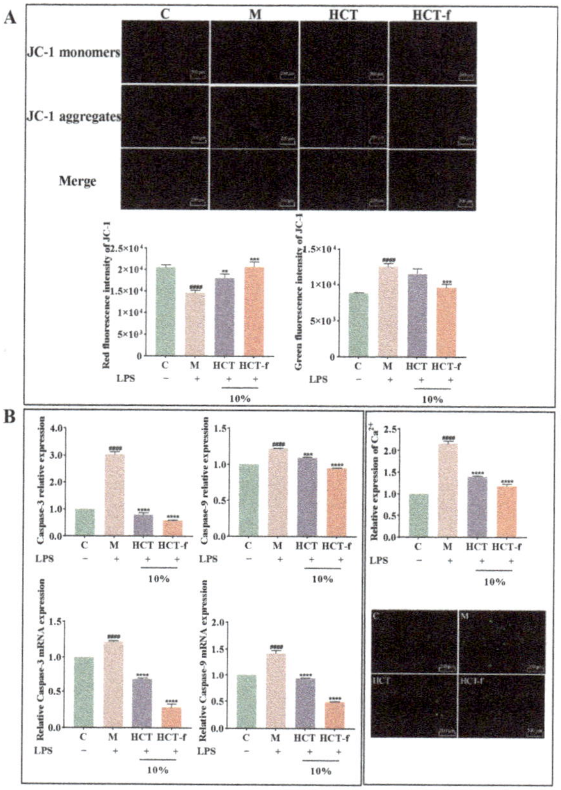

Figure 6. Effects of HCT and HCT-f on the apoptotic profile of HaCaT cells after LPS-induced injury. (**A**) Effects of HCT and HCT-f on the mitochondrial membrane potential in HaCaT cells. (**B**) Effects of HCT and HCT-f on the content and relative mRNA expression of two apoptotic proteins, Caspase-3 and Caspase-9, in HaCaT cells. (**C**) Effects of HCT and HCT-f on the changes in Ca^{2+} content and fluorescence intensity in HaCaT cells. One-way ANOVA was performed to determine statistical significance. #### $p < 0.0001$ vs. the control group, ** $p < 0.01$, *** $p < 0.001$, **** $p < 0.0001$ vs. the model group.

As shown in Figure 6A, in group C, JC-1 aggregated in the mitochondrial matrix to form polymers, emitting strong red fluorescence. After LPS injury (group M), the membrane potential was reduced, and JC-1 existed in the cytosol in the form of monomers emitting strong green fluorescence, which induced an early apoptotic cell response. After 10% HCT and HCT-f treatments, the content of JC-1 polymer significantly increased, while the monomer content decreased significantly compared to group M. The effect of HCT-f was superior to that of HCT, and after 10% HCT-f treatment, it was equal to the undamaged state.

The alteration of mitochondrial permeability leads to the formation of apoptotic vesicles, which in turn promotes the activation of the Caspase family. Caspase-9 plays a crucial role in initiating apoptosis during the apoptotic process. Stimulation of apoptotic signaling activates the Caspase-9 precursor, which then triggers the Caspase-3 activation and initiates the Caspase cascade reaction, ultimately leading to apoptosis [36]. We further confirmed the anti-apoptotic effect of HCT-f by measuring the content and relative mRNA expression of Caspase-3 and Caspase-9, two proteins closely associated with apoptosis. The experimental results in Figure 6B show that the relative expression of both apoptosis-related proteins increased significantly after cells were stimulated with LPS. The up-regulation of the two apoptosis-related proteins was significantly reduced following treatment with HCT and HCT-f. Moreover, the anti-apoptotic effect of HCT-f was found to be superior to that of HCT.

In conclusion, it can be observed that HCT and HCT-f can inhibit LPS injury-induced apoptosis by mitigating the decrease in mitochondrial membrane potential, reducing Ca^{2+} inward flow, and suppressing the relative expression of Caspase-3 and Caspase-9 and their mRNA.

3.6. Effects of HCT-f on Inflammatory Factor Release and Skin Barrier Function

Dysregulation of apoptosis plays an important role in the pathophysiology of chronic inflammation and its associated symptoms. Further development of skin inflammation leads to impairment of the skin's barrier function, making it unable to resist the invasion of foreign harmful substances, thus exacerbating the skin's inflammatory response. As a result, the skin is caught in a vicious cycle of inflammation and barrier damage.

In Figure 7A, following LPS stimulation, the expression of IL-6, IL-8, TNF-α, and iNOS increased. The relative expression of their mRNAs also significantly increased. Subsequently, treatment with HCT and HCT-f resulted in a significant decrease in the expression of all four inflammatory factors. Notably, HCT-f demonstrated more effective inhibition of inflammatory factor expression compared to HCT.

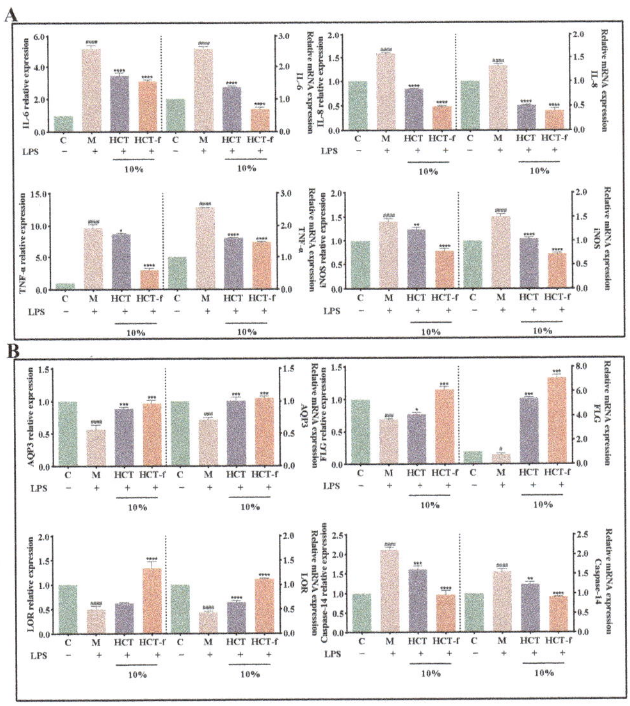

Figure 7. Effects of HCT and HCT-f on the release and transcript levels of inflammatory and barrier factors in LPS-induced injured HaCaT cells. (**A**) Effects of HCT and HCT-f on the release of IL-6, IL-8, TNF-α, and iNOS, as well as their mRNA expression in HaCaT cells with LPS-induced injury. (**B**) Effects of HCT and HCT-f on the release of AQP3, FLG, LOR, and Caspase-14, and their mRNA expression in HaCaT cells with LPS-induced injury. One-way ANOVA was performed to determine statistical significance. # $p < 0.05$, ### $p < 0.001$, #### $p < 0.0001$ vs. the control group; * $p < 0.05$, ** $p < 0.01$, *** $p < 0.001$, **** $p < 0.0001$ vs. the model group.

Figure 7B shows that the content and relative expression of mRNA of three proteins, AQP3, FLG, and LOR, which are closely related to the skin barrier function, decreased significantly after LPS stimulation. Additionally, the content of Caspase-14 shearase and its relative expression of mRNA increased significantly. After treating the cells with both HCT and HCT-f samples, there was a significant increase in the content and relative expression of three proteins (AQP3, FLG, and LOR) and in the relative expression of their mRNAs. The effect of HCT-f was superior. After treating the cells with both HCT and HCT-f samples, there was a decrease in the content of Caspase-14 and the relative expression of its mRNA. The effect of HCT-f was superior.

3.7. Effects of HCT-f on the Transcriptional Activity of Genes in the MAPK/NF-κB Pathway

MAPKs are a group of signaling proteins that include p38 mitogen-activated protein kinase, c-Jun nh2-terminal kinase (JNK), and extracellular signal-regulated kinase (ERK) [37]. MAPKs and phosphorylated MAPK proteins can regulate NF-κB activation [38]. The activation of this signaling pathway is closely related to the development of inflammatory responses [39].

Figure 8 illustrates that following LPS-induced injury, the relative expression of mRNAs for all four genes (JNK, ERK, p38, and NF-κB) increased. Subsequently, after treatment with HCT and HCT-f, the relative expression of mRNA for all four genes decreased significantly. The inhibition of MAPK/NF-κB pathway activation by HCT-f was found to be superior to that of HCT. Our experimental results tentatively suggest that HCT-f may exert anti-inflammatory effects by inhibiting the LPS-induced activation of the MAPK/NF-κB pathway.

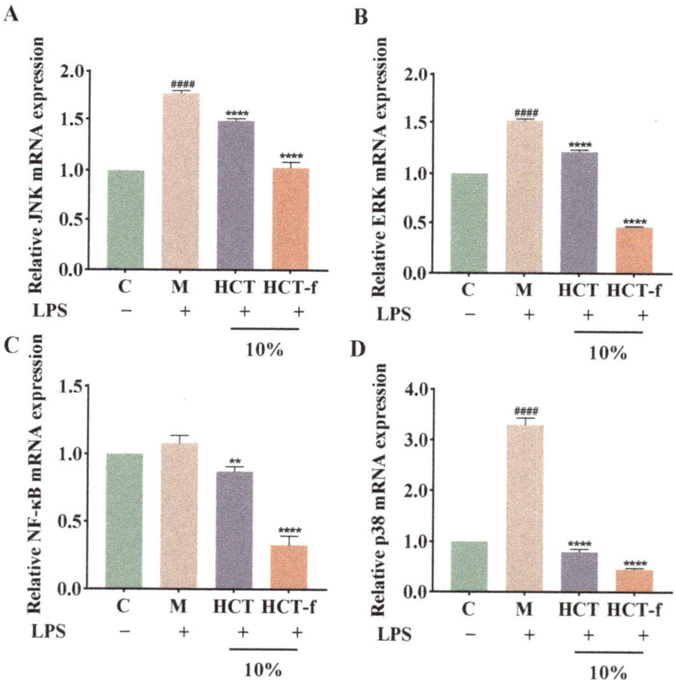

Figure 8. Effects of HCT and HCT-f on gene expression of the JNK, ERK, NF-κB, and p38. (**A**) JNK; (**B**) ERK; (**C**) NF-κB; (**D**) p38. One-way ANOVA was performed to determine statistical significance. #### $p < 0.0001$ vs. the control group; ** $p < 0.01$, **** $p < 0.0001$ vs. the model group.

3.8. Safety of HCT-f

In order to test whether the samples were irritating, RBC test and HCT-CAM test were carried out, and the results of the experiments are shown in Figure 9. The results (Figure 9A) indicate that the hemolysis rate of HCT and HCT-f with volume fractions ranging from 0.625% to 10% was extremely low. Neither of them exhibited significant hemolysis, demonstrating the high safety of the samples. The grading of vascular irritation by HCT and HCT-f is specifically shown in Table 9. ES = 12 for the positive control of 0.1 mol/L NaOH, while ES = 0 for HCT and HCT-f. Based on the experimental results in Figure 9B, it can be observed that neither HCT nor HCT-f caused damage such as hemorrhage or coagulation of blood vessels when applied to the CAM. This indicates that the samples did not cause eye irritation. All of the experimental results mentioned above indicate that HCT-f is safe and has promising application prospects.

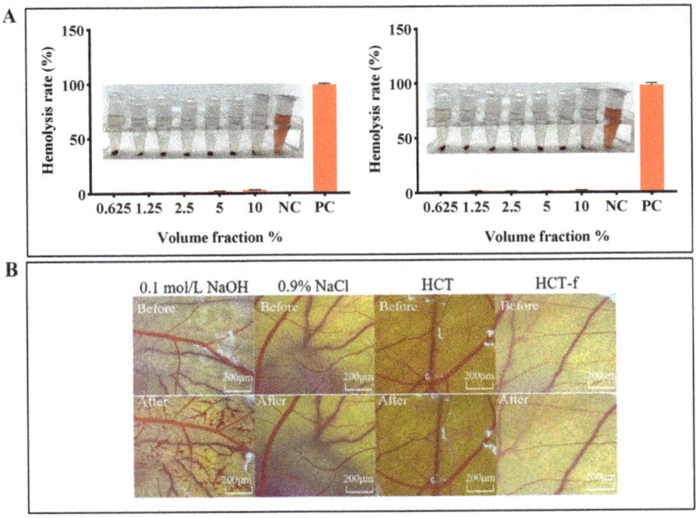

Figure 9. An exploration of the safety of HCT and HCT-f. (**A**) Effects of HCT and HCT-f on hemolysis of rabbit erythrocytes. (**B**) Effects of HCT and HCT-f on vascular damage in the chorioallantoic membrane of the chick embryo.

Table 9. ES scale for CAM bleeding after HCT and HCT-f treatment.

Sample	Bleeding Type	Rating/Point						Effective Score (ES)
		1	2	3	4	5	6	
0.9% NaCl solution	Hemorrhage	0	0	0	0	0	0	0
	Coagulation	0	0	0	0	0	0	
	Dissolution	0	0	0	0	0	0	
0.1 mol/L NaOH	Hemorrhage	2	2	2	2	2	2	12
	Coagulation	2	1	1	1	2	1	
	Dissolution	1	1	1	1	1	1	
HCT	Hemorrhage	0	0	0	0	0	0	0
	Coagulation	0	0	0	0	0	0	
	Dissolution	0	0	0	0	0	0	
HCT-f	Hemorrhage	0	0	0	0	0	0	0
	Coagulation	0	0	0	0	0	0	
	Dissolution	0	0	0	0	0	0	

4. Discussion

It has been shown that *Houttuynia cordata* Thunb possesses a variety of biological effects such as antioxidant, antimicrobial, anti-tumor, and anti-inflammatory effects. Experiments in mice have shown that *Houttuynia cordata* Thunb polysaccharides can alleviate LPS-induced pneumonic injury by inhibiting macrophage hyperactivation [40]. It has been shown that the *Clavispora lusitaniae* produces extracellular polysaccharides during metabolism. These polysaccharides have a variety of physiological and pharmacological activities, such as anti-tumor, anti-aging and anti-infection [41]. The advantage of using microbial fermentation to prepare fermentation broth for direct applications in food and cosmetics is that no organic reagents are required in the fermentation process that may produce substances in their metabolism that are beneficial to the fermentation substrate in terms of its efficacy. The investigation of the efficacy and safety of *Houttuynia cordata* Thunb fermentation broth is a worthwhile research direction. Based on these findings, we hypothesized that *Clavispora lusitaniae Houttuynia cordata* Thunb fermentation broth (HCT-f) could play a significant role in repairing LPS-induced skin inflammation through antioxidant and anti-inflammatory pathways.

The physical properties of HCT-f were investigated using GPC and SEM. The experimental results showed that the molecular weight of HCT-f was higher than that of HCT. The difference could be attributed to the enzymes produced during the microbial fermentation process, which destroyed the plant cell wall and promoted the release of the active substances. Additionally, the surface morphology and structure of HCT-f exhibited more pores, roughness, and better structural homogeneity. These characteristics suggest that HCT-f might offer improved absorption and adhesion effects.

The results of biochemical experiments showed that a significant increase in total sugars, polysaccharides, total proteins, and total phenols in HCT-f compared to HCT. This increase is likely attributed to the production of enzymes that aid in breaking down the plant cell wall during the fermentation process. The results of the in vitro antioxidant assay showed that HCT-f exhibited better scavenging ability than HCT for DPPH radicals, hydroxyl radicals, and ABTS radicals. Additionally, its reduction of Fe^{2+} was also more effective. There is a positive correlation between the content of active substances and the physiological and pharmacological activities of plants in general. The results of the above experiments suggest that HCT-f may have a better application prospect compared to HCT.

Excessive generation of ROS, abnormal increase in Ca^{2+} concentration, disruption of mitochondrial membrane permeability, and aberrant activation of the Caspase family all contribute to the induction of apoptosis in cells [42]. Oxidative stress is a condition characterized by an imbalance between oxidative and antioxidant processes in the body, primarily triggered by the overproduction of ROS. Under normal conditions, the antioxidant enzymes in the body eliminate excessive ROS to maintain a normal ROS concentration. When cells are stimulated by LPS, they generate a significant amount of ROS, leading to the accumulation of free radicals. Excessive free radicals can directly induce DNA and organelle damage, trigger apoptosis, and stimulate the production and release of inflammatory factors [43]. It has been shown that abnormal elevation of Ca^{2+} concentration promotes apoptosis [44]. Mitochondria are essential for cellular energy metabolism and production of ROS, and are involved in various processes such as steroid metabolism, Ca^{2+} homeostasis, and free radical scavenging [45]. Mitochondrial membrane permeability transition is closely related to apoptosis. The opening of the mitochondrial permeability transition pores is induced by elevated Ca^{2+} concentration and overproduction of reactive oxygen species [46]. Apoptosis is closely related to the activation of the Caspase family. After apoptosis is induced, the initiating Caspases are recruited to specific complexes for their activation. Subsequently, the apoptotic signal is amplified in a cascade through the hydrolysis of Caspases downstream of heterologous activation [47].

We measured the changes in mitochondrial membrane potential, Ca^{2+} expression, the content of Caspase-3 and Caspase-9, and the relative expression of mRNA in HaCaT cells after LPS stimulation. The experimental results showed that after LPS stimulation, there was a significant increase in the content of ROS occurring in the cells. The sample treatment was effective at reducing the content of ROS. Following LPS-induced injury, there was a significant increase in the Ca^{2+} content and a significant decrease in the mitochondrial membrane potential in the HaCaT cells. These changes indicated that the cells were gradually progressing towards apoptosis after LPS stimulation. Subsequent to HCT-f treatment, the Ca^{2+} content was significantly reduced, the mitochondrial membrane potential was significantly reduced, and the mitochondrial membrane potential was significantly restored. This suggests that the samples may inhibit apoptosis by suppressing ROS and excessive Ca^{2+} production, as well as by repairing the reduction of mitochondrial membrane potential. The decrease in the secretion of two pro-apoptotic factors, Caspase-3 and Caspase-9, and the reduction in their relative mRNA expression following HCT-f treatment further support our experimental results.

LPS is a common inducer used to construct models of inflammation. It promotes the release of a variety of inflammatory factors and triggers a severe inflammatory response. When the skin is stimulated by external irritants, it promotes the expression of inflammatory factors, which further damages the skin's barrier function. In this study, we found that HCT-f could attenuate the response of LPS-induced inflammation by inhibiting the expression of four pro-inflammatory cytokines (IL-6, IL-8, TNF-α, and iNOS). AQP3, FLG, LOR, and Caspase-14 are closely related to the barrier function of the skin. AQP3, which is mainly expressed in the plasma membrane of the keratinocytes of the epidermis, is known for its role in the transport of water and glycerol [48]. The cutaneous membrane sleeve forms the basis of the skin. It is created by the transmembrane proteins of keratinocytes (loricrin, filaggrin) cross-linking with each other under the action of transglutaminase [49]. Caspase-14 has been closely associated with the development of skin diseases such as photoaging and psoriasis [50]. Our study showed that HCT-f significantly promoted the expression of three proteins (AQP3, FLG, and LOR) and their mRNA, while also significantly down-regulating the expression of Caspase-14 and its mRNA. Collectively, these findings suggest that treating cells with HCT-f can significantly alleviate the inflammatory response and impaired barrier function of the skin caused by LPS injury.

To investigate the mechanism by which HCT-f exerts anti-inflammatory and barrier repair functions, we further explored the MAPK/NF-κB pathway, as this pathway is closely related to the occurrence of inflammatory responses. When stimulated by LPS, the activated IκB kinase complex induces IκB-α phosphorylation and degradation, leading to the activation of nuclear translocation of NF-κB [51]. Our experimental results showed that stimulation with LPS activated the MAPK and NF-κB pathways. Treatment of cells with HCT-f effectively inhibited the expression of JNK, ERK, p38, and NF-κB mRNA. From this, we hypothesized that HCT-f may play an anti-inflammatory role by inhibiting the activation of the MAPK/NF-κB pathway. However, the specific effect of HCT-f on the MAPK/NF-κB pathway needs to be further verified by Western blotting experiments. The results of the RBC test and HCT-CAM test indicated that HCT-f did not cause eye irritation and was deemed safe.

Although our experimental results demonstrated that fermented *Houttuynia cordata* Thunb has a higher content of active substances, better antioxidant efficacy, and can alleviate the inflammatory damage caused by LPS stimulation by inhibiting the activation of the MAPK/NF-κB pathway, we did not exclude the influence of the *Clavispora lusitaniae* itself and the metabolites produced during the fermentation process on the experimental results. Additionally, we did not clarify the specific substances in HCT-f that exert antioxidant and anti-inflammatory effects. Although experimental results have shown that enzymes released during microbial fermentation can promote the rupture of plant cell walls, facilitating the release of active ingredients from the plant, the specific enzymes responsible for this effect during the fermentation process in our study remain unknown. Recent studies have demonstrated that the alkaloids found in *Houttuynia cordata* Thunb can exhibit anti-inflammatory effects by inhibiting nitric oxide (NO) production in RAW 264.7 cells under LPS-induced injury [52]. *Houttuynia cordata* Thunb oil could exert anti-inflammatory effects by reducing the synthesis of iNOS and TNF-α proteins in macrophages induced by LPS stimulation [53]. Our study only explored the effect of HCT-f on LPS-induced HaCaT cell damage and did not investigate its impact on other cells. We plan to conduct further research to explore the effects of HCT-f on other cell types in the future. The results of the chicken embryo test and the erythrocyte hemolysis test provide preliminary evidence that the samples are highly safe. However, this does not guarantee the same level of safety when applied to humans. In future research, we will further clarify the specific antioxidant and anti-inflammatory substances in HCT-f. We will also conduct animal and human studies to further demonstrate its efficacy and safety, aiming to realize the high-value utilization of *Houttuynia cordata* Thunb.

5. Conclusions

In conclusion, our experimental results indicated that compared with HCT, there was a significant increase in the content of active substances and in vitro antioxidant activity occurred in HCT-f. HCT-f could alleviate apoptosis and inflammatory responses caused by LPS. In addition, our study also indicated that HCT-f might exert the above effects by inhibiting the activation of the MAPK/NF-κB signaling pathway. The experimental results of the safety evaluation indicated that HCT-f is safe and has promising prospects for application.

Author Contributions: Conceptualization, Z.S. and J.F.; methodology, C.W.; software, Y.T. and Y.X.; validation, Z.S. and Z.W.; formal analysis, Z.S. and J.F.; investigation, D.W. and Y.T.; resources, D.W. and M.L.; data curation, D.W.; writing—original draft preparation, Z.S. and J.F.; writing—review and editing, Z.S.; visualization, J.G.; supervision, C.W.; project administration, C.W. All authors have read and agreed to the published version of the manuscript.

Funding: This research received no external funding.

Institutional Review Board Statement: Not applicable.

Informed Consent Statement: Not applicable.

Data Availability Statement: The original contributions presented in the study are included in the article, further inquiries can be directed to the corresponding author.

Conflicts of Interest: Author Yuncai Tian was employed by the company Shanghai AZ Science & Technology Co., Ltd. Author Yuhua Xu was employed by the company Shanghai AZ Science & Technology Co., Ltd. The remaining authors declare that the research was conducted in the absence of any commercial or financial relationships that could be construed as a potential conflict of interest.

References

1. He, B.W.; Wang, F.F.; Qu, L.P. Anti-Inflammatory and Antioxidant Properties of Physalis alkekengi L. Extracts In Vitro and In Vivo: Potential Application for Skin Care. *Evid.-Based Complement. Altern. Med.* **2022**, *2022*, 1–16. [CrossRef] [PubMed]
2. Lin, C.-k.; Chen, B.Y.; Ting, J.U.; Rogio, K.G.G.; Tsai, P.W.; Liu, C. Deciphering *Houttuynia cordata* extract as electron shuttles with anti-COVID-19 activity and its performance in microbial fuel cells. *J. Taiwan Inst. Chem. Eng.* **2023**, *145*, 104838. [CrossRef] [PubMed]
3. Subhawa, S.; Chewonarin, T.; Banjerdpongchai, R. The Effects of *Houttuynia cordata* Thunb and Piper ribesioides Wall Extracts on Breast Carcinoma Cell Proliferation, Migration, Invasion and Apoptosis. *Molecules* **2020**, *25*, 1196. [CrossRef] [PubMed]
4. Oh, S.Y. An Effective Quality Control of Pharmacologically Active Volatiles of *Houttuynia cordata* Thunb by Fast Gas Chromatography-Surface Acoustic Wave Sensor. *Molecules* **2015**, *20*, 10298–10312. [CrossRef] [PubMed]
5. Ghosh, A.; Ghosh, B.; Parihar, N.; Ilaweibaphyrnai, M.; Panda, S.R.; Alexander, A.; Chella, N.; Murty, U.S.N.; Naidu, V.G.M.; Jagadeesh, K.G.; et al. Nutraceutical prospects of *Houttuynia cordata* against the infectious viruses. *Food Biosci.* **2022**, *50*, 101977. [CrossRef] [PubMed]
6. Luo, Q.; Meng, P.H.; Jiang, D.W.; Han, Z.M.; Wang, Z.H.; Tan, G.F.; Zhang, J. Comprehensive Assessment of *Houttuynia cordata* Thunb., an Important Medicinal Plant and Vegetable. *Agronomy* **2022**, *12*, 2582. [CrossRef]

7. Rafiq, S.; Hao, H.; Ijaz, M.; Raza, A. Pharmacological Effects of *Houttuynia cordata* Thunb (H. cordata): A Comprehensive Review. *Pharmaceuticals* **2022**, *15*, 1079. [CrossRef] [PubMed]
8. Kim, J.H.; Cho, Y.J.; Pan, J.H.; Park, T.S.; Shim, S.M.; Kim, Y.J. Antiatherogenic and antioxidative effects of *Houttuynia cordata* extracts in rats fed a high-fat diet. *Food Sci. Biotechnol.* **2014**, *23*, 2069–2074. [CrossRef]
9. Fang, H.; Yang, Z.; Yang, L. Protective effect of *Houttuynia cordata* extract on propofol-induced injury of rat hippocampal neurons by regulating PI3K/Akt and Toll-like receptor 4/NF-κB signaling pathway. *NeuroReport* **2021**, *32*, 577–582. [CrossRef]
10. Yubin, J.; Junjun, Y.; Miao, Y.; Yue, C.; Guo, S.Z.; Qiao, A.N. Study on Medicinal Plant Active Substances Extraction and Antibacterial Activity of Houttuynia Cordata. IOP conference series. *Earth Environ. Sci.* **2017**, *100*, 12056. [CrossRef]
11. Fan, G.S.; Liu, P.X.; Chang, X.; Yin, H.; Cheng, L.J.; Teng, C.; Gong, Y.; Li, X.T. Isolation and Identification of a High-Yield Ethyl Caproate-Producing Yeast from Daqu and Optimization of Its Fermentation. *Front. Microbiol.* **2021**, *12*, 663744. [CrossRef] [PubMed]
12. Yildiran, H.; Kiliç, G.B.; Çakmakçi, A.G.K. Characterization and comparison of yeasts from different sources for some probiotic properties and exopolysaccharide production. *Food Sci. Technol.* **2019**, *39* (Suppl. S2), 646–653. [CrossRef]
13. Cissé, H.; Kagambèga, B.; Sawadogo, A.; Tankoano, A.; Sangaré, G.; Traoré, Y.; Ouoba, I.I.L.; Savadogo, A. Molecular characterization of *Bacillus*, lactic acid bacteria and yeast as potential probiotic isolated from fermented food. *Sci. Afr.* **2019**, *6*, e175. [CrossRef]
14. Shukla, R.; Hemansi Kumar, G.; Shukla, S.; Saini, J.K.; Kuhad, R.C. Biorefinery potential of newly isolated yeast *Clavispora lusitaniae* for co-production of erythritol and ethanol. *Biomass Convers. Biorefinery* **2023**, *13*, 8061–8073. [CrossRef]
15. Sun, W.L.; Shahrajabian, M.H.; Lin, M. Research Progress of Fermented Functional Foods and Protein Factory-Microbial Fermentation Technology. *Fermentation* **2022**, *8*, 688. [CrossRef]
16. Chen, G.Q.; Liu, X.Y. On The Future Fermentation. *Microb. Biotechnol.* **2020**, *14*, 18–21. [CrossRef]
17. Zhou, X.R.; Zhao, Y.M.; Xu, G.F. *Bacillus subtilis* and *Bifidobacteria bifidum* Fermentation Effects on Various Active Ingredient Contents in *Cornus officinalis* Fruit. *Molecules* **2023**, *28*, 1032. [CrossRef]
18. Liang, Y.K.; Duan, H.B.; Zhang, P.; Han, H.; Gao, F.X.; Li, Y.X.; Xu, Z.Y. Extraction and isolation of the active ingredients of dandelion and its antifungal activity against *Candida albicans*. *Mol. Med. Rep.* **2020**, *21*, 229–239. [CrossRef] [PubMed]
19. Costa, M.; Cardoso, C.; Afonso, C.; Bandarra, N.M.; Prates, J.A.M. Current knowledge and future perspectives of the use of seaweeds for livestock production and meat quality: A systematic review. *J. Anim. Physiol. Anim. Nutr.* **2021**, *105*, 1075–1102. [CrossRef]
20. Fang, J.X.; Sun, Q.R.; Wang, Z.W.; Song, Z.X.; Geng, J.M.; Wang, C.T.; Li, M.; Wang, D.D. Enhancement of Human Epidermal Cell Defense against UVB Damage by Fermentation of *Passiflora edulis Sims* Peel with *Saccharomyces cerevisiae*. *Nutrients* **2023**, *15*, 501. [CrossRef]
21. Shi, J.H.; Shi, S.; Xie, W.B.; Zhao, M.; Li, Y.; Zhang, J.; Li, N.; Bai, X.Z.; Cai, W.X.; Hu, X.L.; et al. IL-10 alleviates lipopolysaccharide-induced skin scarring via IL-10R/STAT3 axis regulating TLR4/NF-κB pathway in dermal fibroblasts. *J. Cell. Mol. Med.* **2021**, *25*, 1554–1567. [CrossRef] [PubMed]
22. Sp, N.; Kang, D.Y.; Kim, H.D.; Rugamba, A.; Jo, E.S.; Park, J.C.; Bae, S.W.; Lee, J.M.; Jang, K.J. Natural Sulfurs Inhibit LPS-Induced Inflammatory Responses through NF-κB Signaling in CCD-986Sk Skin Fibroblasts. *Life* **2021**, *11*, 427. [CrossRef] [PubMed]
23. Kim, C.; Ji, J.; Baek, S.H.; Lee, J.H.; Ha, I.J.; Lim, S.S.; Yoon, H.J.; Nam, Y.J.; Ahn, K.S. Fermented dried *Citrus unshiu* peel extracts exert anti-inflammatory activities in LPS-induced RAW264.7 macrophages and improve skin moisturizing efficacy in immortalized human HaCaT keratinocytes. *Pharm. Biol.* **2019**, *57*, 392–402. [CrossRef] [PubMed]
24. Yi, L.X.; Weng, T.J.; Nie, P.H.; Zhu, L.; Gao, M.M.; Jia, H.X.; Yang, S.H.; Li, X.B.; Zhang, L.; Xu, Y.; et al. Overexpression of interleukin-10 in engineered macrophages protects endothelial cells against LPS-induced injury in vitro. *FEBS Open Bio* **2022**, *12*, 605–615. [CrossRef] [PubMed]
25. Wang, F.; Liu, J.T.; An, Q.; Wang, Y.M.; Yang, Y.; Huo, T.; Yang, S.M.; Ju, R.J.; Quan, Q.H. Aloe Extracts Inhibit Skin Inflammatory Responses by Regulating NF-κB, ERK, and JNK Signaling Pathways in an LPS-Induced RAW264.7 Macrophages Model. *Clin. Cosmet. Investig. Dermatol.* **2023**, *16*, 267–278. [PubMed]
26. Liu, P.Q.; Li, Y.; Wang, W.L.; Bai, Y.Z.; Jia, H.M.; Yuan, Z.W. Role and mechanisms of the NF-κB signaling pathway in various developmental processes. *Biomed. Pharmacother.* **2022**, *153*, 113513. [CrossRef] [PubMed]
27. Serra, G.; Incani, A.; Serreli, G.; Porru, L.; Melis, M.P.; Tuberoso, C.I.G.; Rossin, D.; Biasi, F.; Deiana, M. Olive oil polyphenols reduce oxysterols -induced redox imbalance and pro-inflammatory response in intestinal cells. *Redox Biol.* **2018**, *17*, 348–354. [CrossRef] [PubMed]
28. Wu, Q.; Wang, Y.H.; Li, Q.F. Matairesinol exerts anti-inflammatory and antioxidant effects in sepsis-mediated brain injury by repressing the MAPK and NF-κB pathways through up-regulating AMPK. *Aging* **2021**, *13*, 23780–23795. [CrossRef]
29. Chun, J.M.; Nho, K.J.; Kim, H.S.; Lee, A.Y.; Moon, B.C.; Kim, H.K. An ethyl acetate fraction derived from *Houttuynia cordata* extract inhibits the production of inflammatory markers by suppressing NF-κB and MAPK activation in lipopolysaccharide-stimulated RAW 264.7 macrophages. *BMC Complement Altern. Med.* **2014**, *14*, 234. [CrossRef]

30. Sun, Q.R.; Fang, J.X.; Wang, Z.W.; Song, Z.X.; Geng, J.M.; Wang, D.D.; Wang, C.T.; Li, M. Two *Laminaria japonica* Fermentation Broths Alleviate Oxidative Stress and Inflammatory Response Caused by UVB Damage: Photoprotective and Reparative Effects. *Mar. Drugs* **2022**, *20*, 650. [CrossRef]
31. Su, Y.F.; Zhang, Y.T.; Fu, H.; Yao, F.F.; Liu, P.P.; Mo, Q.T.; Wang, D.D.; Zhao, D.; Wang, C.T.; Li, M. Physicochemical and Anti-UVB-Induced Skin Inflammatory Properties of *Lacticaseibacillus paracasei* Subsp. *paracasei* SS-01 Strain Exopolysaccharide. *Fermentation* **2022**, *8*, 198. [CrossRef]
32. Shin, M.J.; Lee, C.S.; Kim, S.H. Screening for Lactic Acid Bacterial Strains as Probiotics Exhibiting Anti-inflammatory and Antioxidative Characteristic Via Immune Modulation in HaCaT Cell. *Probiotics Antimicrob. Proteins* **2023**, *15*, 1665–1680. [CrossRef] [PubMed]
33. Shoshan-Barmatz, V.; De, S.; Meir, A. The Mitochondrial Voltage-Dependent Anion Channel 1, Ca^{2+} Transport, Apoptosis, and Their Regulation. *Front. Oncol.* **2017**, *7*, 60. [CrossRef]
34. Sun, W.Y.; Zheng, Y.; Lu, Z.Y.; Wang, H.; Feng, Z.H.; Wang, J.; Xiao, S.X.; Liu, F.; Liu, J.K. LL-37 attenuates inflammatory impairment via mTOR signaling-dependent mitochondrial protection. *Int. J. Biochem. Cell Biol.* **2014**, *54*, 26–35. [CrossRef]
35. David, E.C. Calcium Signaling. *Cell* **2007**, *12*, 131. [CrossRef]
36. Bai, L.N.; Yang, J.H.; Zhang, H.; Liao, W.; Cen, Y.G. PTB domain and leucine zipper motif 1 (APPL1) inhibits myocardial ischemia/hypoxia-reperfusion injury via inactivation of apoptotic protease activating factor-1 (APAF-1)/Caspase9 signaling pathway. *Bioengineered* **2021**, *12*, 4385–4396. [CrossRef] [PubMed]
37. Alam, M.B.; Kwon, Y.G.; Simu, S.Y.; Shahriyar, S.A.; Lee, S.H. Attenuation of Inflammatory Symptoms by Icariside B2 in Carrageenan and LPS-Induced Inflammation Models via Regulation of MAPK/NF-κB Signaling Cascades. *Biomolecules* **2020**, *10*, 1037. [CrossRef]
38. Ge, L.L.; Chen, L.; Mo, Q.G.; Zhou, G.; Meng, X.S.; Wang, Y.W. Total phenylethanoid glycosides and magnoloside Ia from Magnolia officinalis var. biloba fruits inhibit ultraviolet B-induced phototoxicity and inflammation through MAPK/NF-κB signaling pathways. *R. Soc. Chem.* **2018**, *8*, 4362–4371.
39. Jeong, H.J.; Park, M.; Kim, D.W.; Ryu, E.J.; Yong, J.I.; Cha, H.J.; Kim, S.J.; Yeo, H.J.; Jeong, J.H.; Kim, D.S.; et al. Down-regulation of MAPK/NF-κB signaling underlies anti-inflammatory response induced by transduced PEP-1-Prx2 proteins in LPS-induced Raw 264.7 and TPA-induced mouse ear edema model. *Int. Immunopharmacol.* **2014**, *23*, 426–433. [CrossRef]
40. Xu, Y.Y.; Zhang, Y.Y.; Ou, Y.Y.; Lu, X.X.; Pan, L.Y.; Li, H.; Lu, Y.; Chen, D.F. *Houttuynia cordata* Thunb. polysaccharides ameliorates lipopolysaccharide-induced acute lung injury in mice. *J. Ethnopharmacol.* **2015**, *173*, 81–90. [CrossRef]
41. Jeewon, R.; Aullybux, A.A.; Puchooa, D.; Nazurally, N.; Alrefaei, A.F.; Zhang, Y. Marine Microbial Polysaccharides: An Untapped Resource for Biotechnological Applications. *Mar. Drugs* **2023**, *21*, 420. [CrossRef] [PubMed]
42. Verfaillie, T.; Rubio, N.; Garg, A.D.; Bultynck, G.; Rizzuto, R.; Decuypere, J.P.; Piette, J.; Linehan, C.; Gupta, S.; Samali, A.; et al. PERK is required at the ER-mitochondrial contact sites to convey apoptosis after ROS-based ER stress. *Cell Death Differ.* **2012**, *19*, 1880–1891. [CrossRef] [PubMed]
43. Pi, J.; Zhang, Q.; Fu, J.Q.; Woods, C.G.; Hou, Y.Y.; Corkey, B.E.; Collins, A.; Andersen, M.E. ROS signaling, oxidative stress and Nrf2 in pancreatic beta-cell function. *Toxicol. Appl. Pharmacol.* **2010**, *244*, 77–83. [CrossRef]
44. Saleem, M.Z.; Alshwmi, M.; Zhang, H.; Din, S.R.U.; Nisar, M.A.; Khan, M.; Alam, S.; Alam, G.; Jin, L.L.; Ma, T.H. Inhibition of JNK-Mediated Autophagy Promotes Proscillaridin A-Induced Apoptosis via ROS Generation, Intracellular Ca^{+2} Oscillation and Inhibiting STAT3 Signaling in Breast Cancer Cells. *Front. Pharmacol.* **2020**, *11*, 1055. [CrossRef]
45. Chauhan, L.K.; Varshney, M.; Pandey, V.; Pandey, V.; Sharme, P.; Verma, V.K.; Kumar, P.; Goel, S.K. ROS-dependent genotoxicity, cell cycle perturbations and apoptosis in mouse bone marrow cells exposed to formulated mixture of cypermethrin and chlorpyrifos. *Mutagenesis* **2016**, *31*, 635–642. [CrossRef] [PubMed]
46. Zhang, H.C.; Shangguan, F.; Zhang, L.; Ma, N.F.; Song, S.L.; Ma, L.; Liu, C.T.; Liu, M.K.; An, J.; Li, H.; et al. A novel mechanism of 6-methoxydihydroavicine in suppressing ovarian carcinoma by disrupting mitochondrial homeostasis and triggering ROS/MAPK mediated apoptosis. *Front. Pharmacol.* **2023**, *14*, 1093650. [CrossRef] [PubMed]
47. Zhao, Y.X.; Lei, M.X.; Wang, Z.Y.; Qiao, G.L.; Yang, T.L.; Zhang, J. TCR-induced, PKC-θ-mediated NF-κB activation is regulated by a caspase-8-caspase-9-caspase-3 cascade. *Biochem. Biophys. Res. Commun.* **2014**, *450*, 526–531. [CrossRef] [PubMed]
48. Agren, J.; Zelenin, S.; Svensson, L.B.; Nejsum, L.N.; Nielsen, S.; Aperia, A.; Sedin, G. Antenatal Corticosteroids and Postnatal Fluid Restriction Produce Differential Effects on AQP3 Expression, Water Handling, and Barrier Function in Perinatal Rat Epidermis. *Dermatol. Res. Pract.* **2010**, *2010*, 789729. [CrossRef]
49. Yoneda, K.; Demitsu, T.; Manabe, M.; Igarashi, J.; Kosaka, H.; Inagaki, N.; Takahashi, H.; Kon, A.; Kakurai, M.; Kubota, Y. Expression of wild-type, but not mutant, loricrin causes programmed cell death in HaCaT keratinocytes. *J. Dermatol.* **2010**, *37*, 956–964. [CrossRef]
50. Jiang, Y.J.; Kim, P.; Uchida, Y.; Elias, P.M.; Bikle, D.D.; Grundeld, C.; Feingold, K.R. Ceramides stimulate caspase-14 expression in human keratinocytes. *Exp. Dermatol.* **2013**, *22*, 113–118. [CrossRef]
51. Ouyang, Q.Q.; Li, Y.C.; Mei, S.; Zhang, Q.; Li, X.J.; Luo, H.; Zhu, Y.Z.; Wu, K.F. Protective effects of GLHP from Gracilaria lemaneiformis against UVB-induced photodamage in human immortalized keratinocytes cells and BALB/c mice. *Exp. Gerontol.* **2021**, *155*, 111550. [CrossRef] [PubMed]

52. Ahn, J.; Chae, H.S.; Chin, Y.W.; Kim, J. Alkaloids from aerial parts of *Houttuynia cordata* and their anti-inflammatory activity. *Bioorganic Med. Chem. Lett.* **2017**, *27*, 2807–2811. [CrossRef] [PubMed]
53. Li, W.F.; Fan, T.; Zhang, Y.M.; Fan, T.; Zhou, P.; Niu, X.F.; He, L.C. *Houttuynia cordata* Thunb. Volatile Oil Exhibited Anti-inflammatory Effects In Vivo and Inhibited Nitric Oxide and Tumor Necrosis Factor-α Production in LPS-stimulated Mouse Peritoneal Macrophages In Vitro. *Phytother. Res.* **2013**, *27*, 1629–1639. [CrossRef] [PubMed]

foods

MDPI

Article

Microalgae Biomass Production from Rice Husk as Alternative Media Cultivation and Extraction of Phycocyanin Using 3D-Printed Ohmic Heating Reactor

Gabriela Cid-Ibarra [1], Rosa M. Rodríguez-Jasso [1], Gilver Rosero-Chasoy [1], Ruth Belmares [2], Juan Carlos Contreras-Esquivel [3], Samanta Machado-Cepeda [1], Alejandra Cabello-Galindo [1] and Héctor A. Ruiz [1,*,†]

[1] Biorefinery Group, Food Research Department, School of Chemistry, Autonomous University of Coahuila, Saltillo 25280, Coahuila, Mexico; rrodriguezjasso@uadec.edu.mx (R.M.R.-J.); samanta.machado@uadec.edu.mx (S.M.-C.)

[2] Functional Foods and Nutrition Group, Food Research Department, School of Chemistry, Autonomous University of Coahuila, Saltillo 25280, Coahuila, Mexico; ruthbelmares@uadec.edu.mx (R.B.)

[3] Laboratory of Applied Glycobiotechnology, Food Research Department, School of Chemistry, Autonomous University of Coahuila, Saltillo 25280, Coahuila, Mexico

* Correspondence: hector_ruiz_leza@uadec.edu.mx

† webpage: www.biorefinerygroup.com.

Abstract: Phycocyanin is a highly valued pigment present in *Spirulina platensis* biomass with applications in the food industry in terms of biorefinery concepts; specifically, its antioxidant and antimicrobial capacity are an advantage that could be incorporated into a food matrix. This study aims to use rice husk as an alternative culture medium for *S. platensis* biomass growth and phycocyanin extraction by ohmic heating processing using a 3D-printed reactor. *S. platensis* was cultivated in rice husk extract (RHE) from 0–100% (v/v). The highest content of microalgal biomass was 1.75 ± 0.01 g/L, with a specific growth rate of 0.125 ± 0.01 h^{-1}. For the phycocyanin extraction under an ohmic heating process, a 3D-printed reactor was designed and built. To optimize phycocyanin extraction, a central composite rotatable design (CCDR) was evaluated, with three factors: time (min), temperature (°C), and pH. The highest phycocyanin content was 75.80 ± 0.98 mg/g in *S. platensis* biomass grown with rice husk extract. Ohmic heating is a promising method for rapid phycocyanin extraction, and rice husk as a culture medium is an alternative for the growth of *S. platensis* biomass in the integration of second- and third-generation biorefineries.

Keywords: biorefinery; *Spirulina platensis*; pigment; bioeconomy; high added-value compounds; fermentation

1. Introduction

The intensity of color is a crucial aspect when it comes to the appearance of food products, and it plays a significant role in consumer perception. This makes it attractive to the consumer and increases their potential to choose a particular product, hence its economic relevance [1,2]. Artificial colorants are used because they are cheap, stable, and bright; however, health complications are found when they are consumed in a food matrix [3]; some of the possible complications are allergies and neurological disorders. Synthetic dyes have been restricted and investigated by the U.S. Food and Drug Administration (FDA) and California Office of Environmental Health Hazard Assessment (OEHHA), as well as the European Food Safety Authority (EFSA) [4–7].

For this reason, natural dyes are gaining interest, as they are not harmful and have no undesirable or adverse effects on human health. Natural pigments can be beneficial due to their bioactive compounds [8,9]. Among natural dyes, microalgal biomass can be used as a natural resource to obtain interesting pigments. Some intracellular pigments of industrial interest in microalgae biomass are chlorophylls, carotenoids, xanthophylls, and phycobiliproteins (phycocyanin) [10]. Phycocyanin is part of a complex protein pigment called a phycobiliprotein (phycoerythrin, phycocyanin, and allophycocyanin), which gives a blue-green color to the microalgal biomass, and this pigment represents about 40–60% of the total protein content of the biomass [11]. These pigments are found in aquatic organisms such *Acanthophora specifera, Bangia atropurea, Gracilaria corticata, Laurencia papillosa,* and *S. platensis* [12,13], and due to the phycocyanin content in these microorganisms, they have a blue-green color.

Phycocyanin is used extensively in the pharmaceutical and food industries due to its antioxidant, anti-inflammatory, and anti-microbial properties. Food-grade phycocyanin costs approximately USD 0.13 per milligram, and analytical grade phycocyanin costs USD 15.00 per milligram [14]. Due to its widespread application, phycocyanin has generated USD 250 million in the last decade [15]. Some examples of phycocyanin use in the food industry include ice cream, beverages, and confectionery [13]. One of the biggest challenges in the obtention of phycocyanin is the extraction process; the most common methods to hydrolyze or break down the cell wall and release the pigment are freezing and thawing cycles, bead milling, ultra-sonication, and ohmic heating [16–18].

Among the various methods for disrupting cell wall of *S. platensis*, freezing and thawing have been reported to be the highest-yielding methods for obtaining phycobiliproteins [19]. However, the cycles of freezing and thawing may affect the stability of the compounds from the aggressive cell wall disruption. Another alternative is the ohmic heating process, in which an electrical field is applied to a conductive sample [9]. In the case of cellular material, such as *S. platensis*, ohmic heating causes electroporation of the cell membrane. This occurs when the electrical field applied in the sample has a higher dielectric strength, and the lipids present in the membrane are disrupted, leading to permeabilization of the membrane and releasing intracellular compounds such as phycocyanin [20]. Figure 1 shows the mechanisms and effects of ohmic heating processing on *S. platensis*. Ohmic heating has disadvantages, including possible corrosion of the electrodes and the complex conductivity of the sample [21]. Nevertheless, ohmic heating offers interesting advantages such as short time of extraction and homogeneity of heating [10]. In addition, ohmic heating allows the implementation and design of a reactor, depending on the requirements of the sample, the operational conditions, and the possibility of using different materials for construction.

These characteristics lead to ohmic heating being an alternative method for the extraction of phycocyanin. The phycocyanin present in *S. platensis* is close to 20% of the total biomass [22]; however, this extraction process gives a low yield of these pigments. Although ohmic heating is an interesting alternative for phycocyanin extraction, various conditions need to be studied in detail to achieve better yields of phycocyanin using the method. Factors such as temperature, time, and pH must be studied in detail.

In order to implement an extraction method for phycocyanin, improvements in the cultivation of *S. platensis* have been sought; some desirable characteristics for the growth of a culture are higher yield, ease of implementation and operation, and time of culture growth. An alternative growth medium involves the use of agro-waste [23]. Some examples are tofu wastewater [24], cheese whey [25], and fish waste [26]. A viable alternative is the use of rice husk to grow *S. platensis* for the subsequent extraction of phycocyanin. Rice husk is a residue produced from rice milling [27] and accounts for about 20% of the whole rice grain. This agro-waste is used to produce biofuels, livestock feed, and bio-fertilizers [28]. Due to the composition of rice husk, it can be used for the growth of photosynthetic organisms [29]. The main components of rice husk compounds are SiO_2, SiC, Si_3N_4, Si, Mg_2Si, hemicellulose 24.3%, cellulose 34.4%, lignin 19.2%, starch, and protein [30]. To date, the utilization of rice husk extract as a culture medium has not been explored for the growth of *S. platensis*; additionally, the starch content and trace metals present in the waste can be used for such growth. This work aims to produce *S. platensis* biomass from rice husk as an alternative for cultivation and to extract phycocyanin using a 3D-printed ohmic heating reactor in terms of a biorefinery concept.

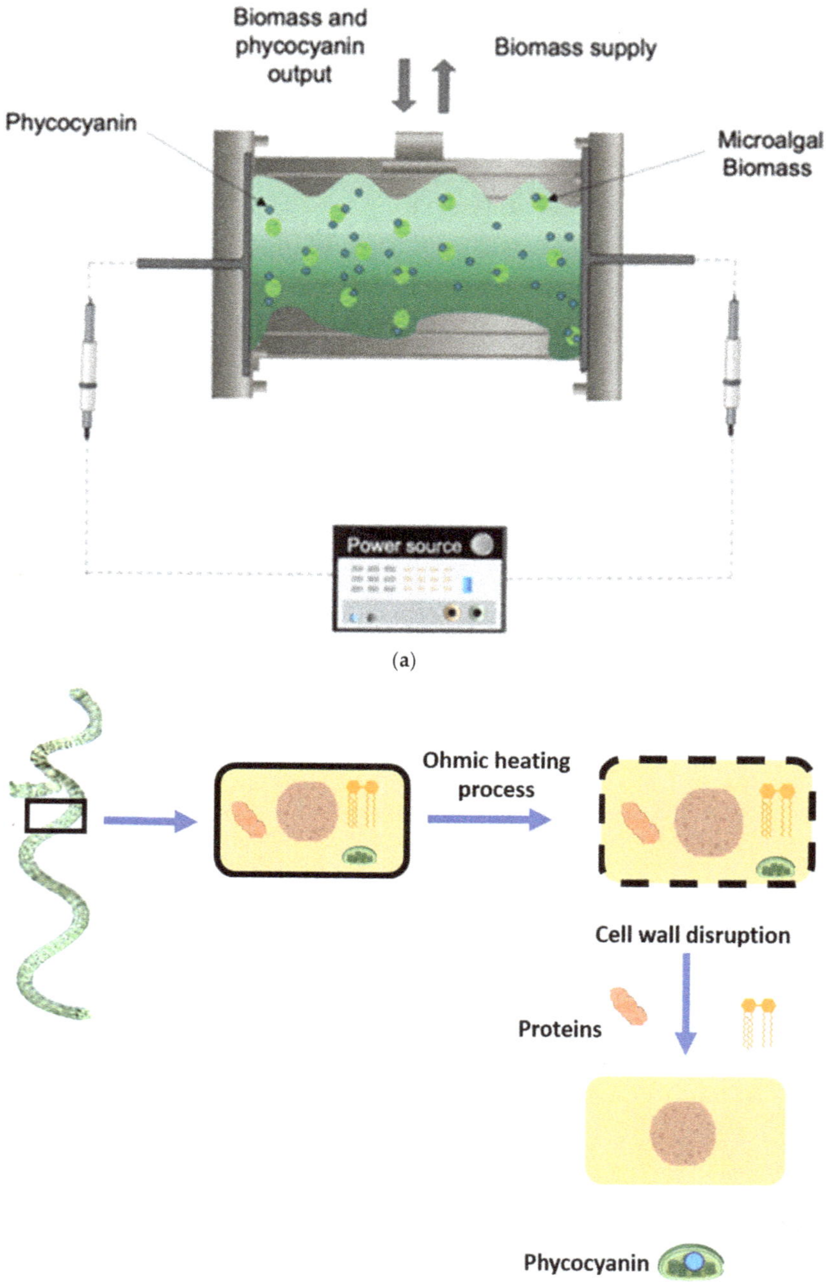

(a)

(b)

Figure 1. (**a**) Ohmic heating reactor. (**b**) Diagram of cell wall disruption by ohmic heating process.

2. Materials and Methods

2.1. Pre-Inoculum and Growth Conditions

S. platensis LEB-52 strain was kindly provided by the Passo Fundo University, Brazil (Prof. Luciane Colla). *S. platensis* was cultivated in previously sterilized Zarrouk medium [31].

The growth conditions for the pre-inoculum were photo periods of 13 h of light and 11 h of darkness, a light intensity of 73 μmol/($m^2 \times$), and room temperature for 10 days. The working volume was 250 mL in an Erlenmeyer flask with a capacity of 500 mL [31].

2.2. Rice Husk Extract as a Cultivation Medium

The rice husk (RH) was obtained from Gea (Jalisco, Mexico). The water-soluble compounds present in the rice husk were extracted. Briefly, 40 g of rice husk was weighed and placed in a Schott flask to which 1 L of distilled water was added. The extraction of rice husk components was carried out for 1 h in a water bath (100 °C) at an internal temperature of 76 °C (Schott flask). After the extraction step, the solids from the rice husk were removed by filter paper using a vacuum pump, and the liquid extract of the rice husk was sterilized with UV light for 15 min. The final pH of the rice husk extract was 7.5, and the rice husk was stored at 4 °C until use [32].

2.3. Experimental Setup for the Growth of *S. platensis* Using Rice Husk as a Culture Medium

The *S. platensis* growth was evaluated in the rice husk extract at different concentrations. Table 1 shows the different treatments to determine the effect of rice husk extract on the growth of *S. platensis*. All experiments were batch-run with an inoculum content of 15% (*v*/*v*). All the treatments were sterilized with UV light for 15 min prior to inoculation. The working volume was 300 mL in an Erlenmeyer flask of 500 mL, and the experimental conditions were the same as mentioned above; all treatments had a constant nitrogen source of 0.7 g/L $NaNO_3$. The treatments T1–T6 contained different percentages of rice husk extract (0–100%) supplemented with a nitrogen source. The control (T7) was Zarrouk medium.

Table 1. Treatments with different concentrations of rice husk as a culture medium for the growth of *S. platensis*.

Treatments	Rice Husk Extract */Medium (%, *v*/*v*)	Distilled Water (%, *v*/*v*)
T1	0	100
T2	10	90
T3	20	80
T4	40	60
T5	80	20
T6	100	0
T7 (blank)	Zarrouk	NA

* Initial concentration 40 g/L; NA = not applicable.

The sampling was carried out for 2 days during the growth kinetics over 14 days. A sample of 10 mL was taken to determine the biomass concentration. The protein and phycocyanin contents were evaluated at the end of the period [33]. For microalgal kinetics, the phase of highest yield, known as the exponential phase, was used, in which the highest amount of biomass produced (g/L) is related to time (days) in an exponential function [34].

2.4. Analytical Methods for the Characterization of Rice Husk Extract and *S. platensis* Biomass

2.4.1. Carbohydrate Determination in Rice Husk Extract

The carbohydrate content in the rice husk extract (RHE) was determined using the anthrone method. To prepare the anthrone reagent, 0.2 g of anthrone was dissolved in 5 mL of 96% ethanol and 75% H_2SO_4. For the total carbohydrate assay, 200 μL of rice husk extract was mixed with 800 μL of the anthrone reagent in a test tube. A glucose solution (1000 ppm) was used in the calibration curve for a standard solution. The samples were vortexed and heated at 100 °C for 5 min, followed by cooling for 5 min. Then, 200 μL of each sample was transferred to a microplate and measured using a spectrophotometer (SunriseTM A-5082, Grödig, Austria) at 530 nm [31].

2.4.2. Fourier-Transform Infrared Spectroscopy (FTIR) for the Analysis of Rice Husk and Phycocyanin

For the identification of functional groups from starch present in the rice husk extract an FTIR analysis was performed; starch can be used for the growth of microalgae as a carbon source. The rice husk (RH) and the rice husk extract (RHE) (lyophilized) were ground, sieved, and analyzed by FTIR using a Perkin Elmer Frontier FTIR spectrometer (Waltham, MA, USA) under the following conditions: resolution at 4 cm^{-1} with 16 scans per test. Spectra were collected from 4000 to 500 cm^{-1} in absorbance mode, normalized, and baseline corrected [34].

2.4.3. Elemental Characterization by X-ray Fluorescence of Rice Husk

For determination of the elemental composition of the rice husk extract (RHE), milling and subsequent screening of the husk were performed, while to obtain the solids contained in the rice husk extract, lyophilization was used (Labconco company, Kansas City, MO, USA). The resulting solid was weighed (0.2 g) and analyzed by X-ray fluorescence (Panalytical, Epsilon 1, Almelo, The Netherlands), before being integrated by a spectrometer with the use of the Omniam software [35]. For the total solids content in the rice husk extract, the liquid part was evaporated, and the rest of the solids were weighed and heated at 600 °C for 3 h for the de-ashed content.

2.4.4. Characterization of the Rice Husk by HPLC

For the characterization of RHE, high-performance liquid chromatography (HPLC) was performed. For hydrolysis, 0.5 g of the rice husk was weighed (previously ground and sifted) and placed in a flask to which 5 mL of H_2SO_4 at 72% was added. The samples were agitated manually for 1 h at 30 °C. After the agitation of the samples, water was added to dilute the acid to 4%. Post-hydrolysis, the sample content in the flasks was autoclaved for 1 h at 121 °C. After the treatment, the samples were filtered, and the solid part of the samples was dried for 24 h and weighed for the determination of the amount of Klason lignin present in the rice husk. The liquid phase of the sample was filtered through a 0.45 μm nylon filter for the injection of the sample into the HPLC system (Agilent 1260 Infinity II, Ratingen, Germany). The analysis was performed on Metacarb 87H (300 mm × 7.8 mm, Agilent, Ratingen, Germany) at temperature of 60 °C with a 0.7 mL/min flow rate of 0.005 mol/L sulfuric acid as the mobile phase. Xylose, arabinose, glucose, and acetic acid were determined and related to the polysaccharides present in the rice husk [36].

2.5. Determination of the Biomass Growth S. platensis

A representative sample of the culture medium (10 mL) was filtered using vacuum filtration, and the filter with the biomass was dried at 50 °C for 24 h. Subsequently, the biomass (content in the filter) was weighed to determine the dry weight of the biomass. During the 14-day growth period, a sample was taken every 2 days to obtain the growth kinetics [37].

2.6. Determination of Protein Content in Microalgal Biomass

For the release of the proteins present in the *S. platensis* biomass, the freezing and thawing method in three cycles was applied for cell wall breaking; after this step, the biomass (10 mL) was centrifuged at 6000 rpm for 15 min at 0 °C. After the centrifugation, the supernatant was discarded, and the pellet (biomass) was mixed with 1.5 mL of 25% (*v*/*v*) ethanol in NaOH (1 N). The samples were centrifuged again and sonicated for 15 min (110 W). For the quantification of proteins, 100 μL of the samples was used and placed in a microplate, and a standard curve was elaborated using bovine serum albumin; the regents of the Lowry method [38,39] were used, and the analysis required the use of the colorimetric method in a microplate spectrophotometer (Sunrise M A-5082, Grödig, Austria) at 660 nm.

Determination of Phycocyanin Content

A culture volume of 45 mL of the microalgal biomass was harvested at 14 days. The harvested sample was centrifugated at 6000 rpm for 15 min. The liquid phase was discarded, and the pellet was re-suspended with a phosphate buffer at pH 6.6 [16]. The pellet with buffer was mixed by vortexing. To evaluate the phycocyanin content present in the *S. platensis* biomass from the different treatments of rice husk extract, the samples were subjected to five cycles of freezing and thawing as an extraction method to release the intracellular phycocyanin. The phycocyanin was separated from the biomass by decantation. The phycocyanin concentration was determined by absorbance at 620 and 652 nm. The phycocyanin concentration in mg/mL was calculated from Equation (1) [16].

$$CPC = (OD620 - 0.47OD\ 652)/5.34 \quad (1)$$

where CPC is the C-phycocyanin concentration in mg/mL and OD is the optical density, reading at 620 and 652 nm. The remaining values correspond to constants for phycocyanin.

2.7. Design of Ohmic 3D Reactor for the Phycocyanin Extraction

For the extraction of phycocyanin from *S. platensis*, the ohmic heating processing technology was used, and a 3D-printed reactor was designed and built. The software employed for this purpose was Solidworks 2018, the software provided an effective means of design and development of images, and these files are compatible with the 3D printer.

The design of the reactor was created with the following dimensions: 15 mm of radium, 50 mm of height, 6 mm of thickness, and a nominal volume of 50 mL. The reactor also has a hole in the middle with a radius of 3.5 mm. The reactor has two caps made of polylactic acid with the following dimensions: 21 mm radius, 15 mm of height, 3 mm of thickness, and a hole in the middle of each cap with a 3.5 mm radius. Figure 2 shows the reactor design. The reactor was printed on a 3D printer (Creality Ender-3) (Shenzhen, China) under the following conditions: temperature in the extruder of 220 °C, bed temperature of 60 °C, 0.12 mm between layers, smoothed fill pattern for the top infill, and monotonic fill pattern for the bottom infill using polylactic acid (PLA) as filling. For the thermal conductivity inside the reactor, two electrodes of 316L steel were added to the caps.

The biomass growth in the rice husk extract (40%) was filtered and weighed (1000 ppm); the wet biomass was added to a phosphate buffer in a range of pH 5.6–7.3. The samples were mixed and added into the 3D-printed reactor with a working volume of 20 mL. During the phycocyanin extraction, 40 V was applied in the reactor to reach the temperatures used in the experimental design (please see Table 2). The temperature control in the holding time was achieved manually by turning the power source on and off once the assay temperature was reached.

Table 2. Coded levels and variables of the CCDR for experimental design.

Variables	Symbols	−1.68 (α)	−1	0	+1	1.68 (α)
Temperature (°C)	X_1	29.77	40	55	70	80.23
Time (min)	X_2	1.59	50	10	15	18.41
pH	X_3	5.66	6	6.5	7	7.34

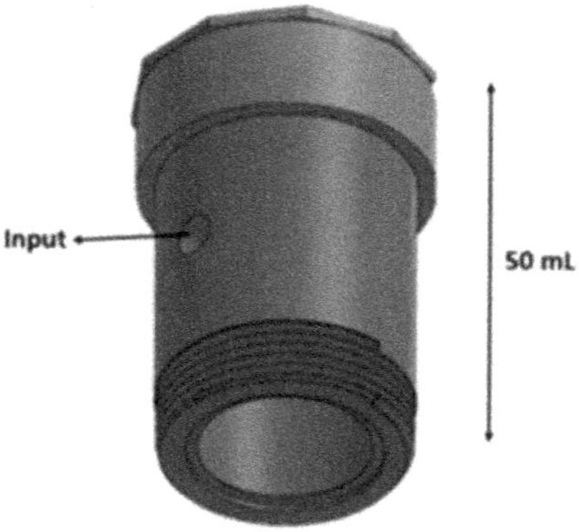

(**a**)

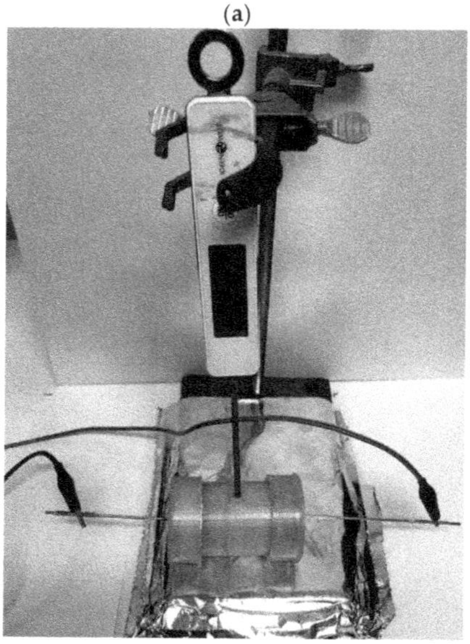

(**b**)

Figure 2. (**a**) 3D reactor for ohmic heating processing. (**b**) Photo of the ohmic heating reactor.

2.8. Experimental Design to Optimize the Phycocyanin Extraction by Ohmic Heating

Central compound rotatable design (CCRD) was applied. Temperature, time, and pH were studied as factors related to phycocyanin extraction and its optimization. The temperature applied was in the range of 30–80 °C. For the extraction of cyano-bacteria pigments, the temperature used in ohmic heating is 70 °C [40]; this temperature is related to a greater electroporation phenomenon and, consequently, permeation of the cell membrane for the release of pigments; lower temperatures in the 40–50 °C range are related to the extraction of phycocyanin [41]. The pH in the extraction of phycocyanin is in the range of above 5 and below 8, and the stability of these pigments is best in the range of 6–7 [42]. The time used for the extraction of cyanobacteria pigments is found in the range of 5–60 min [40,43]. Table 2 shows the levels for factors present in the CCRD with the coded levels. A total of 18 experimental runs were performed in random order. The response surface methodology (RSM) was used to optimize the studied factor from the second-order polynomic Equation (2).

The independent variables and levels (coded and un-coded) presented in Table 2 were used for the central compound design (Equation (2)).

$$Y = \beta_0 + \sum_{i=1}^{k} \beta_i X_i + \sum_{i<j}^{k} \beta_{ij} X_i^2 + \sum\sum \beta_{ij} X_i X_j + \varepsilon \quad (2)$$

where Y represents the predicted response (phycocyanin content) and the independent variables correspond to X_1, X_2, and X_3 [44]. The regression coefficients of variables for the intercept are as follows: β_0 is a constant, β_i are linear coefficients, β_{ii} are quadratic coefficients, and β_{ij} are the coefficients of the interaction and $p < 0.05$. All the experimental assays were analyzed by the MATLAB r20217a software.

2.9. Statistical Analysis for the Evaluation of Rice Husk Extract as a Culture Medium

For the selection of the best treatment for the growth of *S. platensis* using rice husk as a culture medium and the analysis of the optimization of the phycocyanin extraction, a statistical analysis was applied. The analysis was performed using ANOVA followed by an analysis of the comparison of means by the Tukey method. All the analyses were performed considering 95% confidence and $p \leq 0.05$. The software used was developed by Prof. Emilio Olivares-Sáenz at the School of Agronomy, Universidad Autónoma de Nuevo León (UANL, San Nicolás de las Garza, Mexico). The treatments used were duplicated using the mean of the replicates, with the standard deviation in the error bars.

3. Results

3.1. Rice Husk and Phycocyanin Composition Analysis by FTIR

Fourier-transform analysis (FTIR) was used to detect the vibration of molecules present in the rice husk extract (carbohydrates, mainly starch) and phycocyanin (signals of amine I, II, and III). Please see Figure 3.

(a) I. The FTIR of the biomass of *S. platensis* presents different signals corresponding to different vibrations as follows: groups formed by stretching vibrations of -OH and -NH (3265 cm^{-1}); stretching vibrations of -CH (2800–2900 cm^{-1}); stretches of -CO corresponding to ketones, aldehydes mainly (1700–1400 cm^{-1}); stretching vibrations of -C-O and -C-C (1300–1000 cm^{-1}); and some signals corresponding to -*p*-O and -S-O (750–900 cm^{-1}) [45]. *S. platensis* biomass shows a very similar composition; no evidence of difference is shown between the biomass obtained using the different cultures (rice husk extract and Zarrouk media).

II. The structural composition of phycocyanin remains consistent regardless of the growth conditions. This assertion applies consistently to phycocyanin extracted from biomass cultivated under control conditions (utilizing Zarrouk medium), as well as from biomass cultivated using rice husk extract. The composition of the phycocyanin corresponds to the following vibration signals: N-H amide I band (1640 cm^{-1}), amide II (1500 cm^{-1}), amide III (1460 cm^{-1}), carbohydrates such as starch (1200 cm^{-1}), and some signs of carbonates (700–850 cm^{-1}) [46].

(b) I. Starch shows the typical trace of carbohydrates in the zones of 900–1200 cm^{-1} and 1200 cm^{-1} [37]. The rice husk presents signals in the carbohydrates zone (900–1200 cm^{-1}), mainly starch, a protein composition with vibration signals of -NH close to 1500–1600 cm^{-1}, and signals corresponding to silica (780–800 cm^{-1}) [47]. The rice husk has components such as a source of nitrogen (presence of amine) and a source of carbohydrate (mainly starch) necessary for the growth of *S. platensis* biomass [38]. Enzymes commonly present in cyanobacteria are amylase, cellulases, and β-glucosidases; these types of enzymes could hydrolyze the glycosidic β bonds 1–4 of the chains of amylose and amylopectin present in the starch and could explain how *S. platensis* obtains glucose as a source of carbon. The presence of glucose can increase the yield of *S. platensis* biomass and increase the production of phycocyanin [48,49].

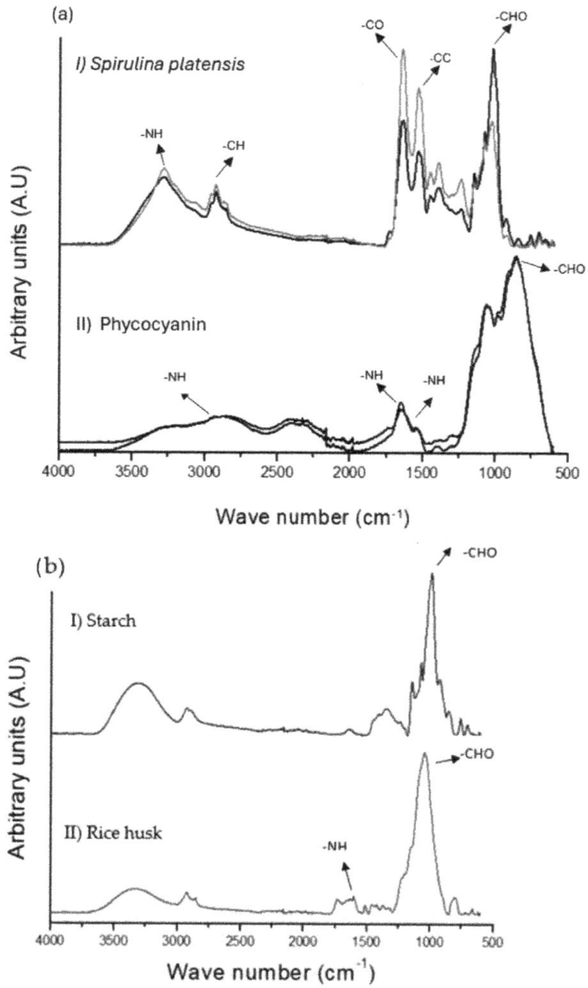

Figure 3. (**a**) FTIR analysis of the *S. platensis* biomass (II) and phycocyanin structure present in *S. platensis* biomass (II). (**b**) FTIR analysis of (I) the starch and (II) rice husk.

3.2. Ash Content and Elemental Composition of Rice Husk Extract by X-ray Fluorescence Spectrometry

The total solids content present in the rice husk extract was 0.90 ± 0.03 g/L. The elemental composition of rice husk extract (RHE) shows important compounds such as potassium at 22.48 ± 0.06%, silica at 1.95 ± 0.06%, magnesium at 2.17 ± 0.06%, phosphorous at 1.95 ± 0.06%, calcium at 1.83 ± 0.05%, iron at 0.03 ± 0.00%, and metal traces. The heating extraction of the components present in the rice husk extract was effective for the solubilization of nutrients necessary for the growth of *S. platensis*.

The presence of components such as nitrogen and a carbon source are essential for microalgae growth [48]. The trace metals are related to microalgae growth, especially the Fe^{2+}, Cu^{2+}, and Zn^{2+} present in the RHE. These metals are also associated with the production of pigments in microalgae [49]. $NaNO_3$ has an important role in protein production, including that of phycobiliproteins. To produce phycocyanin, the presence of Fe^{2+} is fundamental; even though this element is not a structural part of phycocyanin, it is an important precursor of phycocyanin and co-factor in several enzymes. The enzyme ferrochelatase inserts a ferrous ion to produce a stable structure of heme. These structures are opened with an oxygenase enzyme, resulting in biliverdin IX α, a precursor of the phycobiliproteins [44].

3.3. *Total Carbohydrates Present in Rice Husk Extract*

The total carbohydrate content of the rice husk extract determined by the Anthrone method was 143 ± 0.07 mg/L, representing 31.8% ± 0.01 of the total solids. These carbohydrates in the rice husk are primarily composed of glucose and xylose.

3.4. *Effect of the Concentrations of Rice Husk Extract on the Concentration of* S. platensis *Biomass*

Seven treatments were performed for the evaluation of the effect of the rice husk extract (RHE) on the concentration of the *S. platensis* biomass (carried out by duplicate); all the treatments had similar growth behavior, with marked growth on the eighth day (please see Figure 4). The range of biomass for the treatments of rice husk extract was 1.25–1.75 g/L. The highest concentration corresponds to treatment 4 (rice husk extract 40%) with a biomass content of 1.75 ± 0.71 g/L. The lowest yield corresponds to treatment 1 (0% rice husk) with a total yield of 0.9 ± 0.141 g/L. The treatments with rice husk show a better or equal yield in comparison with the Zarrouk medium; the control treatment has a content of biomass of 1.55 ± 0.03. When using Zarrouk as a culture medium, usually the biomass is in the range of 0.6–2.5 g/L, depending on the modified nutrients present in the Zarrouk medium [18]. The culture that contained different concentrations of rice husk extract and $NaNO_3$ was efficient for the growth of *S. platensis* biomass, which is an alternative to growing microalgae. This is because the rice husk has nutrients that are present in the extract, such as sources of SiO_2, sources of C (starch), and traces of elements such as Ca, Mg, Fe, and Zn, mainly [48,49]. At an appropriate pH of 7.5, all the present conditions promote the growth of *S. platensis* using rice husk extract as a culture medium.

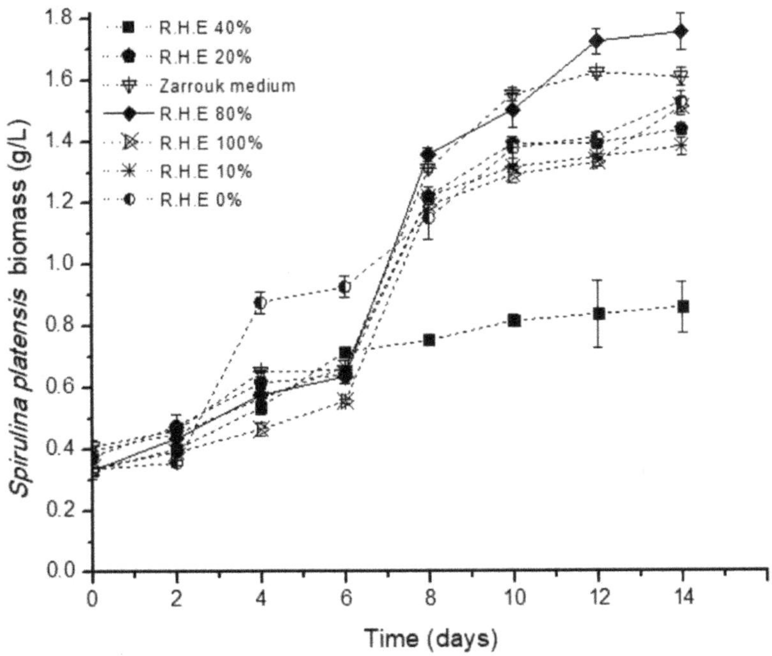

Figure 4. Effect of the different concentrations of rice husk extract (RHE) on the growth of *S. platensis* biomass, and the control Zarrouk's medium.

3.5. Effect of the Concentrations of Rice Husk Extract on the Growth Rate, Protein Content, and Phycocyanin Content of S. platensis

To evaluate the effect of the rice husk extract as a culture medium, the protein content, phycocyanin content, and growth rate of *S. platensis* were analyzed. Table 3 shows the treatments analyzed. The highest specific growth rates are present in treatments 4 (40% *v/v* RHE) and 3 (20% *v/v*, RHE), with 0.125 ± 0.000 and 0.115 ± 0.001, respectively, followed by treatment 1 corresponding to Zarrouk medium. The use of rice husk as a culture medium proves to be suitable for the growth of *S. platensis* biomass compared to the synthetic Zarrouk medium.

Table 3. Effect of the rice husk extract (RHE) on the specific growth, protein content, and phycocyanin content in the biomass of *S. platensis*.

Treatments/Rice Husk (%)	Specific Growth Rate μ (h^{-1})	Protein Content (mg/g)	Phycocyanin Content (mg/g)
T1 Zarrouk	0.109 ± 0.003 ^{c}	608.940 ± 12.720 ^{b}	25.25 ± 2.600 ^{ab}
T2 R.H 10	0.107 ± 0.002 ^{e}	499.000± 24.030 ^{c}	31.46 ± 0.750 ^{a}
T3 R.H 20	0.115 ± 0.001 ^{b}	544.550± 22.690 ^{c}	29.31 ± 0.240 ^{a}
T4 R.H 40	0.125 ± 0.000 ^{a}	672.270 ± 0.260 ^{a}	28.47 ± 3.040 ^{a}
T5 R.H 80	0.108 ± 0.002 ^{d}	502.330 ± 8.810 ^{c}	21.55 ± 1.890 ^{bc}
T6 R.H 100	0.099 ± 0.002 ^{f}	591.220 ± 17.100 ^{b}	23.24 ± 0.260 ^{c}
T7 Water *	0.055 ± 0.038 ^{g}	326.770 ± 17.100 ^{d}	19.83 ± 6.380 ^{ab}

Notes: values are presented as the means ± standard deviations. Values in the same column with different letters are significantly different (Tukey's analysis, $p \leq 0.05$). * RH (0%) = water = control.

The highest specific rates show that rice husk extract as a culture medium for *S. platensis* growth is better than the Zarrouk medium. The protein contents present in all the treatments show that the highest content is present in treatment 4, at 672.270 ± 0.260 mg/g, followed by treatment 1 corresponding to Zarrouk medium. The content of protein expected for *S. platensis* biomass in the literature is reported to be in the range of 600–700 mg/g [22].

Finally, phycocyanin is present in all the treatments. The highest phycocyanin content is 25–31 mg/g in treatments 2, 3, and 4, followed by treatment 1. The phycocyanin extracted by the freezing and thawing method is in the range of 19–73 mg/g of *S. platensis* biomass [50]. The use of rice husk extract as a culture medium for the growth of *S. platensis* is a viable method with some advantages such as easy preparation of the culture medium and lack of need to adjust the pH (7.6). The rice husk extract is an interesting alternative for the growth of *S. platensis* biomass; the composition of the different polysaccharides leads to the opportunity for the use of rice husk for the extraction of high-value compounds such as oligosaccharides and fermentable sugars to produce bioethanol for second-generation biorefineries, as well as culture media for microalgae.

3.6. Central Composite Rotatable Design Model for the Optimization of Phycocyanin Extraction Using Ohmic Heating

The effect of temperature, time, and pH on phycocyanin extraction using ohmic heating was investigated using CCRD. The levels of the independent variables used in the experimental design were determined based on preliminary experiments. Table 4 shows all the experimental runs (18), and the response (Y) of phycocyanin extraction was analyzed. The highest concentration of phycocyanin was extracted when the variables were in the central point as follows: pH 6.5, 10 min, and 55 °C. The highest concentrations of phycocyanin were in the range of 52–68 mg/g.

Table 4. Effects of the different experimental runs on phycocyanin extraction.

Run	X_1	X_2	X_3	Phycocyanin (mg/g)
1	0	0	0	64.70
2	0	0	−1.68	33.70
3	−1	−1	1	39.80
4	−1.68	0	0	44.20
5	0	0	0	68.40
6	1.68	0	0	7.90
7	−1	−1	−1	24.30
8	−1	1	1	43.10
9	1	−1	−1	7.70
10	−1	1	−1	20.80
11	0	0	0	52.50
12	0	−1.68	0	55.00
13	1	−1	1	12.30
14	1	1	−1	12.00
15	1	1	1	7.80
16	0	0	1.68	8.00
17	0	0	0	52.15
18	0	1.68	0	15.20

Coded variables: X_1 = temperature (°C), X_2 = time (min), X_3 = (pH).

The linearity and quadratic effect of the treatments on the response variables were obtained by ANOVA and are shown in Table 5. The independent variables (temperature b2, time b3, and pH b4) are significant ($p < 0.05$). However, the interaction of the factors was not significant ($p > 0.05$), as is presented in Table 5. The mathematical model obtained from the coefficients in Table 5 has an R^2 of 0.83 and a root mean square error (RMSE) of 12.6. Such values are acceptable for the extraction of phycocyanin by ohmic heating.

Table 5. Analysis of variance (ANOVA) of the second-order polynomial model for the extraction of phycocyanin using ohmic heating.

Source	Estimated	Stnd. Err	tStat	*p*-Value
b1	59.700	6.257	9.541	0.000
b2	−21.915	6.826	−3.210	0.012
b3	−9.741	6.938	−1.403	0.197
b4	−28.787	7.041	−4.088	0.003
b5	−25.494	7.134	−3.573	0.007
b6	−20.710	7.617	−2.718	0.02
b7	−0.724	6.803	−0.106	0.917
b8	0.016	8.885	0.001	0.99
b9	−9.344	8.885	−1.05	0.323
b10	−0.513	8.885	−0.057	0.955

The predictive equation for the yield of phycocyanin (Equation (3)) is as follows:

$$\text{Phycocyanin (mg/g)} = 59.65 - 10.92\,X_1 - 4.93\,X_2 - 0.36\,X_3 - 12.75\,X_1^2 - 9.55\,X_2^2 - 14.59\,X_3^2 - 0.25\,X_1X_3 - 4.67\,X_2X_3 \quad (3)$$

A three-dimensional response surface plot is presented in Figure 5. The effects of the variables on the extraction of phycocyanin are found in Figure 5a–c. To date, several reports exist on the use of ohmic heating for the extraction of different compounds; however, the implementation of these technologies for phycocyanin extraction has been limited. For the use of ohmic heating technology, the conditions reported were a temperature of 44 °C/ 30 min and 37 °C/60 min [40,50]. Under these operational conditions, the highest phycocyanin contents were 45 and 40 mg/g, respectively. In the present study, the experimental runs that were close to the reported values were runs 1, 3, 4, and 8. They present yields of phycocyanin in the range of 33–43 mg/g.

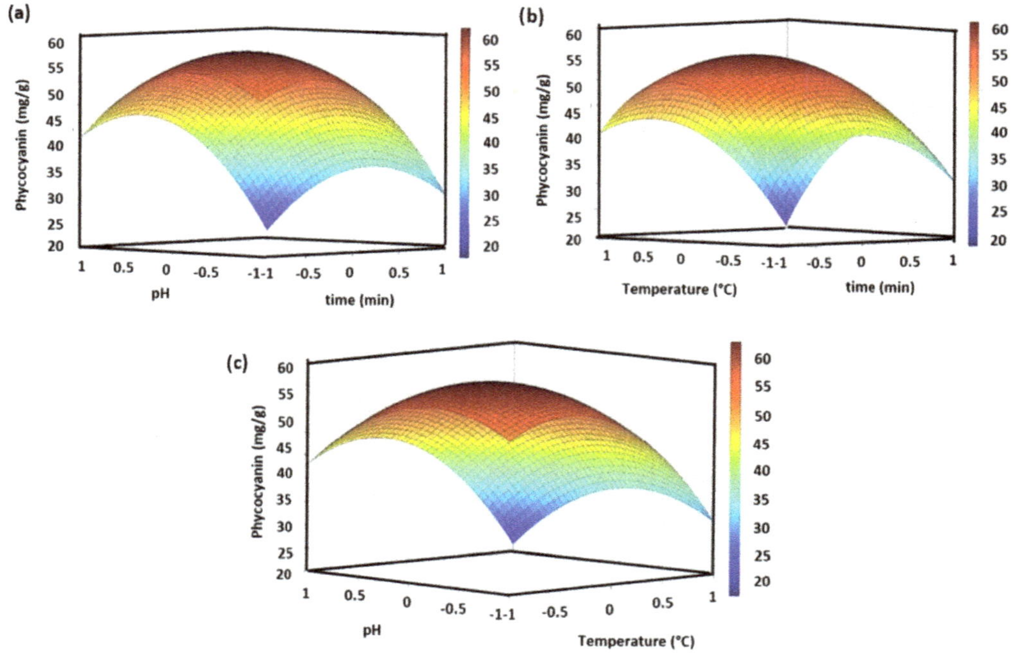

Figure 5. Effect of the variables (**a**–**c**) pH, time, and temperature on the ohmic heating for the content of phycocyanin present in the *S. platensis* biomass.

In runs 2, 3, and 8, similarities were found regarding high temperature values (≥40 °C) and the same time (10 min); however, the three runs presented high pH values of ≥7. A basic pH value can affect the stability of the pigment [45].

Although the phenomenon of electroporation and cellular permeability (due to heat and time) may be present, the pigment may have difficulties being released because there are no differences in the isoelectric point of phycocyanin (4.6–5.2) [46] and the phosphate buffer at ≥7 pH [45]. Experimental run 4 presents an adequate pH (6.5) and time (10 min); however, the low temperature (≤40 °C) presented difficulty for the electroporation phenomenon and membrane permeability. It is important to mention that the temperature values of experimental runs 3, 4, and 8 are very similar to those reported in the literature [39,50].

This indicates that low temperatures (≤40 °C) are inadequate for achieving higher yields of phycocyanin. Experimental runs 6, 9, and 15 were exposed to a high temperature of 70 °C. Under this temperature, although membrane permeability and electroporation are favored, pigment degradation was observed, with a yield close to 7 mg/g. These treatments obtained the lowest amounts of pigment and lost the characteristic blue color of phycocyanin.

For the highest content of phycocyanin, the variables are close to the central points, as follows: time of 10 min, temperature of 55 °C, and pH of 6.5. The highest concentrations of phycocyanin are in the range of 52–68 mg/g.

The content of phycocyanin increased when using the optimization process, reaching a total of 75.80 ± 0.98 mg/g. The results indicate that under these conditions, it is possible to extract a greater amount of phycocyanin. The predictive equation indicated that the expected content of phycocyanin was 62.54 mg/g. The difference between the theoretical and actual values is mainly due to the adjustment of the model used.

3.7. Optimization of the Extraction of Phycocyanin Using Ohmic Heating

The independent variables influence the extraction of phycocyanin present in the *S. platensis* biomass ($p < 0.05$). The predicted model showed an optimal concentration of phycocyanin of 62.64 (mg/g) when the temperature, time, and pH were 50.71 °C, 8.60 min, and 6.52, respectively. However, when the phycocyanin concentration was evaluated, the optimal yield was 75.79 ± 0.98 (mg/g).

The results present in the experimental design showed a higher concentration of phycocyanin in comparison with the predicted model. The recovery yield of phycocyanin present in the ohmic heating process is higher in comparison with the freezing and thawing method using rice husk extract as a culture medium in both cases, with the freezing and thawing method showing a phycocyanin concentration in the range of 25–31 mg/g. The optimal conditions of the CCDR showed that extraction needs to be carried out at 50 °C. To maintain the structure of the proteins present, it is recommended that the temperature does not exceed 60 °C, as higher temperatures have been reported to denature proteins. It is important to mention that the exposure time at 50 °C in this study was separated into short extraction periods (8 min), so critical temperatures were not reached [47]. Temperatures above 60 °C can denature the secondary, tertiary, and quaternary structures of allophycocyanin, affecting the color and stability of phycocyanin [18,50].

The temperature used for the phycocyanin extraction showed an impact on the cell wall membrane, and the electroporation under these conditions was higher in comparison with that at lower temperatures; the consequence of membrane permeability causes the release of the pigments present in the biomass of *S. platensis*. The appropriate temperature for this purpose is 50 °C, and factors such as a pH of 6.5 are essential for the ionic interaction of phycocyanin and the buffer related to the release of this protein pigment. On the other hand, ohmic heating processing is a promising alternative to use at other scales of operation in the extraction of compounds. However, scaling-up strategies need to be studied. Furthermore, several advantages can be attributed to this technology, such as the construction materials for the reactor and the possible energy consumption, which, by using relatively low temperatures, could be low. Finally, phycocyanin extraction showed good yields compared to conventional methods; electroporation and permeabilization of the *S. platensis* membrane under optimized conditions allowed the release of phycocyanin into the liquid medium. Furthermore, the ohmic heating process is an alternative for the release of other intracellular compounds, not only phycocyanin, in short periods of operation.

4. Conclusions

This work showed the development of a bioprocess for the cultivation and growth of *S. platensis* in an alternative medium such as rice husk extract. The use of this agro-industrial waste allows the integration of the second and third generation of biorefineries. In addition, the design and construction of a 3D-printed reactor was efficient in the extraction of phycocyanin from the cultivated *S. platensis* and had good performance under the operating conditions studied. Furthermore, the electroporation phenomenon in addition to membrane permeability allows compounds to be extracted without altering their properties, as they are released into the medium (buffer) through ionic interaction under the study conditions present in ohmic heating. This development will provide the basis for scaling up the process and obtaining compounds of industrial interest, such as phycocyanin, under a circular bioeconomy concept and incorporating them into a food matrix.

Author Contributions: Conceptualization, G.C.-I., R.M.R.-J., A.C.-G. and H.A.R.; methodology, G.C.-I., G.R.-C., A.C.-G., R.B. and S.M.-C.; software, G.C.-I., G.R.-C., R.M.R.-J., S.M.-C. and H.A.R.; validation, G.R.-C., R.M.R.-J. and H.A.R.; data analysis, G.C.-I., R.M.R.-J. and H.A.R.; investigation, G.C.-I., J.C.C.-E., A.C.-G., G.R.-C., R.M.R.-J. and H.A.R.; resources, R.M.R.-J., R.B., J.C.C.-E. and H.A.R.;

writing—original draft preparation, G.C.-I.; writing—review and editing, G.C.-I., R.M.R.-J. and H.A.R.; visualization, R.M.R.-J. and H.A.R.; supervision, R.M.R.-J. and H.A.R.; project administration, R.M.R.-J. and H.A.R.; funding acquisition, R.M.R.-J. and H.A.R. All authors have read and agreed to the published version of the manuscript.

Funding: This research was supported by the National Council of Humanities, Sciences, and Technologies (CONAHCYT, Mexico) Infrastructure Project—FOP02-2021-04, grant number 317250, and the Consejo Estatal de Ciencia y Tecnología (COECYT, Coahuila, Mexico) with the "Fondo Destinado a Promover el Desarrollo de la Ciencia y la Tecnología en el Estado de Coahuila (FONCYT)" Project "Sustainable cultivation of *S. platensis* on a large scale for use in livestock feed"—COAH-2022-C19-C123.

Institutional Review Board Statement: Not applicable.

Informed Consent Statement: Not applicable.

Data Availability Statement: The original contributions presented in the study are included in the article, further inquiries can be directed to the corresponding author.

Acknowledgments: The authors Gabriela Cid-Ibarra, Samanta Machado-Cepeda, and Alejandra Cabello-Galindo thank the National Council of Humanities, Sciences, and Technologies (CONAHCYT, Mexico) for their Ph.D. fellowship support (grant numbers 863516, 1154370, and 711463, respectively). The authors thank Luciane M. Colla (University of Passo Fundo, Brazil) for kindly providing the *S. platensis* microalgae strain. Also, the authors thank Claudia López Badillo (Autonomous University of Coahuila) for the elemental composition analysis of rice husk.

Conflicts of Interest: The authors declare no conflict of interest.

References

1. Sen, T.; Barrow, C.J.; Deshmukh, S.K. Microbial Pigments in the Food Industry—Challenges and the Way Forward. *Front. Nutr.* **2019**, *6*, 7. [CrossRef] [PubMed]
2. Dilrukshi, P.G.T.; Munasinghe, H.; Silva, A.B.G.; De Silva, P.G.S.M. Identification of Synthetic Food Colours in Selected Confectioneries and Beverages in Jaffna District, Sri Lanka. *J. Food Qual.* **2019**, 7453169. [CrossRef]
3. Mota, I.G.C.; Neves, R.A.M.D.; Nascimento, S.S.D.C.; Maciel, B.L.L.; Morais, A.H.D.A.; Passos, T.S. Artificial Dyes: Health Risks and the Need for Revision of International Regulations. *Food Rev. Int.* **2023**, *39*, 1578–1593. [CrossRef]
4. Martelli, G.; Folli, C.; Visai, L.; Daglia, M.; Ferrari, D. Thermal Stability Improvement of Blue Colorant C-Phycocyanin from *Spirulina Platensis* for Food Industry Applications. *Process Biochem.* **2014**, *49*, 154–159. [CrossRef]
5. Roda-Serrat, M.C.; Christensen, K.V.; El-Houri, R.B.; Fretté, X.; Christensen, L.P. Fast Cleavage of Phycocyanobilin from Phycocyanin for Use in Food Colouring. *Food Chem.* **2018**, *240*, 655–661. [CrossRef] [PubMed]
6. Chappell, G.A.; Britt, J.K.; Borghoff, S.J. Systematic Assessment of Mechanistic Data for FDA-Certified Food Colors and Neurodevelopmental Processes. *Food Chem. Toxicol.* **2020**, *140*, 111310. [CrossRef] [PubMed]
7. Leuschner, R.G.K.; Robinson, T.P.; Hugas, M.; Cocconcelli, P.S.; Richard-Forget, F.; Klein, G.; Licht, T.R.; Nguyen-The, C.; Querol, A.; Richardson, M. Qualified Presumption of Safety (QPS): A Generic Risk Assessment Approach for Biological Agents Notified to the European Food Safety Authority (EFSA). *Trends Food Sci. Technol.* **2010**, *21*, 425–435. [CrossRef]
8. Nabi, B.G.; Mukhtar, K.; Ahmed, W.; Manzoor, M.F.; Ranjha, M.M.A.N.; Kieliszek, M.; Bhat, Z.F.; Aadil, R.M. Natural Pigments: Anthocyanins, Carotenoids, Chlorophylls, and Betalains as Colorants in Food Products. *Food Biosci.* **2023**, *52*, 102403. [CrossRef]
9. Di Salvo, E.; Lo Vecchio, G.; De Pasquale, R.; De Maria, L.; Tardugno, R.; Vadalà, R.; Cicero, N. Natural Pigments Production and Their Application in Food, Health and Other Industries. *Nutrients* **2023**, *15*, 1923. [CrossRef] [PubMed]
10. Nwoba, E.G.; Ogbonna, C.N.; Ishika, T.; Vadiveloo, A. Microalgal Pigments: A Source of Natural Food Colors. In *Microalgae Biotechnology for Food, Health and High Value Products*; Springer: Berlin/Heidelberg, Germany, 2020; pp. 81–123.
11. Prabakaran, G.; Sampathkumar, P.; Kavisri, M.; Moovendhan, M. Extraction and Characterization of Phycocyanin from *Spirulina platensis* and Evaluation of Its Anticancer, Antidiabetic and Antiinflammatory Effect. *Int. J. Biol. Macromol.* **2020**, *153*, 256–263. [CrossRef] [PubMed]
12. Yuan, B.; Li, Z.; Shan, H.; Dashnyam, B.; Xu, X.; McClements, D.J.; Zhang, B.; Tan, M.; Wang, Z.; Cao, C. A Review of Recent Strategies to Improve the Physical Stability of Phycocyanin. *Curr. Res. Food Sci.* **2022**, *5*, 2329–2337. [CrossRef] [PubMed]
13. Imchen, T.; Singh, K.S. Marine Algae Colorants: Antioxidant, Anti-Diabetic Properties and Applications in Food Industry. *Algal. Res.* **2023**, *69*, 102898. [CrossRef]
14. Hsieh-Lo, M.; Castillo, G.; Ochoa-Becerra, M.A.; Mojica, L. Phycocyanin and Phycoerythrin: Strategies to Improve Production Yield and Chemical Stability. *Algal. Res.* **2019**, *42*, 101600. [CrossRef]
15. Ashaolu, T.J.; Samborska, K.; Lee, C.C.; Tomas, M.; Capanoglu, E.; Tarhan, Ö.; Taze, B.; Jafari, S.M. Phycocyanin, a Super Functional Ingredient from Algae; Properties, Purification Characterization, and Applications. *Int. J. Biol. Macromol.* **2021**, *193*, 2320–2331. [CrossRef] [PubMed]

16. Gorgich, M.; Passos, M.L.C.; Mata, T.M.; Martins, A.A.; Saraiva, M.L.M.F.S.; Caetano, N.S. Enhancing Extraction and Purification of Phycocyanin from Arthrospira Sp. with Lower Energy Consumption. *Energy Rep.* **2020**, *6*, 312–318. [CrossRef]
17. Ferreira-Santos, P.; Genisheva, Z.; Pereira, R.N.; Teixeira, J.A.; Rocha, C.M.R. Moderate Electric Fields as a Potential Tool for Sustainable Recovery of Phenolic Compounds from Pinus Pinaster Bark. *ACS Sustain. Chem. Eng.* **2019**, *7*, 8816–8826. [CrossRef]
18. Pez Jaeschke, D.; Rocha Teixeira, I.; Marczak, D.F.; Domeneghini Mercali, G. Phycocyanin from *Spirulina:* A Review of Extraction Methods and Stability. *Food Res. Int.* **2021**, *143*, 110314. [CrossRef] [PubMed]
19. Tan, H.T.; Khong, N.M.H.; Khaw, Y.S.; Ahmad, S.A.; Yusoff, F.M. Optimization of the Freezing-Thawing Method for Extracting Phycobiliproteins from *Arthrospira* sp. *Molecules* **2020**, *25*, 3894. [CrossRef] [PubMed]
20. Alkanan, Z.T.; Altemimi, A.B.; Al-Hilphy, A.R.S.; Watson, D.G.; Pratap-Singh, A. Ohmic Heating in the Food Industry: Developments in Concepts and Applications during 2013–2020. *Appl. Sci.* **2021**, *11*, 2507. [CrossRef]
21. Patel, A.; Singh, M. Ohmic Heating for Food Products—A Review. *Curr. Appl. Sci. Technol.* **2018**, *27*, 1–7. [CrossRef] [PubMed]
22. Kumar, A.; Ramamoorthy, D.; Verma, D.K.; Kumar, A.; Kumar, N.; Kanak, K.R.; Marwein, B.M.; Mohan, K. Antioxidant and Phytonutrient Activities of *Spirulina Platensis. Energy Nexus* **2022**, *6*, 100070. [CrossRef]
23. Khan, S.; Das, P.; Thaher, M.I.; Abdul Quadir, M.; Mahata, C.; Al Jabri, H. Utilization of Nitrogen-Rich Agricultural Waste Streams by Microalgae for the Production of Protein and Value-Added Compounds. *Curr. Opin. Green Sustain. Chem.* **2023**, *41*, 100797. [CrossRef]
24. Nur, M.M.A.; Rahmawati, S.D.; Sari, I.W.; Achmad, Z.; Setyoningrum, T.M.; Jaya, D.; Murni, S.W.; Djarot, I.N. Enhancement of Phycocyanin and Carbohydrate Production from *Spirulina platensis* Growing on Tofu Wastewater by Employing Mixotrophic Cultivation Condition. *Biocatal. Agric. Biotechnol.* **2023**, *47*, 102600. [CrossRef]
25. Girard, J.-M.; Roy, M.-L.; Hafsa, M.B.; Gagnon, J.; Faucheux, N.; Heitz, M.; Tremblay, R.; Deschênes, J.-S. Mixotrophic Cultivation of Green Microalgae *Scenedesmus Obliquus* on Cheese Whey Permeate for Biodiesel Production. *Algal. Res.* **2014**, *5*, 241–248. [CrossRef]
26. Shanthi, G.; Premalatha, M.; Anantharaman, N. Potential Utilization of Fish Waste for the Sustainable Production of Microalgae Rich in Renewable Protein and Phycocyanin-*Arthrospira platensis/Spirulina. J. Clean Prod.* **2021**, *294*, 126106. [CrossRef]
27. Lim, J.S.; Abdul Manan, Z.; Wan Alwi, S.R.; Hashim, H. A Review on Utilisation of Biomass from Rice Industry as a Source of Renewable Energy. *Renew. Suste Energy Rev.* **2012**, *16*, 3084–3094. [CrossRef]
28. Goodman, B.A. Utilization of Waste Straw and Husks from Rice Production: A Review. *J. Bioresour. Bioprod.* **2020**, *5*, 143–162. [CrossRef]
29. Badar, R.; Qureshi, S.A. Composted Rice Husk Improves the Growth and Biochemical Parameters of Sunflower Plants. *J Bot.* **2014**, *2014*, 427648. [CrossRef]
30. Soltani, N.; Bahrami, A.; Pech-Canul, M.I.; González, L.A. Review on the Physicochemical Treatments of Rice Husk for Production of Advanced Materials. *J. Chem. Eng.* **2015**, *264*, 899–935. [CrossRef]
31. Rosero-Chasoy, G.; Rodríguez-Jasso, R.M.; Aguilar, C.N.; Buitrón, G.; Chairez, I.; Ruiz, H.A. Hydrothermal Kinetic Modeling for Microalgae Biomass under Subcritical Condition Cultivated in a Close Bubble Tubular Photobioreactor. *Fuel* **2023**, *334*, 126585. [CrossRef]
32. Wanyo, P.; Kaewseejan, N.; Meeso, N.; Siriamornpun, S. Bioactive Compounds and Antioxidant Properties of Different Solvent Extracts Derived from Thai Rice By-Products. *Appl. Biol. Chem.* **2016**, *59*, 373–384. [CrossRef]
33. Darvehei, P.; Bahri, P.A.; Moheimani, N.R. Model Development for the Growth of Microalgae: A Review. *Renew. Sustain. Energy Rev.* **2018**, *97*, 233–258. [CrossRef]
34. Vázquez-Vuelvas, O.F.; Chávez-Camacho, F.A.; Meza-Velázquez, J.A.; Mendez-Merino, E.; Ríos-Licea, M.M.; Contreras-Esquivel, J.C. A Comparative FTIR Study for Supplemented Agavin as Functional Food. *Food Hydrocoll.* **2020**, *103*, 105642. [CrossRef]
35. García-García, J.D.; Anguiano-Cabello, J.C.; Arredondo-Valdés, R.; Candido del Toro, C.A.; Martínez-Hernández, J.L.; Segura-Ceniceros, E.P.; Govea-Salas, M.; González-Chávez, M.L.; Ramos-González, R.; Esparza-González, S.C. Phytochemical Characterization of *Phoradendron Bollanum* and *Viscum album* subs. *austriacum* as Mexican Mistletoe Plants with Antimicrobial Activity. *Plants* **2021**, *10*, 1299. [CrossRef] [PubMed]
36. Shiva; Rodríguez-Jasso, R.M.; Rosero-Chasoy, G.; López-Sandin, I.; Morais, A.R.C.; Ruiz, H.A. Enzymatic Hydrolysis, Kinetic Modeling of Hemicellulose Fraction, and Energy Efficiency of Autohydrolysis Pretreatment Using Agave Bagasse. *Bioenergy Res.* **2023**, *16*, 75–87. [CrossRef]
37. Rosero-Chasoy, G.; Rodríguez-Jasso, R.M.; Aguilar, C.N.; Buitrón, G.; Chairez, I.; Ruiz, H.A. Growth Kinetics and Quantification of Carbohydrate, Protein, Lipids, and Chlorophyll of *Spirulina platensis* under Aqueous Conditions Using Different Carbon and Nitrogen Sources. *Bioresour. Technol.* **2022**, *346*, 126456. [CrossRef] [PubMed]
38. Lowry, O.H.; Rosebrough, N.J.; Farr, A.L.; Randall, R.J. Protein measurement with the Folin phenol reagent. *Renew. Sustain. Energy Rev.* **1951**, *193*, 265–275. [CrossRef]
39. Pagels, F.; Pereira, R.N.; Amaro, H.M.; Vasconcelos, V.; Guedes, A.C.; Vicente, A.A. Continuous Pressurized Extraction versus Electric Fields-Assisted Extraction of Cyanobacterial Pigments. *J. Biotechnol.* **2021**, *334*, 35–42. [CrossRef] [PubMed]
40. Ferreira-Santos, P.; Miranda, S.M.; Belo, I.; Spigno, G.; Teixeira, J.A.; Rocha, C.M.R. Sequential Multi-Stage Extraction of Biocompounds from *Spirulina platensis*: Combined Effect of Ohmic Heating and Enzymatic Treatment. *Innov. Food Sci. Emerg. Technol.* **2021**, *71*, 102707. [CrossRef]

41. Kassama, L.S.; Shi, J.; Mittal, G.S. Optimization of Supercritical Fluid Extraction of Lycopene from Tomato Skin with Central Composite Rotatable Design Model. *Sep. Purif. Technol.* **2008**, *60*, 278–284. [CrossRef]
42. Ganim, Z.; Chung, H.S.; Smith, A.W.; DeFlores, L.P.; Jones, K.C.; Tokmakoff, A. Amide I Two-Dimensional Infrared Spectroscopy of Proteins. *Chem. Res.* **2008**, *41*, 432–441. [CrossRef] [PubMed]
43. Bhayani, K.; Mitra, M.; Ghosh, T.; Mishra, S. C-Phycocyanin as a Potential Biosensor for Heavy Metals like Hg 2+ in Aquatic Systems. *RSC Adv.* **2016**, *6*, 111599–111605. [CrossRef]
44. Brasil, B.D.S.A.F.; de Siqueira, F.G.; Salum, T.F.C.; Zanette, C.M.; Spier, M.R. Microalgae and Cyanobacteria as Enzyme Biofactories. *Algal Res.* **2017**, *25*, 76–89. [CrossRef]
45. Chen, T.; Zheng, W.; Wong, Y.-S.; Yang, F.; Bai, Y. Accumulation of Selenium in Mixotrophic Culture of *Spirulina platensis* on Glucose. *Bioresour. Technol.* **2006**, *97*, 2260–2265. [CrossRef] [PubMed]
46. Chittapun, S.; Jonjaroen, V.; Khumrangsee, K.; Charoenrat, T. C-Phycocyanin Extraction from Two Freshwater Cyanobacteria by Freeze Thaw and Pulsed Electric Field Techniques to Improve Extraction Efficiency and Purity. *Algal. Res.* **2020**, *46*, 101789. [CrossRef]
47. Pereira, R.N.; Rodrigues, R.M.; Ramos, Ó.L.; Malcata, X.; Teixeira, J.A.; Vicente, A.A. Production of Whey Protein-Based Aggregates Under Ohmic Heating. *Food Biop. Technol.* **2015**, *9*, 576–587. [CrossRef]
48. Ogbuefi, P.S.; Nwaokafor, P.; Njoku, I.J.; Uzuegbunam, O.J. Elemental Characterization of Rice Husk Ash from Local Rice Species in South Eastern Nigeria. *Chem. Afr.* **2020**, *3*, 1081–1085. [CrossRef]
49. Stunda-Zujeva, A. Review of *Spirulina/Arthrospira*. Growth Medium Modifications. *Key Eng. Mater.* **2020**, *850*, 153–158. [CrossRef]
50. Ferreira-Santos, P.; Nunes, R.; De Biasio, F.; Spigno, G.; Gorgoglione, D.; Teixeira, J.A.; Rocha, C.M.R. Influence of Thermal and Electrical Effects of Ohmic Heating on C-Phycocyanin Properties and Biocompounds Recovery from *Spirulina platensis*. *Lebensm.-Wiss. Technol.* **2020**, *128*, 109491. [CrossRef]

foods

MDPI

Article

Screening of a *Saccharomyces cerevisiae* Strain with High 3-Methylthio-1-Propanol Yield and Optimization of Its Fermentation Conditions

Qi Sun [1], Jinghao Ma [1], Rana Abdul Basit [1], Zhilei Fu [2], Xiaoyan Liu [1] and Guangsen Fan [1,3,4,*]

1 Key Laboratory of Geriatric Nutrition and Health, Beijing Technology and Business University, Ministry of Education, Beijing 100048, China; s17321914603l@163.com (Q.S.); 15032049018@163.com (J.M.); basit98ft@gmail.com (R.A.B.); liuxiaoyan@btbu.edu.cn (X.L.)
2 School of Biology and Food Science, Hebei Normal University for Nationalities, Chengde 067000, China; 13833415159@163.com
3 China Food Flavor and Nutrition Health Innovation Center, School of Food and Health, Beijing Technology and Business University, Beijing 100048, China
4 Sweet Code Nutrition & Health Institute, Zibo 256306, China
* Correspondence: fanguangsen@btbu.edu.cn

Abstract: 3-Methylthio-1-propanol (3-Met) is an important flavor compound in various alcoholic beverages such as *Baijiu* and *Huangjiu*. To maintain the content of 3-Met in these alcoholic beverages, it is necessary to screen a micro-organism with high yield of 3-Met from the brewing environment. In this study, the ability of yeast strains from the *Baijiu* brewing to produce 3-Met was analyzed, aiming to obtain yeast with high-yield 3-Met, and its fermentation conditions were optimized. Firstly, 39 yeast strains were screened using 3-Met conversion medium. The results showed that the majority of the strains from *Baijiu* brewing sources could produce 3-Met, and nearly half of the strains produced more than 0.5 g/L of 3-Met. Among these, yeast F10404, Y03401, and Y8#01, produced more than 1.0 g/L of 3-Met, with yeast Y03401 producing the highest amount at 1.30 g/L. Through morphological observation, physiological and biochemical analysis, and molecular biological identification, it was confirmed that yeast Y03401 was a *Saccharomyces cerevisiae*. Subsequently, the optimal fermentation conditions for 3-Met production by this yeast were obtained through single-factor designs, Plackett–Burman test, steepest ascent path design and response surface methodology. When the glucose concentration was 60 g/L, yeast extract concentration was 0.8 g/L, L-methionine concentration was 3.8 g/L, initial pH was 4, incubation time was 63 h, inoculum size was 1.6%, shaking speed was 150 rpm, loading volume was 50 mL/250 mL, and temperature was 26 °C, the content of 3-Met produced by *S. cerevisiae* Y03401 reached a high level of 3.66 g/L. It was also noteworthy that, in contrast to other study findings, this yeast was able to create substantial amounts of 3-Met even in the absence of L-methionine precursor. Based on the clear genome of *S. cerevisiae* and its characteristics in 3-Met production, *S. cerevisiae* Y03401 had broad prospects for application in alcoholic beverages such as *Baijiu*.

Keywords: 3-methylthio-1-propanol; *Saccharomyces cerevisiae*; strain identification; fermentation condition optimization; response surface methodology

1. Introduction

Pineapple alcohol, or 3-Methylthio-1-propanol (3-Met), is a sulfur-containing compound that imparts pleasant meaty, grilled cheese, sweet, onion, or vegetable flavors to alcoholic beverages such as *Baijiu*, *Huangjiu*, wine, rice wine, beer, brandy, and whiskey [1–7]. It is an important flavor compound in these alcoholic beverages. For example, in soy-flavored *Baijiu*, 3-Met imparts a ripe potato aroma [3]; in *Magli* wine, it imparts a unique vegetable aroma [4]; in beer and wine, it contributes a unique cooked potato or vegetable flavor [8,9]; in fruit wines such as cider, it is a major flavor compound that has a significant impact on product quality [6,10]. Furthermore, *Huangjiu* has a strong 3-Met fragrance, particularly in the refreshing variety, which gives *Huangjiu* a cooked vegetable odor [11]. In conclusion, 3-Met has a significant impact on the quality of alcoholic beverages due to its concentration.

Studies have shown that the major source of 3-Met in alcoholic beverages is the metabolic activity of relevant micro-organisms during the brewing process. The principal source was yeast, which produced 3-Met by metabolizing basic materials via the Ehrlich route or precursor chemicals from other microbial metabolisms [12]. Unfortunately, at present, these alcohol products generally have a low content of 3-Met during the brewing process, which affects the quality of the related products to a certain extent [12]. The lack or almost complete absence of critical strains that generated significant amounts of 3-Met during the brewing process was one of the main causes of the aforementioned process. In reported studies, except for *Kluyveromyces lactis* KL71, *Hyphopichia burtonii* YHM-G, *S. cerevisiae* SC408, *S. cerevisiae* SC57, *S. cerevisiae* S288C and *Saccharomycopsis fibuligera* Y1402, other natural strains produced less than 0.5 g/L of 3-Met under optimal fermentation conditions [13–21]. Modern genetic technology and genetic modification have made it entirely possible to increase natural strains' capacity to produce high levels of 3-Met; however, Chinese laws and regulations currently prohibit the use of genetically modified strains in the production of traditional brewing food, which raises concerns among consumers regarding the safety of brewing food [22–24]. Therefore, screening high-yield 3-Met strains from the production environment of brewing food and applying them appropriately to the brewing of alcoholic beverages was an effective way to address the above issues.

Currently, multiple yeast strains that produced 3-Met had been screened from the brewing environment of traditional brewing foods, such as *S. cerevisiae* [25], *Zygosaccharomyces rouxii* [1], *K. lactis* [16], *Geotrichum candidum* [26], *H. burtonii* [21] and *S. fibuligera* [20]. However, except for *H. burtonii* and *S. fibuligera*, which had a higher ability to produce 3-Met, the levels of 3-Met produced by other strains were generally low, which cannot meet the needs of products for 3-Met content. Therefore, it was necessary to further screen relevant functional micro-organisms from brewing foods, especially the production environment of alcoholic beverage products, in order to use these micro-organisms to increase the content of 3-Met in alcoholic beverages and improve their quality.

2. Materials and Methods

2.1. Materials

The experiment's yeast strains were selected from several aroma-type *Baijiu* brewing environments and then preserved by our research team. L-methionine (L-Met), 3-Met standard, and chromatographic grade methanol were purchased from Sigma (St. Louis, MO, USA). Unless otherwise stated, all other chemicals were analytical grade and commercially available.

2.2. Medium

Yeast peptone dextrose (YPD) medium: yeast extract 10 g/L, peptone 20 g/L, glucose 20 g/L, sterilized at 121 °C for 20 min.

3-Met conversion medium: glucose 30 g/L, yeast extract 0.8 g/L, L-Met 4 g/L, KH_2PO_4 8 g/L, K_2HPO_4 6 g/L, NaCl 2 g/L, $ZnSO_4$ 0.03 g/L, MgCl 0.01 g/L, $FeCl_2$ 0.02 g/L (through a 0.22 μm membrane), pH 5.0, sterilized at 121 °C for 20 min.

Wallerstein laboratory nutrient (WL) medium: yeast extract 5 g/L, tryptone 5 g/L, glucose 50 g/L, agar 20 g/L, KH_2PO_4 0.55 g/L, KCl 0.425 g/L, $CaCl_2$ 0.125 g/L, $FeCl_3$ 0.0025 g/L, $MgSO_4$ 0.125 g/L, $MnSO_4$ 0.0025 g/L, bromocresol green 0.022 g/L, pH 6.5, sterilized at 121 °C for 20 min.

Hydrogen sulfide test medium: NaCl 5 g/L, tryptone 25 g/L, gelatin 120 g/L, beef extract 7.5 g/L, 10% $FeSO_4 \cdot 7H_2O$ (sterilized through a 0.22 μm microporous membrane) 5 mL, pH 7.0, sterilized at 115 °C for 20 min.

Indole test medium: NaCl 5 g/L, tryptone 10 g/L, pH 7.8, sterilized at 121 °C for 20 min.

Methyl red test medium: K_2HPO_4 2 g/L, glucose 5 g/L, tryptone 5 g/L, pH 7.0–7.2, sterilized at 112 °C for 30 min.

Voges–Proskauer test medium: NaCl 5 g/L, tryptone 10 g/L, pH 7.8, sterilized at 121 °C for 20 min.

Citrate test medium: $NH_4H_2PO_4$ 1 g/L, NaCl 5 g/L, K_2HPO_4 1 g/L, sodium citrate 2 g/L, agar 20 g/L, $MgSO_4$ 0.2 g/L, pH 6.8, sterilized at 121 °C for 20 min.

Starch hydrolysis test medium: NaCl 5 g/L, beef extract 5 g/L, agar 20 g/L, tryptone 10 g/L, soluble starch 2 g/L, sterilized at 121 °C for 20 min.

Urea medium: tryptone 1 g/L, agar 15 g/L, NaCl 5 g/L, KH_2PO_4 2 g/L, sterilized at 121 °C for 20 min. 20 g/L urea (sterilized through a 0.22 μm microporous membrane).

Gelatin hydrolysis experimental medium: NaCl 5 g/L, tryptone 10 g/L, gelatin 120 g/L, beef extract 3 g/L, pH 7.2–7.4, sterilized at 121 °C for 20 min.

Skim milk test medium: Dissolve 10.167 g of commercial skim milk medium in 100 mL of water, sterilize at 115 °C for 15 min.

Sugar fermentation medium: NaCl 5 g/L, tryptone 10 g/L, 0.2% bromocresol purple, pH 7.8, sterilized at 115 °C for 20 min.

Carbon assimilation medium: beef extract 5 g/L, tryptone 10 g/L, NaCl 3 g/L, Na_2HPO_4 2 g/L, sugar or alcohol 10 g/L, bromothymol blue 0.03 g/L, sterilized at 115 °C for 20 min.

Nitrogen assimilation medium: $MgSO_4 \cdot 7H_2O$ 0.05 g/L, glucose 2 g/L, KH_2PO_4 0.1 g/L, pH 5–6, nitrogen source 1 g/L, sterilized at 115 °C for 15 min.

2.3. *Screening of High-Yielding 3-Met Yeast Strains*

The yeast strains preserved in glycerin tube at −20 °C were inoculated into YPD medium and activated for 48 h at 28 °C and 180 rpm. The activated yeast was then inoculated into 3-Met conversion medium with a 1% inoculum size and incubation for 48 h at 28 °C and 180 rpm. After centrifugation at 4 °C and 13,751× *g*, the supernatant was collected, and the 3-Met content was determined using high-performance liquid chromatography (HPLC).

2.4. *Identification and Performance Testing of Yeast Strains*

The high-yielding 3-Met yeast strains were identified according to Ma et al., Buchana et al. and Dong et al. [21,27,28], and their growth temperature range (20, 25, 30, 35, 40, 45, and 50 °C), pH range (1–14), glucose tolerance (4.8, 9.1, 13.0, 16.7, 20, 23.1, 25.9, 28.6, 31.0, 33.3, 35.5, 37.5, 39.4, 41.2, 42.9, 44.4, 45.9, 47.4, 48.7, and 50.0%, *w*/*w*), NaCl tolerance (0, 3, 6, 8, 11, 13, 15, 17, 19, and 21%, *w*/*v*), ethanol tolerance (0, 3, 6, 9, 12, 15, 18, and 21%, *v*/*v*), and 3-Met tolerance (1, 2, 3, 4, 5, 6, 7, 8, 9, 10, 11, 12, 13, 14, and 15 g/L) were determined using liquid culture methods. The cell density after 36 h of culture was measured using turbidity, and the absorbance values at 560 nm of different samples were determined with the corresponding blank control of the culture medium without inoculation.

2.5. *Optimization of 3-Met Production by Single-Factor Design (SFD)*

SFD was conducted to explore the effects of medium components (glucose concentration, yeast extract concentration, L-Met concentration, surfactant type and concentration) and fermentation conditions (initial pH, temperature, incubation time, inoculum size, shaking speed, and loading volume) on 3-Met production by *S. cerevisiae* Y03401. The important factors affecting 3-Met yield and their corresponding optimal levels were determined (Table 1).

Table 1. Factors and levels of SFD and their optimization conditions.

Factor	Level/Type	Optimization
Glucose concentration (g/L)	10, 20, 30, 40, 50, 60, 70, 80, 90, 100 and 110	60–90
Yeast extract concentration (g/L)	0, 0.4, 0.8, 1.2, 1.6 and 2	0.4
L-Met concentration (g/L)	0, 2, 4, 6, 8 and 10	2–4
Surfactant types	Control, Tween-20, Tween-40, Tween-60, Tween-80, Triton-100 and Glycerol	Control
Initial pH	3, 4, 5, 6 and 7	4
Temperature (°C)	22, 26, 30, 34 and 38	26
Incubation time (h)	0, 24, 36, 48, 60 and 72	36–60
Inoculum size (%)	0.2, 0.4, 0.8, 1.6 and 3.2	1.6
Shaking speed (rpm)	50, 100, 150, 200 and 250	150–200
Loading volume (mL/250 mL)	25, 50, 75, 100 and 125	50

2.6. Optimization of 3-Met Production by Plackett–Burman (PB) Test

Based on the results of SFD, nine factors including glucose concentration (X_1), yeast extract concentration (X_2), L-Met concentration (X_3), temperature (X_4), initial pH (X_5), incubation time (X_6), shaking speed (X_7), loading volume (X_8), and inoculum size (X_9) were selected for the PB test to further determine the main factors influencing 3-Met yield. This was performed in order to subsequently conduct response surface methodology (RSMD) to obtain the optimal fermentation conditions (Table 2).

Table 2. Plackett–Burman test design and results.

Test Number	X_1 (g/L)	X_2 (g/L)	X_3 (g/L)	X_4 (°C)	X_5	X_6 (h)	X_7 (rpm)	X_8 (mL/250 mL)	X_9 (%)	3-Met Concentration (g/L)
1	50	0.1	4	28	5	24	200	75	0.8	0.94
2	60	0.4	2	26	4	36	150	50	1.6	1.73
3	50	0.7	4	28	3	48	200	25	2.4	3.1
4	60	0.4	2	26	4	36	150	50	1.6	1.55
5	70	0.1	4	24	3	24	200	75	2.4	1.19
6	70	0.7	4	24	5	48	100	75	0.8	2.37
7	70	0.7	0	28	5	24	200	25	0.8	1.52
8	50	0.1	0	28	5	48	100	75	2.4	1.02
9	70	0.1	0	24	5	48	200	25	2.4	1.12
10	50	0.7	4	24	5	24	100	25	2.4	2.21
11	50	0.7	0	24	3	48	200	75	0.8	1.51
12	60	0.4	2	26	4	36	150	50	1.6	1.74
13	50	0.1	0	24	3	24	100	25	0.8	0.77
14	70	0.1	4	28	3	48	100	25	0.8	1.30
15	70	0.7	0	28	3	24	100	75	2.4	1.48

2.7. Optimization of 3-Met Production by Steepest Ascent Path Design (SAPD)

Based on the results of the PB test, factors significantly affecting 3-Met production were selected for the SAPD to obtain a reasonable range for the design of RSMD. The direction of each variable in the experiment was determined based on the regression results of the PB test, and the step size for each variable was determined based on the regression model of the PB test and actual experience. The specific design was shown in Table 3.

Table 3. Experimental designs and the results of SAPD.

Groups	1	2	3	4	5
Incubation time (h)	24	36	48	60	72
L-Met concentration (g/L)	0	1	2	3	4
Yeast extract concentration (g/L)	0.2	0.4	0.6	0.8	1
3-Met concentration (g/L)	1.08	1.78	2.37	3.29	1.03

2.8. Optimization of 3-Met Production by Response Surface Methodology (RSMD)

Based on the determination of the center point from the SAPD, an RSMD was conducted to optimize three factors: incubation time (A), L-Met concentration (B), and yeast extract concentration (C). According to the Box–Behnken experimental design (BBD, Design-Expert Software 11.0, StatEase Inc., Minneapolis, MN, USA), a total of 17 experiments were designed (including five repetitions of the center point) (Table 4).

Table 4. The RSMD and the responses of the dependent variables.

Test Number	Incubation Time (h)		L-Met Concentration (g/L)		Yeast Extract Concentration (g/L)		3-Met Concentration (g/L)
	A	Code A	B	Code B	C	Code C	Y
1	60	0	3	0	0.8	0	3.37
2	72	1	3	0	0.6	−1	2.24
3	60	0	3	0	0.8	0	3.35
4	72	1	3	0	1	1	2.74
5	60	0	3	0	0.8	0	3.45
6	48	−1	3	0	0.6	−1	1.46
7	48	−1	1	−1	0.8	0	1.69
8	60	0	1	−1	0.6	−1	1.94
9	48	−1	3	0	1	1	1.97
10	72	1	5	1	0.8	0	3.02
11	60	0	3	0	0.8	0	3.45
12	72	1	1	−1	0.8	0	2.46
13	60	0	1	−1	1	1	2.49
14	60	0	5	1	0.6	−1	2.54
15	48	−1	5	1	0.8	0	2.24
16	60	0	3	0	0.8	0	3.54
17	60	0	5	1	1	1	3.00

2.9. Determination of 3-Met Content

After centrifugation at 13,751× g for 10 min, the supernatant of the fermentation broth was filtered through a 0.22 μm membrane, and then appropriately diluted with pure water for HPLC detection. The HPLC method followed the parameters described by Ma et al. [21], as follows: C-18 reverse-phase column (ZORBAX Eclipse Plus C-18, 4.6 × 250 mm, 5 μm); mobile phase of methanol:water = 10:90, with pure phosphoric acid to adjust the pH to 3; flow rate: 0.7 mL/min; detection wavelength 215 nm; column temperature 30 °C; injection volume 10 μL.

2.10. Data Analysis

A one-way ANOVA ($p < 0.05$) with Tukey's test was used to analyze statistical differences in the assessed methodologies. Each experiment was conducted in triplicate, and the experimental data were processed and plotted using Excel 2019, SPSS 24.0, and Design-Expert 11.

3. Results and Discussion

3.1. Screening of High-Yielding 3-Met Strain

A total of 39 strains that were kept in our lab were separately activated and inoculated into a 3-Met conversion medium. Their 3-Met production was assessed during a 48 h fermentation period. The findings are shown in Table 5, which demonstrates that most of the strains were capable of producing 3-Met. Only eight of the examined strains failed to produce any discernible 3-Met. Moreover, several strains of the yeasts that produce 3-Met showed differing capacities for the manufacture of 3-Met. Nearly half of the strains produced 3-Met at concentrations exceeding 0.5 g/L, which was unexpectedly high compared to previous research reports. This might be related to the fact that most of the selected strains originated from brewing environments of sesame-flavored *Baijiu*. Notably, among the tested strains, three strains produced significantly higher levels of 3-Met, exceeding 1.0 g/L, namely the strains F10404, Y03401, and Y8#01, with yield of 1.21 g/L, 1.30 g/L, and 1.10 g/L, respectively. Remarkably, strain Y03401 was chosen for further investigation because it showed the highest 3-Met output among all the strains examined.

3.2. Identification and Performance Testing of Strain Y03401

Strain Y03401 produced blue–green colonies that were elevated and had smooth edges, but the colonies' centers remained blue–green after growing on WL medium (Figure 1a). The surface of the colony was smooth, sticky, wet, and pickable. Under a microscope, the cells showed conventional yeast features: they were round or elliptical, budding at one end, and devoid of mycelium (Figure 1b).

Table 5. Yield of 3-Met by various strains.

Strain	3-Met Concentration (g/L)	Strain	3-Met Concentration (g/L)	Strain	3-Met Concentration (g/L)
1021	0.78 ± 0.06	F12502	0.31 ± 0.09	F1918	0.37 ± 0.06
1213	0.31 ± 0.11	F13004	0.58 ± 0.04	F1915	0.55 ± 0.07
1273	0.46 ± 0.03	F13011	0.00 ± 0.00	F2405	0.00 ± 0.00
1344	0.21 ± 0.07	F13017	0.00 ± 0.0	Y03401	1.30 ± 0.06
1701	0.51 ± 0.03	F1504	0.77 ± 0.08	Y1401	0.00 ± 0.00
32704	0.57 ± 0.10	F1701	0.21 ± 0.05	Y1405	0.66 ± 0.13
32708	0.48 ± 0.08	F1702	0.00 ± 0.00	Y1511	0.27 ± 0.06
32714	0.43 ± 0.01	F1904	0.36 ± 0.06	Y1517	0.00 ± 0.00
33197	0.59 ± 0.07	F1905	0.00 ± 0.00	Y2#09	0.38 ± 0.02
33260	0.36 ± 0.02	F1912	0.00 ± 0.00	Y41114	0.45 ± 0.06
33298	0.76 ± 0.11	F1913	0.33 ± 0.08	Y8#01	1.10 ± 0.10
33319	0.92 ± 0.06	F1914	0.77 ± 0.11	Y8#03	0.96 ± 0.08
F10404	1.21 ± 0.08	F1916	0.52 ± 0.04	YX3307	0.39 ± 0.05

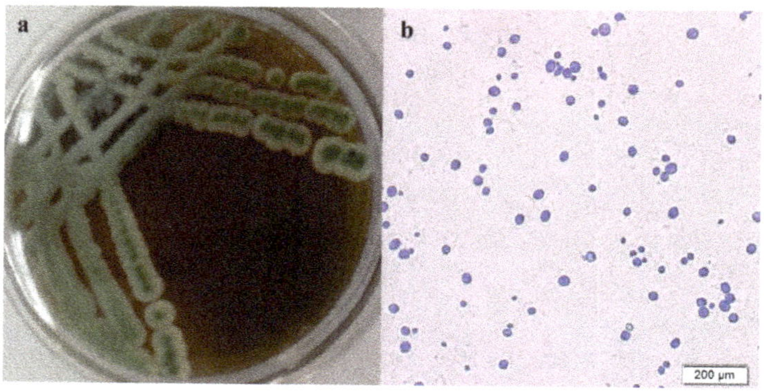

Figure 1. Colony morphology (**a**) on WL medium and morphological characteristics of yeast colonies under microscope (**b**).

Analysis of its physiological and biochemical characteristics (Table 6) revealed that yeast Y03401 could grow when D-xylose, fructose, glucose, maltose, lactose, and sucrose were used as the sole carbon sources. It produced acid and gas when fructose, glucose, maltose, and sucrose were the sole carbon sources, while only producing gas, not acid, when D-xylose and lactose were the sole carbon sources. In carbon source assimilation tests, yeast Y03401 could grow when ethanol, glycerol, and inulin were used as the sole carbon sources but not when D-ribose was used. In nitrogen assimilation tests, urea, potassium nitrate, ammonium sulfate, L-Met, and L-tyrosine all promoted the growth of yeast Y03401 as sole nitrogen sources. Additionally, yeast Y03401 showed poor results in the tests for hydrogen sulfide, indole, Voges–Proskauer, urea hydrolysis, and gelatin liquefaction, but it showed positive results in the tests for methyl red, citrate utilization, and starch hydrolysis. This showed that yeast Y03401 could thrive when sodium citrate was the only carbon source and created amylase to break down starch. Y03401 may ferment glucose and produce acidic chemicals such as pyruvic acid, lactic acid, succinic acid, acetic acid, or formic acid. After extracting the whole DNA from yeast Y03401, the 26s rDNA gene fragment was successfully amplified using specific primers, NL1 (5′-GCATATCAATAAGCGGAGGAAAAG-3′) and

NL4 (5′-GGTCCGTGTTTCAAGACGG-3′). This yeast's identity was verified by comparing and analyzing the gene sequence in the NCBI database, which revealed a 100% similarity to the *S. cerevisiae* source. It was noteworthy, nevertheless, that this particular yeast's physiological and biochemical traits did not match those of *S. cerevisiae* Y1914 and Y3401, which could be because of variations in their living conditions leading to partial variations in their physiological and biochemical traits [29,30].

Table 6. Physiological, biochemical characteristics and performance test results of *S. cerevisiae* Y03401.

Tests	Substrate Type	Results	Tests	Substrate Type	Results
Nitrogen source assimilation tests	Urea	+	Sugar fermentation tests	D-Xylose	Gas production, no acid
	Ammonium sulfate	+		Fructose	Acid and gas production
	Potassium nitrate	+		Glucose	Acid and gas production
	L-Met	+		Maltose	Acid and gas production
	L-Tyrosine	+		Lactose	Gas production, no acid
Carbon source assimilation tests	Ethanol	+		Sucrose	Acid and gas production
	Glycerol	+	Other tests	Hydrogen sulfide test	−
	Inulin	+		Indole test	−
	D-Ribose	−		Methyl red test	+
Tolerance range	Temperature (°C)	20–40		Voges–Proskauer test	−
	NaCl (%, *w*/*v*)	0–6		Citrate test	+
	Ethanol (%, *v*/*v*)	0–6		Urea test	−
	Glucose (%, *w*/*w*)	0–45.9		Starch hydrolysis test	+
	pH	2–10		Gelatin liquidized test	−
	3-Met (g/L)	1–12			

Note: "+", positive response; "−", negative response.

Experimental results indicated that the highest growth temperature for *S. cerevisiae* Y03401 was 40 °C, and its optimal growth temperature was 25 °C, suitable for the brewing of *Baijiu*, *Huangjiu*, and other wine products [31]. *S. cerevisiae* Y03401 had low tolerance to NaCl, only 6%, although it cannot be used in some high-salt-concentration foods, it adapted well to the production of alcohol-containing products. *S. cerevisiae* Y03401 had moderate ethanol tolerance and grew well when the ethanol volume fraction is 6% or lower, which made it possible to adjust to the 6%-or-less ethanol content that is typically produced in the brewing of *Baijiu* and some other alcoholic beverages [32,33]. *S. cerevisiae* Y03401 tolerated a maximum glucose concentration of 45.9%, slightly higher than *H. burtonii* YHM-G and *Clavispora lusitaniae* YX3307, showing relatively high sugar tolerance [21,34]. It had broad pH tolerance and grew at pH 2–10, making it an excellent strain that can adapt to different environmental pH levels, with its optimal growth pH being 4.0, which might be related to its origin in the brewing of *Baijiu* [31]. In addition, *S. cerevisiae* Y03401 had a tolerance of 12 g/L to 3-Met, lower than *H. burtonii* YHM-G and *S. cerevisiae* CEN.PK113–7D [21,25]. In conclusion, *S. cerevisiae* Y03401 was quite suitable for use in *Baijiu*, *Huangjiu*, and other wine products.

3.3. Optimization of 3-Met Production by SFD

3.3.1. Effect of Medium Components on 3-Met Production by *S. cerevisiae* Y03401

Glucose is a vital component of microbial growth, metabolism, and reproduction. It serves as both the energy supply and the carbon skeleton for microbial cell creation. However, since various microbes have varying glucose tolerances, higher glucose concentration was not always preferable for the development and metabolism of microbes; rather, it was helpful in specific circumstances [35]. The influence of glucose concentration on the production of 3-Met by *S. cerevisiae* Y03401 is clearly depicted in Figure 2a. The figure showed that when glucose concentration increased, the demand for energy and carbon sources necessary for the growth, metabolism, and reproduction of *S. cerevisiae* Y03401 escalated progressively. This increased the concentration of 3-Met that was produced. The concentration of 3-Met peaked between 60 and 90 g/L as the glucose content rose. However, further increasing the glucose concentration led to a decrease in 3-Met yield, as the high osmotic pressure caused by high glucose concentration affected the growth and metabolism of *S. cerevisiae* Y03401. Considering the yield and cost, the glucose concentration was chosen as 60 g/L, higher than the optimal glucose concentration reported in previous studies, which might be related to the higher glucose tolerance of this strain [17,21].

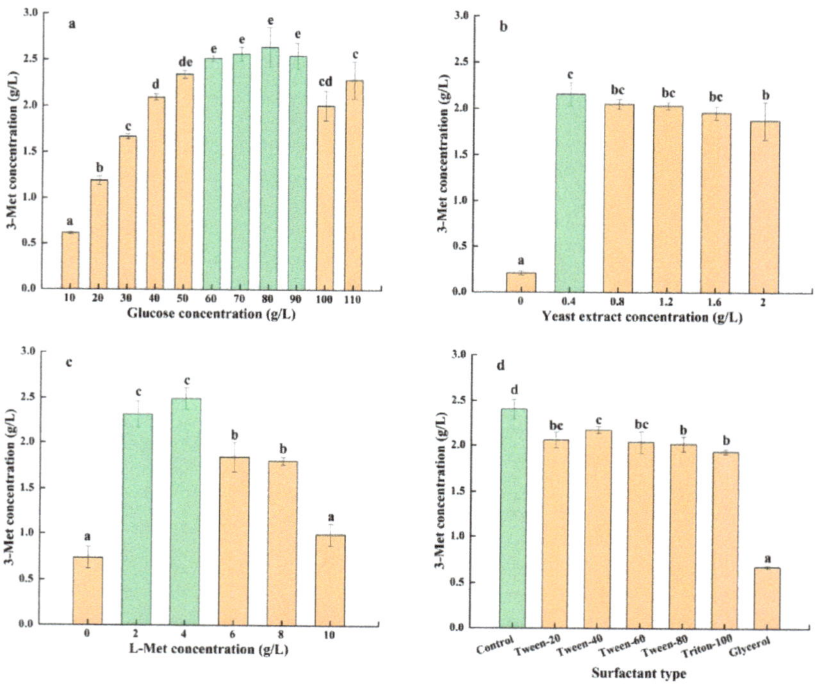

Figure 2. (**a**) Glucose concentration (10, 20, 30, 40, 50, 60, 70, 80, 90, 100 and 110 g/L), (**b**) yeast extract concentration (0, 0.4, 0.8, 1.2, 1.6 and 2 g/L), (**c**) L-Met concentration (0, 2, 4, 6, 8 and 10 g/L) and (**d**) surfactant types (Control, Tween-20, Tween-40, Tween-60, Tween-80, Triton-100 and Glycerol). The same letters in the column indicate that the data do not differ significantly at 5% probability using Tukey's test. The green bar means the best conditions on 3-Met concentration.

Previous studies have shown that the Ehrlich path was used by yeast to generate 3-Met. The presence of additional nitrogen sources in the medium would affect the yeast's ability to convert the substrate L-Met into 3-Met. However, earlier studies showed that yeast could make 3-Met more effectively when it was exposed to an appropriate concentration of yeast extract. This might be due to the fact that yeast extract, in addition to serving as a nitrogen source, also contained numerous other nutritional components such as various vitamins, amino acids, and biotin, which were important growth factors for yeast's growth and reproduction, thereby promoting its high yield of 3-Met [14,21]. As Figure 2b illustrates, the experimental groups with varying yeast extract quantities produced greater amounts of 3-Met in comparison to the control group that did not have any yeast extract added. *S. cerevisiae* Y03401 generated the highest concentration of 3-Met, particularly at a yeast extract concentration of 0.4 g/L. However, excessively high yeast extract concentrations would inhibit the production of 3-Met, which fully confirmed the conclusions reported in previous studies. Lower concentrations of yeast extract provided the necessary growth factors for yeast, while higher concentrations of yeast extract would exert a strong inhibitory effect on the Ehrlich pathway of yeast [21]. In contrast to other studies, *S. cerevisiae* Y03401 needed a considerably low optimum concentration of yeast extract in order to produce 3-Met. This might be because various strains of the yeast have somewhat different associated metabolic pathways [17,21].

L-Met is an essential precursor for yeast to produce 3-Met through the Ehrlich pathway, and its concentration is crucial for the synthesis of 3-Met by yeast [36]. The concentration of 3-Met produced by *S. cerevisiae* Y03401 initially increased and subsequently decreased in response to an increase in L-Met concentration, as shown in Figure 2c. The highest concentration of 3-Met was achieved at L-Met concentrations of 2–4 g/L. On the other hand, 3-methylthio-1-propanoic acid production might be excessive in yeast due to too-high L-Met concentrations, which would prevent yeast development and metabolism. The yeast would desulfurate L-Met to lower its effective concentration to counteract this inhibition, which would lower the amount of 3-Met it produced [37,38]. Contrary to earlier study findings, *S. cerevisiae* Y03401 was able to synthesize significant amounts of 3-Met with a yield more than 0.5 g/L even in the absence of L-Met. The reason for this might be because the strain can either manufacture 3-Met from glucose de novo or it can synthesize L-Met from glucose and utilize the Ehrlich route to make 3-Met. Additionally, for this reason, this strain's optimal L-Met concentrations for generating 3-Met were somewhat lower than those of other strains [16,21].

According to earlier research, adding appropriate surfactants to the medium, including Tween and Triton, may aid in the production of 3-Met by the yeast. These surfactants may have enhanced the permeability of the cell membrane, the dissolved oxygen concentration, and the expression of enzymes involved in metabolism while the cells were being incubated [21,39]. However, in this study, the expected effect was not achieved (Figure 2d). Among the tested surfactants, none of them could promote the production of 3-Met by *S. cerevisiae* Y03401. On the contrary, they all had a certain inhibitory effect, especially glycerol and this might be due to differences in the composition of the cell membrane of yeast strains, leading to different effects of surfactants on their growth and metabolism.

3.3.2. Effect of Fermentation Conditions on 3-Met Production by *S. cerevisiae* Y03401

The growth, reproduction, and metabolic pathways of micro-organisms are influenced by pH, which also plays a crucial regulatory role in the accumulation of essential metabolic products. pH influences the presence of nutrients in the microbial growth environment, the membrane potential and permeability of the microbial cells, and membrane protein function [40]. The results of the initial pH's effect on the yield of 3-Met by *S. cerevisiae* Y03401 are shown in Figure 3a. It was observed that at an initial pH of 4.0, *S. cerevisiae* Y03401 produced the highest level of 3-Met, while lower or higher initial pH levels would affect its yield. This might be attributed to the fact that, under these initial pH conditions, the capacity of *S. cerevisiae* Y03401 to absorb nutrients or the functionality of enzymes and proteins involved in the Ehrlich pathway was relatively impaired, as evidenced by its optimal growth pH of 4.0. Additionally, this was similar to the optimal pH for 3-Met production by *H. burtonii* YHM-G, which also originated from *Baijiu*, and this might be related to the slightly acidic environment in which both were used for *Baijiu* production [21,29].

Temperature is an important parameter that affects the growth, reproduction, and metabolic pathways of micro-organisms. Suitable temperatures are conducive to the growth, reproduction, and accumulation of metabolic products in micro-organisms, while lower temperatures will affect the diffusion of nutrients and the fluidity of microbial cell membranes, leading to a slower rate of nutrient intake and lower enzymatic activity within the micro-organism, which is not conducive to its growth and reproduction. On the other hand, higher temperatures cause damage to the microbial cell membrane and loss of enzymatic activity within the organism, which ultimately hinders its growth and reproduction [41]. Based on the growth temperature range of *S. cerevisiae* Y03401, the production of 3-Met was investigated within the range of 22–38 °C. The results (Figure 3b) showed that with an increase in the temperature, the amount of 3-Met produced initially increased and then decreased. When the temperature was 26 °C, the yield reached its maximum, which was consistent with its optimal growth temperature and slightly lower than the reported optimal temperature for other yeast strains producing 3-Met [14,21].

The large accumulation of microbial metabolic products generally begins during the logarithmic growth phase, while the peak production mainly occurs during the stable growth phase and the pre-death phase [21]. Different micro-organisms often have different optimal fermentation periods for accumulating metabolic products. Therefore, in order to accumulate more 3-Met, it was necessary to study the fermentation process of *S. cerevisiae* Y03401 for producing 3-Met. The results (Figure 3c) showed that with the prolongation of incubation time, the content of 3-Met produced by *S. cerevisiae* Y03401 gradually increased. The yield reached its highest point at 48 h of fermentation, and then decreased slightly. This might be because the high concentration of 3-Met can have a toxic effect on yeast cells, and the yeast converted it into other metabolic products with weaker toxic effects for self-protection. Alternatively, it might be due to the lack of nutrients in the late stage of fermentation, and the yeast used 3-Met as a nutrient and metabolized it into other substances. It was worth noting that compared with *H. burtonii* YHM-G, this yeast accumulated a large amount of 3-Met in a shorter time (within 24 h), which might be related to the higher enzymatic activity of the Ehrlich pathway, the higher incubation time, or the ability of de novo synthesis [21].

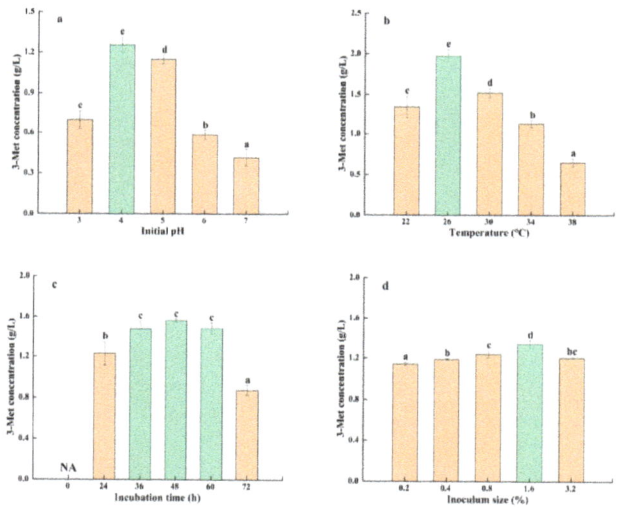

Figure 3. (**a**) initial pH (3, 4, 5, 6 and 7), (**b**) temperature (22, 26, 30, 34 and 38 °C), (**c**) incubation time (0, 24, 36, 48, 60 and 72 h), (**d**) inoculum size (0.2, 0.4, 0.8, 1.2 and 3.2%), (**e**) shaking speed (50, 100, 150, 200 and 250 rpm), (**f**) loading volume (25, 50, 75, 100 and 125 mL/250 mL). According to Tukey's test, the identical letters in the column indicate there is no significant difference in the data with a 5% probability. The optimal 3-Met concentration levels are shown by the green bar. The letter NA stands for "no 3-Met detection".

In the fermentation process, the growth rate of the strain and the length of the fermentation period are closely related to the inoculum size. Generally, a properly larger inoculum size can improve the growth rate, shorten the fermentation period, and reduce production costs [21]. The quantity of 3-Met generated by *S. cerevisiae* Y03401 somewhat increased as the inoculum size increased, as Figure 3d illustrates. The highest yield occurred at 1.6% inoculum size, and when inoculum size increased, the output declined marginally. This may be the result of a larger inoculum size causing the fermentation system's oxygen concentration to be inadequate and nutrients to be used more quickly, which hindered the yeast's ability to produce the necessary metabolic products and undergo necessary transformations. However, overall, *S. cerevisiae* Y03401 produced a high yield of 3-Met under different inoculum sizes, which might be related to its good environmental adaptability and its ability to rapidly convert L-Met into 3-Met.

It has been shown in earlier research that substantial yields of 3-Met by yeast were possible in an oxygen-rich environment [16]. The two most direct elements influencing the oxygen concentration in a shaken flask fermentation system were the loading volume and the shaking speed. Therefore, it was crucial to investigate how these two variables influenced the capability of *S. cerevisiae* Y03401 to produce 3-Met. At lower shaking speeds (50 rpm), the flask's oxygen level was insufficient, adversely affecting the growth and fermentation of *S. cerevisiae* Y03401, and subsequently reducing the production of 3-Met, as depicted in Figure 3e. The yield of 3-Met increased as the shaking speed increased because the oxygen concentration grew steadily. The maximum yield was obtained at 150–200 rpm shaking rates but when the shaking speed increased, the tremendous shear force brought on by these high speeds damaged the yeast cells to some degree, reducing the yeast's capacity to ferment even while the oxygen concentration was rising. These results were consistent with the SFD optimization of shaking speed for *H. burtonii* YHM-G [21]. Furthermore, the yield of 3-Met produced by yeast also rose initially, then decreased, as loading volume increased (Figure 3f). It is probable that the combined effects of oxygen concentration and physical shear force led to the greatest production of 3-Met, which was seen at a loading volume of 50 mL/250 mL.

3.4. Optimization of 3-Met Production by PB Test

The concentration of glucose, yeast extract, L-Met, temperature, initial pH, incubation time, shaking speed, loading volume, and inoculum size were among the nine factors that were chosen for the PB test based on the SFD results. These factors were intended to shed additional light on the significant factors influencing the yield of 3-Met by *S. cerevisiae* Y03401. The fitting regression equation linking 3-Met yield to each element was derived by statistical analysis of the experimental data, and it looks like this:

$$Y = 0.09 - 0.00491 \times X_1 + 1.620 \times X_2 + 0.1540 \times X_3 + 0.0079 \times X_4 - 0.0146 \times X_5 + 0.01602 \times X_6 + 0.00041 \times X_7 - 0.00503 \times X_8 + 0.1795 \times X_9$$

Through variance analysis and significance analysis (Table 7), it was found that the F-value of the fitted regression model was 8.46, with a corresponding p-value of 0.015 ($p < 0.05$), indicating the significance of the fitted model. The coefficient of determination R^2 for the regression equation was 0.9384, indicating that the model had high accuracy and was used to predict the yield of 3-Met by *S. cerevisiae* Y03401. Additionally, the regression equation showed that loading volume, initial pH, and glucose concentration had a negative effect on the yield of 3-Met by *S. cerevisiae* Y03401, while the other six factors had a positive effect. The significance analysis indicated that yeast extract concentration, L-Met concentration, and incubation time had a significant impact on the yield of 3-Met, while the other factors had an insignificant impact. Therefore, further optimization was carried out for these three factors, while the levels of the other factors were set to the optimal levels determined by SFD designs, glucose concentration of 60 g/L, initial pH of 4.0, temperature of 26 °C, inoculum size of 1.6%, shaking speed of 150 rpm, and loading volume of 50 mL/250 mL.

Table 7. Levels of the variables and statistical analysis in PB for yield of 3-Met.

Code	Variable	Low Level (−1)	High Level (+1)	Effect (EXi)	*F* Values	*p* Values	Rank	Significance
X_1	Glucose concentration (g/L)	50	70	−0.0491	0.45	0.532	6	-
X_2	Yeast extract concentration (g/L)	0.1	0.7	0.4859	44.1	0.001	1	***
X_3	L-Met concentration (g/L)	0	4	0.3081	17.72	0.008	2	**
X_4	Temperature (°C)	24	28	0.0159	0.05	0.837	8	-
X_5	Initial pH	3	5	−0.0146	0.04	0.849	9	-
X_6	Incubation time (h)	24	48	0.1922	6.9	0.047	3	*
X_7	Shaking speed (rpm)	100	200	0.0203	0.08	0.792	7	-
X_8	Loading volume (mL/250 mL)	25	75	−0.1256	2.95	0.147	5	-
X_9	Inoculum size (%)	0.8	3.2	0.1436	3.85	0.107	4	-
	model				8.46	0.015		*
	$R^2 = 0.9384$			$R^2_{Adj} = 0.8274$				

Note: "-" Not significant ($p > 0.05$); "*" Significant at 5% level ($p < 0.05$); "**" Significant at 1% level ($p < 0.01$); "***" Significant at 0.1% level ($p < 0.001$).

3.5. Optimization of 3-Met Production by SAPD

The results of the SAPD showed that the yield of 3-Met produced by *S. cerevisiae* Y03401 matched the expected values. As the three significant factors progressed in the direction of increasing their influence on the response value, the yield of 3-Met gradually reached its peak and then showed a decreasing trend. This led to the determination of the central values for the optimal optimization range for subsequent RSMD: yeast extract concentration of 0.8 g/L, incubation time of 60 h, and L-Met concentration of 3 g/L. Under the above fermentation conditions, the yield of 3-Met by *S. cerevisiae* Y03401 could reach 3.29 g/L.

3.6. Optimization of 3-Met Production by RSMD

In a multi-component environment, particularly when there are interactions between the variables, the RSMD can efficiently ascertain the ideal degree of impact of each element and the interactions between the factors, swiftly establishing the best circumstances. Based on the principles of Box–Behnken experimental design (BBD), 17 experiments were designed for the three factors at three levels: L-Met concentration (g/L), incubation time (h), and yeast extract concentration (g/L). The experimental results showed that the yield of 3-Met under different fermentation conditions ranged from 1.46 to 3.54 g/L, with the highest yield at the central values, averaging 3.43 g/L. Through multivariate regression analysis and fitting of a multivariate quadratic equation, the regression equation obtained was as follows:

$$Y = -28.76 + 0.6451 \times A + 0.695 \times B + 25.28 \times C - 0.005102 \times A^2 - 0.08619 \times B^2 - 14.869 \times C^2 + 0.00010 \times AB - 0.0010 \times AC - 0.0562 \times BC$$

Through analysis of variance (Table 8), the simulated equation had an *F* value of 239.27, corresponding to a p value < 0.0001, indicating that the model was significant. The corresponding coefficient of determination R^2 for the regression equation was 0.9968, indicating that the established model could explain the response value well. In addition, the lack of fit was not significant, indicating a good correlation between the actual and predicted values. The model could be used to predict the yield of 3-Met. Through significance analysis, it was found that in the regression equation, the first-order terms A (yeast extract concentration), B (L-Met concentration), and C (incubation time) were all significant, indicating a significant linear effect of the three factors on the yield of 3-Met, consistent with the PB test results. However, the order of significance was different, with L-Met concentration > incubation time > yeast extract concentration in the RSMD. This may be due to the different number of factors and calculation methods in the two experimental designs. There was no significant interaction between the three factors, indicating that the interaction effect between the three factors on the yield of 3-Met by *S. cerevisiae* Y03401 was not significant. The quadratic terms A^2, B^2, and C^2 were all significant, indicating that there was a surface effect of yeast extract concentration, L-Met concentration, and incubation time on the yield of 3-Met by *S. cerevisiae* Y03401. Based on the regression equation, a corresponding response surface three-dimensional graph was drawn (Figure 4), which showed that with the increase of the three factors, the yield of 3-Met first increased and then decreased. According to the fitted regression equation, when the yeast extract concentration was 0.8 g/L, L-Met concentration was 3.8 g/L, and incubation time was 63 h, the yield of 3-Met by *S. cerevisiae* Y03401 was the highest, predicted to be 3.57 g/L. This was verified by experiments conducted under this condition, and the yield of 3-Met was 3.66 g/L, which was close to the predicted value of 3.57 g/L, indicating that the model could predict the optimal conditions for the yield of 3-Met well. The amount of 3-Met produced by *S. cerevisiae* Y03401 was higher than that reported for *H. burtonii* YHM-G, but lower than that reported for *S. fibuligera* Y1402 which was currently the highest reported natural strain for the production of 3-Met [20,21]. In addition, it should be noted that compared with *H. burtonii* YHM-G, *S. cerevisiae* Y03401 had relatively weak tolerance to 3-Met, but its yield was higher than that of the former, indicating that there was a certain correlation between the ability of the strain to produce 3-Met and its tolerance to 3-Met, but there was no proportional relationship [21].

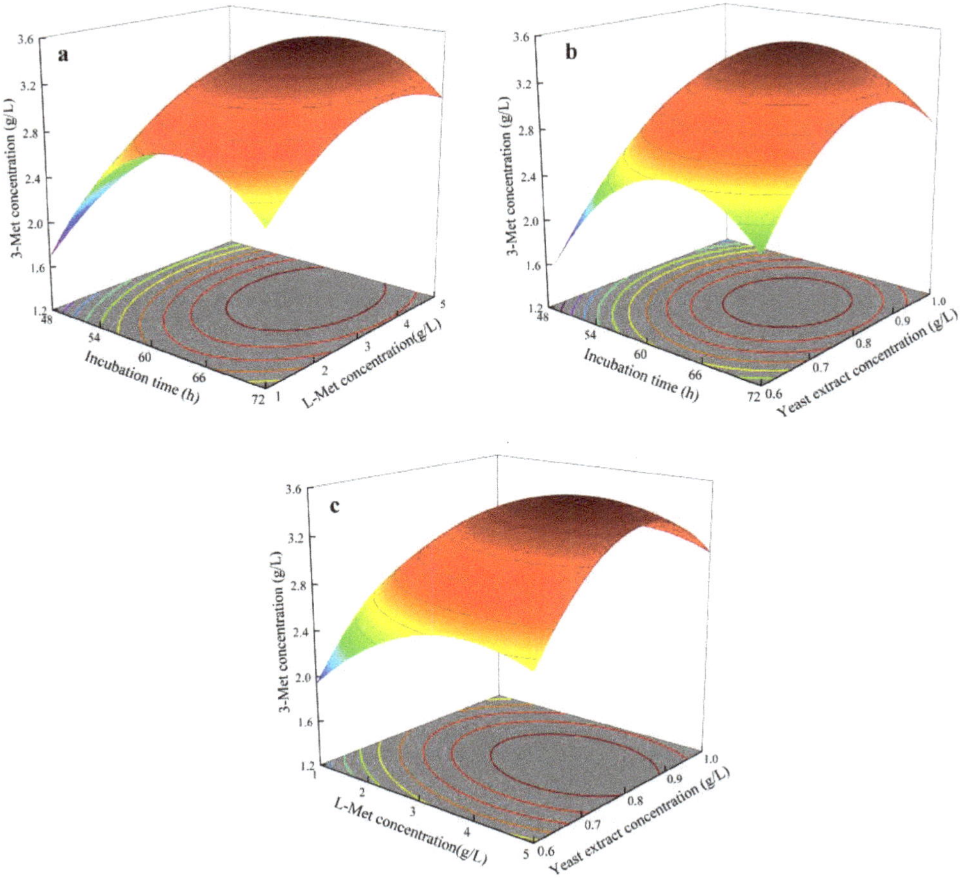

Figure 4. Response surface and contour map of the interaction of various factors affecting the yield of 3-Methylthiopropanol. (**a**) interaction between incubation time and L-Met concentration; (**b**) interaction between incubation time and yeast extract concentration; (**c**) interaction between L-Met concentration and yeast extract concentration. The colour change from blue to red corresponds to increase in responses.

Table 8. Regression coefficients and their significance for 3-Met yield from the results of the BBD.

Source	Sum of Squares	DF	Mean Square	*F* Values	*p* Values
Model	7.03873	9	0.78208	239.27	0.000
A-Yeast extract concentration	0.02282	1	0.02282	6.98	0.033
B-L-Met concentration	0.34501	1	0.34501	105.55	0.000
C-Incubation time	0.09187	1	0.09187	28.11	0.001
AB	0.00203	1	0.00203	0.62	0.457
AC	0.00003	1	0.00003	0.01	0.933
BC	0.00002	1	0.00002	0.01	0.933
A^2	1.48938	1	1.48938	455.67	0.000
B^2	0.50043	1	0.50043	153.10	0.000
C^2	2.27308	1	2.27308	695.44	0.000
Residual	0.02288	7	0.00327		
Lack of Fit	0.00000	3	0.00000	0.00	1.000
Pure Error	0.02288	4	0.00572		
Cor Total	7.06161	16			
	$R^2 = 0.9968$		$R^2_{Adj} = 0.9926$		

4. Conclusions

This article first screened the yeast strains preserved by our research team for the production of 3-Met, and selected a yeast strain Y03401 with high yield of 3-Met, which was identified as a *S. cerevisiae*. Subsequently, SFD, PB tests, SAPD, and RSMD were used to optimize the fermentation conditions for the production of 3-Met. The results showed that the highest yield of 3-Met was produced, reaching 3.66 g/L, when the glucose concentration was 60 g/L, yeast extract concentration was 0.8 g/L, and L-Met concentration was 3.8 g/L. The initial pH was 4, the incubation period was 63 h, the inoculum size was 1.6%, the shaking speed was 150 rpm, the loading volume was 50 mL/250 mL, and the temperature was 26 °C. This was one of the few natural strains that produce 3-Met with a yield of more than 3 g/L at present, providing theoretical reference for its application in alcoholic beverages such as liquor. In addition, it is necessary to conduct further in-depth research on the mechanism of the high yield of 3-Met in the absence of L-Met.

Author Contributions: Providing experimental conditions, G.F. and X.L.; providing experimental ideas and methods, G.F. and Z.F.; methodology, Q.S. and J.M.; formal analysis, Q.S.; investigation, G.F.; resources, Z.F.; data curation, Q.S.; writing—original draft preparation, R.A.B. and G.F.; writing—review and editing, X.L.; visualization, Q.S.; supervision, G.F.; project administration, G.F.; funding acquisition, G.F. All authors have read and agreed to the published version of the manuscript.

Funding: This research was supported by the National Natural Science Foundation of China (31701592), the Beijing Natural Science Foundation (6222003), and the Open Research Fund Program of Henan Key Laboratory of Industrial Microbial Resources and Fermentation Technology (HIMFT20190205).

Institutional Review Board Statement: Not applicable.

Informed Consent Statement: Not applicable.

Data Availability Statement: The original contributions presented in the study are included in the article, further inquiries can be directed to the corresponding authors.

Acknowledgments: The authors thank Zibo Sweet Code Nutrition & Health Institute for the support with instruments and equipment.

Conflicts of Interest: The authors declare no conflicts of interest.

References

1. Liu, B.; Wang, X.F.; Zhao, J.H.; Qin, L.H.; Shi, L.; Zhou, T.; Liu, S.; Wang, C.L. Effects of salinity on the synthesis of 3-methylthiopropanol, 2-phenylethanol, and isoamyl acetate in *Zygosaccharomyces rouxii* and *Z. rouxii* 3-2. *Bioproc. Biosyst. Eng.* **2020**, *43*, 831–838. [CrossRef]
2. Laurence, G.; Philippe, P.; Aurore, T.; Sonia, C. 3-methylthiopropionaldehyde as precursor of dimethyl trisulfide in aged beers. *J. Agric. Food Chem.* **2000**, *48*, 6196–6199. [CrossRef]

3. Fan, H.Y.; Fan, W.L.; Xu, Y. Characterization of key odorants in Chinese Chixiang aroma-type liquor by gas chromatography-olfactometry, quantitative measurements, aroma recombination, and omission studies. *J. Agric. Food Chem.* **2015**, *63*, 3660–3668. [CrossRef]
4. Park, H.J.; Lee, S.M.; Song, S.H.; Kim, Y.S. Characterization of volatile components in *Makgeolli*, a traditional Korean rice wine, with or without pasteurization, during storage. *Molecules* **2013**, *18*, 5317–5325. [CrossRef]
5. Jiang, B.; Xi, Z.M.; Luo, M.J.; Zhang, Z.W. Comparison on aroma compounds in Cabernet Sauvignon and Merlot wines from four wine grape-growing regions in China. *Food Res. Int.* **2013**, *51*, 482–489. [CrossRef]
6. Peng, B.Z.; Li, F.L.; Cui, L.; Guo, Y.D. Effects of fermentation temperature on key aroma compounds and sensory properties of apple wine. *J. Food Sci.* **2015**, *80*, S2937–S2943. [CrossRef]
7. Chen, S.; Xu, Y. The influence of yeast strains on the volatile flavour compounds of Chinese rice wine. *J. Inst. Brew.* **2010**, *116*, 190–196. [CrossRef]
8. Fedrizzi, B.; Magno, F.; Badocco, D.; Nicolini, G.; Versini, G. Aging effects and grape variety dependence on the content of sulfur volatiles in wine. *J. Agric. Food Chem.* **2007**, *55*, 10880–10887. [CrossRef]
9. Liu, J.; Wu, Q.; Wang, P.; Lin, J.C.; Huang, L.; Xu, Y. Synergistic effect in core microbiota associated with sulfur metabolism in spontaneous Chinese liquor fermentation. *Appl. Environ. Microbiol.* **2017**, *83*, e01475-17. [CrossRef]
10. Peng, B.Z.; Yue, T.L.; Yuan, Y.H. Analysis of key aroma components in cider from *Shaanxi* (China) *Fuji* apple. *Int. J. Food Sci. Technol.* **2009**, *44*, 610–615. [CrossRef]
11. Luo, T. Characteristic Aroma of Light Aroma Style Chinese Rice Wines and Effects of Wheat Qu by GC-O Coupled with GC-MS. Master Thesis, Jiangnan University, Wuxi, China, 2008.
12. Du, R.B.; Liu, J.; Jiang, J.; Wang, Y.Q.; Ji, X.A.; Yang, N.; Wu, Q.; Xu, Y. Construction of a synthetic microbial community for the biosynthesis of volatile sulfur compound by multi-module division of labor. *Food Chem.* **2021**, *347*, 129036. [CrossRef]
13. Castillo-Lozano, M.L.D.; Spinnler, A.D.H.E.; Bonnarme, P.; Landaud, S. Comparison of volatile sulphur compound production by cheese-ripening yeasts from methionine and methionine–cysteine mixtures. *Appl. Microbiol. Biotechnol.* **2007**, *75*, 1447–1454. [CrossRef]
14. Seow, Y.X.; Ong, P.K.C.; Liu, S.Q. Production of flavour-active methionol from methionine metabolism by yeasts in coconut cream. *Int. J. Food Microbiol.* **2010**, *143*, 235–240. [CrossRef]
15. Buzzini, P.; Romano, S.; Turchetti, B.; Vaughan, A.; Pagnoni, U.M.; Davoli, P. Production of volatile organic sulfur compounds (VOSCs) by basidiomycetous yeasts. *FEMS Yeast Res.* **2005**, *5*, 379–385. [CrossRef]
16. Koh, M.K.P.; Sun, J.C.; Liu, S.Q. Optimization of L-methionine bioconversion to aroma-active methionol by *Kluyveromyces lactis* using the Taguchi method. *J. Food Res.* **2013**, *2*, 90–100. [CrossRef]
17. Hou, S.; Zhao, L.; Yang, X.L.; Wang, C.T. Study on conditions of bio-transformation of 3-(methylthio)-1-propanol by *Saccharomyces cerevisiae* SC408. *J. Food Sci. Technol.* **2011**, *29*, 18–22. [CrossRef]
18. Liu, L.; Yin, S.; Liang, J.R.; Wang, C.T.; Zhao, J.X. Cloning of the aminotransferase gene *ARO8* and its influence on the production of methionol in *Saccharomyces cerevisiae*. *Sci. Technol. Food Ind.* **2014**, *35*, 132–135. [CrossRef]
19. Wang, C.T.; Hou, S.; Zhao, L. Application of coupling of reaction and separation technique (CRST) to biocatalytic processes of natural 3-methylthiopropanol. In Proceedings of the 2010 First International Conference on Cellular, Molecular Biology, Biophysics and Bioengineering, Qiqihar, China, 25–26 December 2010; p. 4.
20. Zhang, Y.J.; Sun, Q.; Liu, X.Y.; Basit, R.A.; Ma, J.H.; Fu, Z.L.; Cheng, L.J.; Fan, G.S.; Teng, C. Screening, identification, and fermentation condition optimization of a high-yield 3-methylthiopropanol yeast and its aroma-producing characteristics. *Foods* **2024**, *13*, 418. [CrossRef]
21. Ma, J.H.; Cheng, L.J.; Zhang, Y.J.; Liu, Y.C.; Sun, Q.; Zhang, J.; Liu, X.Y.; Fan, G.S. Screening of yeasts isolated from *Baijiu* environments for producing 3-methylthio-1-propanol and optimizing production conditions. *Foods* **2022**, *11*, 3616. [CrossRef]
22. Che, Y.X.; Yin, S.; Wang, H.; Yang, H.Q.; Xu, R.X.; Wang, Q.; Wu, Y.P.; Boutet, J.; Huet, R.; Wang, C.T. Production of methionol from 3-methylthiopropionaldehyde by catalysis of the yeast alcohol dehydrogenase Adh4p. *J. Agric. Food Chem.* **2020**, *68*, 4650–4656. [CrossRef]
23. Yin, S.; Lang, T.D.; Xiao, X.; Liu, L.; Sun, B.G.; Wang, C.T. Significant enhancement of methionol production by co-expression of the aminotransferase gene *ARO8* and the decarboxylase gene *ARO10* in *Saccharomyces cerevisiae*. *FEMS Microbiol. Lett.* **2015**, *362*, fnu043. [CrossRef]
24. Deed, R.C.; Hou, R.; Kinzurik, M.I.; Gardner, R.C.; Fedrizzi, B. The role of yeast *ARO8*, *ARO9* and *ARO10* genes in the biosynthesis of 3-(methylthio)-1-propanol from L-methionine during fermentation in synthetic grape medium. *FEMS Yeast Res.* **2019**, *19*, foy109. [CrossRef]
25. Etschmann, M.M.W.; Kötter, P.; Hauf, J.; Bluemke, W.; Entian, K.D.; Schrader, J. Production of the aroma chemicals 3-(methylthio)-1-propanol and 3-(methylthio)-propylacetate with yeasts. *Appl. Microbiol. Bio.* **2008**, *80*, 579–587. [CrossRef]
26. Arfi, K.; Tache, R.; Spinnler, H.E.; Bonnarme, P. Dual influence of the carbon source and l-methionine on the synthesis of sulphur compounds in the cheese-ripening yeast *Geotrichum candidum*. *Appl. Microbiol. Bio.* **2003**, *61*, 359–365. [CrossRef]
27. Buchana, R.E.; Gibbons, N.E. *Bergey's Manual of Determinative Bacteriology*, 8th ed.; Science Press: Beijing, China, 1984; pp. 382–533.
28. Dong, X.Z.; Cai, M.Y. *Ommon Bacteria Manual System Identification*; Science Press: Beijing, China, 2001; pp. 348–392.

29. Fan, G.S.; Teng, C.; Xu, D.; Fu, Z.L.; Liu, P.X.; Wu, Q.H.; Yang, R.; Li, X.T. Improving ethyl acetate production in Baijiu manufacture by *Wickerhamomyces anomalus* and *Saccharomyces cerevisiae* mixed culture fermentations. *BioMed Res. Int.* **2019**, *2019*, 1470543. [CrossRef]
30. Fan, G.S.; Liu, P.X.; Wu, Q.H.; Fu, Z.L.; Zhu, Y.T.; Cheng, L.J.; Li, X.T. Isolation and identification of a high—Yield ethyl alcohol yeast in aerobic conditions and its biological characteristics. *Sci. Technol. Food Ind.* **2019**, *40*, 138–143. [CrossRef]
31. Wang, X.S.; Wang, B.W.; Sun, Z.G.; Tan, W.; Zheng, J.; Zhu, W.Y. Effects of modernized fermentation on the microbial community succession and ethyl lactate metabolism in Chinese baijiu fermentation. *Food Res. Int.* **2022**, *159*, 111566. [CrossRef]
32. Hong, L.X.; Fu, G.M.; Liu, T.; Chen, Y.R.; Wu, S.W.; Cai, W.Q.; Xie, Z.J.; Wan, Y. Functional microbial agents enhance ethanol contents and regulate the volatile compounds in Chinese *Baijiu*. *Food Biosci.* **2021**, *44*, 101411. [CrossRef]
33. Fan, G.S.; Xu, D.; Fu, Z.L.; Xu, C.Y.; Teng, C.; Sun, B.G.; Li, X.T. Screen of aroma-producing yeast strains from *Gujinggong daqu* and analysis of volatile flavor compounds produced by them. *J. Chin. Inst. Food Sci. Technol.* **2018**, *18*, 220–229. [CrossRef]
34. Fan, G.S.; Liu, P.X.; Chang, X.; Yin, H.; Cheng, L.J.; Teng, C.; Gong, Y.; Li, X.T. Isolation and identification of a high-yield ethyl caproate-producing yeast from *Daqu* and optimization of its fermentation. *Front. Microbiol.* **2021**, *12*, 663744. [CrossRef]
35. Sánchez, S.; Chávez, A.; Forero, A.; García-Huante, Y.; Romero, A.; Sánchez, M.; Rocha, D.; Sánchez, B.; Ávalos, M.; Guzmán-Trampe, S.; et al. Carbon source regulation of antibiotic production. *J. Antibiot.* **2010**, *63*, 442–459. [CrossRef]
36. Perpète, P.; Duthoit, O.; Maeyer, S.D.; Imray, L.; Lawton, A.I.; Stavropoulos, K.E.; Gitonga, V.W.; Hewlins, M.J.E.; Dickinson, J.R. Methionine catabolism in *Saccharomyces cerevisiae*. *FEMS Yeast Res.* **2006**, *6*, 48–56. [CrossRef]
37. Mouret, J.R.; Perez, M.; Angenieux, M.; Nicolle, P.; Farines, V.; Sablayrolles, J.M. Online-based kinetic analysis of higher alcohol and ester synthesis during winemaking fermentations. *Food Bioproc. Technol.* **2014**, *7*, 1235–1245. [CrossRef]
38. Lwa, H.; Sun, J.C.; Liu, S.Q. Optimization of methionol bioproduction by *Saccharomyces cerevisiae* using response surface methodology. *Ann. Microbiol.* **2015**, *65*, 197–205. [CrossRef]
39. Liang, Y.; Zhu, L.; Gao, M.J.; Zheng, Z.Y.; Wu, J.R.; Zhan, X.B. Influence of Tween-80 on the production and structure of water-insoluble curdlan from *Agrobacterium* sp. *J. Biol. Macromol.* **2018**, *106*, 611–619. [CrossRef]
40. Quek, J.M.B.; Seow, Y.X.; Ong, P.K.C.; Liu, S.Q. Formation of volatile sulfur-containing compounds by *Saccharomyces cerevisiae* in soymilk supplemented with l-methionine. *Food Biotechnol.* **2011**, *25*, 292–304. [CrossRef]
41. Chursov, A.; Kopetzky, S.J.; Bocharov, G.; Frishman, D.; Shneider, A. RNAtips: Analysis of temperature-induced changes of RNA secondary structure. *Nucleic Acids Res.* **2013**, *41*, W486–W491. [CrossRef]

foods

MDPI

Article

Assessment of Sourdough Fermentation Impact on the Antioxidant and Anti-Inflammatory Potential of Pearl Millet from Burkina Faso

Morena Gabriele [1,*], Andrea Cavallero [1], Elena Tomassi [1], Nafiou Arouna [1,2], Július Árvay [3], Vincenzo Longo [1] and Laura Pucci [1]

1 Institute of Agricultural Biology and Biotechnology, National Research Council, 56124 Pisa, Italy; andrea.cavallero@ibba.cnr.it (A.C.); elena.tomassi@ibba.cnr.it (E.T.); nafiou.arounaa@gmail.com (N.A.); vincenzo.longo@ibba.cnr.it (V.L.); laura.pucci@ibba.cnr.it (L.P.)
2 Department of Agricultural Sciences, University of Naples Federico II, Portici, 80055 Naples, Italy
3 Institute of Food Sciences, Faculty of Biotechnology and Food Sciences, Slovak University of Agriculture, 949 76 Nitra, Slovakia; julius.arvay@uniag.sk
* Correspondence: morena.gabriele@ibba.cnr.it

Abstract: Millet, a gluten-free cereal, has received attention for its environmental friendliness and higher protein content than other grains. It represents a staple food in many African countries, where fermentation is traditionally used for preserving food products and preparing different cereal-based products. This study aimed to assess the impact of sourdough fermentation on bioactive compounds and antioxidant and anti-inflammatory properties of pearl millet from Burkina Faso. Phenolic compounds were investigated spectrophotometrically and by HPLC-DAD. The antioxidant activity of unfermented (MF) and fermented (FeMF) millet was evaluated in vitro by spectrophotometric and fluorometric assays and ex vivo on oxidized human erythrocytes for hemolysis inhibition. Finally, the potential anti-inflammatory effect of FeMF and MF was evaluated on human adenocarcinoma cell line (HT-29) exposed to TNF-α inflammatory stimulus. Results revealed significantly higher levels of polyphenols, flavonoids, and in vitro antioxidant activity following millet fermentation. Notable differences in phenolic composition between FeMF and MF are observed, with fermentation facilitating the release of bioactive compounds such as gallic acid, quercetin, and rutin. A dose-dependent protection against oxidative hemolysis was observed in both FeMF- and MF-pretreated erythrocytes. Similarly, pretreatment with FeMF significantly reduced the levels of inflammatory markers in TNF-α-treated cells, with effects comparable to those of MF. Fermentation with sourdough represents a simple and low-cost method to improve the bioactive compounds content and in vitro antioxidant activity of millet flour with promising nutraceutical potential.

Keywords: pearl millet; sourdough fermentation; antioxidant capacity; HT-29 cell line; anti-inflammatory capacity

1. Introduction

Millet, a group of small-seeded gluten-free cereal grains, has long been consumed as a staple food in many regions across the world [1]. Recognizing its environmental friendliness and exceptional resilience to drought and pests, the Food and Agriculture Organization (FAO) designated 2023 as the "International Year of Millet" [2]. Particularly in arid and semi-arid regions grappling with the impacts of climate change, millets' intrinsic hardiness, especially their minimal water requirements, positions them as vital contributors to resilient agricultural systems [3].

During the last years, the interest in millet and its potential as a nutritious and functional food has increased [4]; millet grains offer a wealth of essential nutrients, dietary fiber, and bioactive compounds such as phenolic acids and flavonoids, which contribute to their healthy properties [1,5–7]. Indeed, epidemiological studies have consistently demonstrated a lower incidence of diabetes among individuals who regularly consume millets, suggesting a beneficial role of millets in modulating blood glucose release [1,8,9].

Moreover, millet and, in particular, pearl millet have a higher protein content compared to other cereals like sorghum and wheat [10]. Additionally, the application of fermentation techniques to millet flour has been explored to enhance its nutritional value and functional attributes [6].

Fermentation, a traditional food processing technique, has been utilized to ameliorate the nutritional and functional properties of various food commodities. Especially the fermentation with sourdough, due to its diversified microbial composition, increases the content of bioactive substances, improves protein digestibility, increases the bioavailability of minerals, lowers the glycemic index, allows for the enrichment of dietary fiber, and decreases the content of anti-nutritional factors [11,12]. In particular, the use of natural sourdough has shown a remarkable improvement in the content of bioactive molecules with high antioxidant power in many cereals in our previous articles [13,14].

Fermentation of millet flour has gained attention as a means to enhance its digestibility, bioavailability of nutrients, and release of bioactive compounds [15]. During fermentation, microorganisms, such as yeasts and lactic acid bacteria, metabolize the carbohydrates and proteins in millet flour, resulting in the production of organic acids, vitamins, enzymes, and beneficial metabolites [16,17]. These fermentation-derived products contribute to the improved nutritional quality and functional properties of millet flour [18]. Furthermore, during fermentation, enzymatic hydrolysis processes occur, resulting in the liberation of a vast array of bioactive peptides. These peptides have a diverse range of activities, which include anti-thrombotic, antimicrobial, antihypertensive, immunomodulatory, and antioxidative properties [19]. Cereal-fermented foods have been related to a reduced risk of cardiovascular diseases, diabetes, and cancer due to the presence of antioxidant molecules that can counteract the harmful effects of free radicals [20–22]. However, in unfermented cereals, the majority of natural antioxidants, like phenolic compounds, are often bound to cell walls, glycosylated, or in polymeric forms which affects their bio-accessibility [23,24].

Nowadays, the rising attention to a balanced and healthy diet has led to an increased demand for novel functional foods that can mitigate the risk of various pathologies associated with inflammation [25]. While several studies have examined the nutritional and functional changes that occur during millet fermentation, there is a need for an exhaustive understanding of the impact of fermentation on the bioactive compounds, antioxidant, and anti-inflammatory activity of millet flour [24]. Additionally, the potential health benefits of consuming fermented millet flour and its applications in the development of functional foods warrant further investigation.

This work dealt with the impact of fermentation with sourdough on the nutraceutical properties of pearl millet from Burkina Faso. This gluten-free cereal represents a staple food in many African countries, where fermentation is the principal process for preserving food products and is traditionally used to prepare numerous cereal-based products. Therefore, this work aimed to evaluate the effect of sourdough fermentation on the phytochemical content and antioxidant capacity of the pearl millet flour (*Pennisetum glaucum* L.) by in vitro assays and as hemolysis inhibition on oxidized human erythrocytes. Moreover, the anti-inflammatory potential of fermented and unfermented millet flour was evaluated on the human adenocarcinoma cell (HT-29) line under inflammatory stimulus. Through a comprehensive analysis, the work proposes to shed light on the potential benefits and applications of fermented millet flour as a functional ingredient in the food industry.

2. Materials and Methods

2.1. Chemicals and Reagents

All standards and reagents were of analytical grade. Sodium acetate, sodium carbonate, sodium hydroxide, potassium chloride, potassium persulfate, catechin hydrate, gallic acid, fluorescein sodium salt, quercetin dihydrate, 2,2′-azobis (2-amidinopropane) dihydrochloride (AAPH), 2,2-azinobis(3-ethylbenzthiazoline-6-sulphonic acid) diammonium salt (ABTS), 1,1-diphenyl-2-picrylhydrazyl (DPPH), 2′,7′-dichlorofluorescein diacetate (DCFH-DA), Folin-Ciocalteu reagent, 6-hydroxy-2,5,7,8-tetramethylchromane-2-carboxylic acid (Trolox), dimethyl sulfoxide (DMSO), iso-quercitrin, 4-OH benzoic acid, rutin, trans-ferulic acid, trans-p-coumaric acid, vanillic acid, vitexin, acetonitrile (HPLC grade), ethanol, hydrochloric acid, methanol (HPLC grade), phosphoric acid (ACS grade), trichloroacetic acid, cell culture media and supplements were purchased from Sigma-Aldrich (St. Louis, MO, USA). Aluminum chloride and sodium nitrite were purchased from Carlo Erba (Milan, Italy). A Simplicity 185 purification system (Millipore SAS, Molsheim, France) was used to treat double deionized water (ddH$_2$O; 18.2 MΩ/cm, 20 °C).

2.2. Plant Material and Extraction

Pearl millet, i.e., *Pennisetum glaucum* L. based on sample caryopses morphological evaluation, was purchased from markets in Burkina Faso. Millet flour was obtained by grinding millet grains using a laboratory mill. The fermentation of millet flour was carried out for 96 h using a natural sourdough consisting of different species of Lactobacilli, mainly *L. sanfranciscensis* and *L. pentosus*, and yeast strains, as previously described by Balli et al. [24]. Water (60%) was added to moisten the millet flour (40%), and then the sourdough, provided by "Lievitamente s.r.l." (Viareggio-LU, Italy), was added to initiate fermentation. The fermentation temperature was maintained at around 38 °C, while the pH reached a value of 4. Once fermented, the product was frozen and lyophilized for two days using a freeze-dryer Lyovac GT 2 (SRK Systemtechnik, Riedstadt, Germany). The unfermented (MF) and fermented (FeMF) millet flour was suspended with 10% DMSO in distilled water, sonicated (three cycles: 10 s on/10 s off), and shaken gently for 1 h; samples were centrifuged for 10 min at 2300× *g* at 4 °C (Jouan CR3i centrifuge, Newport Pagnell, UK), and the supernatant was collected, filtered (0.2 µm VWR International PBI, Milan, Italy), and kept at 4 °C in the dark until use [26]. The extraction was carried out in triplicate. These extracts were used for all the analyses if not otherwise specified.

2.3. Total Polyphenols and Flavonoids Content

The unfermented and fermented millet extracts were screened for total polyphenol and flavonoid content as described by Kim et al. [27], with some modifications. Total polyphenols, estimated as Folin-Ciocalteu (FC) reducing capacity, were expressed as mg of gallic acid equivalents (GAE)/g dry weight (dw). Extracts or standard (100 µL) were mixed with 0.2 N Folin-Ciocalteu reagent (500 µL) and incubated for 5 min in the dark. Then, 0.7 M sodium carbonate (400 µL) was added, and the absorbance was measured at 760 nm, after 2 h of incubation at room temperature in the dark. Total flavonoids were quantified using the aluminum chloride colorimetric method and expressed as mg catechin equivalents (CE)/g dw. Extracts or standard (200 µL) were mixed with dH$_2$O (800 µL) and 5% sodium nitrite (60 µL) for 5 min at room temperature. Then, 10% aluminum chloride (60 µL) was added, incubated for 6 min, and finally reactions were neutralized with 1M sodium hydroxide (400 µL). Absorbance was measured at 430 nm after 30 min of incubation at room temperature.

2.4. Quantification of Polyphenols by HPLC-DAD

The unfermented and fermented millet samples (3 g) were extracted with 20 mL of 80% methanol at laboratory temperature for 2 h by horizontal shaker Unimax 2010 (Heidolph Instruments, GmbH, Schwabach, Germany). The extract was filtered through Munktell No 390 paper (Munktell & Filtrak GmbH, Bärenstein, Germany) and stored in closed 20 mL PE vial tubes. Before HPLC analysis, the extract was filtered through syringe filter Q-Max (0.22 µm, 25 mm, PVDF) (Frisenette ApS, Knebel, Denmark). All compounds were determined using an Agilent 1260 Infinity HPLC (Agilent Technologies GmbH, Wäldbronn, Germany) with quaternary solvent manager coupled with degasser (G1311B), sampler manager (G1329B), column manager (G1316A) and DAD detector (G1315C). All HPLC analyses were performed on a Purosphere® reverse phase C18 column (250 mm × 4 mm × 5 μm) (Merck KGaA, Darmstadt, Germany). The mobile phase consisted of gradient acetonitrile (A) and 0.1% phosphoric acid in ddH_2O (B). The gradient elution was as follows: first, 0–1 min isocratic elution (20% A and 80% B), 1–5 min linear gradient elution (25% A and 75% B), 5–15 min (30% A and 70% B) and 15–25 min (40% A and 60% B). The post-run was 3 min. The initial flow rate was 1 mL/min, and the injection volume was 5 μL. Column thermostat was set up to 30 °C and the samples were kept at 4 °C in the sampler manager. The detection wavelengths were set up at 265 nm (4-OH benzoic acid, vanillic acid, rutin), 320 nm (gallic acid, vitexin, trans-*p*-coumaric acid, trans-ferulic acid), and 372 nm (isoquercitrin, quercetin). Data were collected and processed using Agilent Open Lab Chem Station software 2.8 for LC 3D systems.

2.5. *In Vitro Antioxidant Activity*

2.5.1. DPPH Radical Scavenging Activity

The DPPH assay was carried out for the evaluation of the radical scavenging activity of MF and FeMF extracts. An aliquot of extract (50 μL) was added to a methanolic solution of DPPH 60 μM (1950 μL) and mixed for 30 min in the dark at 30 °C. The absorbance was recorded at 517 nm by a Perkin-Elmer Lambda 365 spectrophotometer and the extract concentration corresponding to 50% DPPH inhibition (EC_{50}) was determined according to Guimaraes et al. [28].

2.5.2. ABTS Radical Scavenging Activity

The ABTS assay was carried out according to the method described by Chelucci et al. [29]. ABTS radical cations (ABTS+) were produced through the oxidation of ABTS using potassium persulfate and were subsequently reduced by antioxidants that donate hydrogen. Fresh ABTS stock solution (80 μL of 140 mM potassium persulfate added to 5 mL 7 mM ABTS) was shaken overnight in the dark, then diluted with methanol to obtain an absorbance of 0.7 ± 0.02 at 734 nm (Perkin Elmer UV/VIS Lambda 365, Waltham, MA, USA). Lastly, 190 μL of extract were added to 1 mL of ABTS diluted solution, and the absorbance was read at 734 nm after 10 min incubation. Trolox served as the standard antioxidant. The extent of ABTS+ inhibition was determined, and the result was presented as μmol Trolox equivalent/g dw (TEAC).

2.5.3. Ferric Reducing Antioxidant Power (FRAP) Assay

The FRAP assay was carried out according to Colosimo et al. [14] and 85 μL of aqueous extracts were mixed with 2500 μL of freshly prepared solution of FRAP containing TPTZ 10 mM in HCl 40 mM, $FeCl_3{\cdot}6H_2O$ 20 mM, and acetate buffer 300 mM (pH 3.6) at a ratio of 1:1:10. The sample was gently vortexed and incubated at room temperature for 6 min. The absorbance was then recorded at 593 nm, and the results, expressed as Fe^{2+} equivalent (μM), were estimated using a calibration curve of an aqueous solution of $FeSO_4{\cdot}7H_2O$ (100–2000 μM).

2.5.4. Oxygen Radical Absorbance Capacity (ORAC) Assay

The ORAC assay was carried out as previously described [29]. AAPH was used as a peroxyl radical generator and fluorescein as a probe. The final reaction mixture contained 0.04 mM fluorescein sodium salt in 0.075 M phosphate buffer, pH 7.4, diluted sample, or 5 mM Trolox. The control was 0.075 M phosphate buffer, pH 7.4. The fluorescence decay of fluorescein was measured at λ_{ex} 485 nm and λ_{em} 514 nm (VictorTM X3 Multilabel Plate Reader, Waltham, MA, USA), with Trolox utilized as a standard. ORAC values were quantified in μmol Trolox equivalent (TE)/100 g dw.

2.6. Oxidative Hemolysis of Erythrocytes

Human blood samples were centrifuged at 2300× g at 4 °C for 10 min. After discarding the buffy coat and plasma, erythrocytes were washed twice with PBS pH 7.4. The hemolysis of erythrocyte was quantified at 540 nm according to Mikstacka et al. [30], by measuring the hemoglobin (Hb) released in the supernatant of human erythrocyte pre-treated with FeMF and MF extracts (0.2, 0.4 and 0.8 mg mL^{-1}), then exposed to AAPH peroxyl radicals. Results were presented as a percentage of hemolysis inhibition relative to the control (AAPH-only treated cells). Erythrocytes not treated with AAPH served as a blank sample to account for spontaneous hemolysis.

2.7. Human Intestinal Cell Culture

The HT-29 (human colonic adenocarcinoma, DSMZ, Braunschweig, Germany) cell line was cultured in DMEM/F12 medium supplemented with 100 μg mL^{-1} streptomycin, 100 units mL^{-1} penicillin, and 10% fetal bovine serum at 37 °C in a 5% CO_2 humidified incubator. For treatments, DMEM/F12 medium devoid of phenol red and FBS but containing antibiotics was used. Before exposure to extracts or TNFα, cells underwent a serum starvation phase of 1 h. Following this, HT-29 cells were either pre-treated for 1 h with 0.08 and 0.4 mg mL^{-1} of MF and FeMF extract or not, before being stimulated for 24 h with or without 5 ng mL^{-1} of TNFα. HT-29 were seeded into a 96-well plate (10^4 cells/well) and cell viability was evaluated by MTT assay as previously described [26].

2.8. RT-PCR and Quantitative Real-Time RT-PCR

For gene expression analysis, HT-29 were seeded into a 6-well plate and treated once confluent. Total RNA was extracted and reverse-transcribed using the E.Z.N.A.® Total RNA Kit I (OMEGA bio-tek, Norcross, GA, USA) and the iScriptTM cDNA Synthesis Kit (Bio-Rad, Hercules, CA, USA), respectively. Then, the SsoFastTM EvaGreen® Supermix (Bio-Rad, Hercules, CA, USA) and β-actin, ICAM-1, IL-8, BAX, COX-2, and HO-1 primers [26] were used for quantitative real-time RT-PCR analysis (CFX Connect Real-Time PCR Detection System, Bio-Rad, Hercules, CA, USA). Each gene was analyzed in triplicate, and its expression was determined using the $2^{-\Delta\Delta CT}$ method for relative quantification [26]. Results are presented as the fold change in expression levels relative to the control samples.

2.9. Statistical Analysis

Statistical analysis was carried out using GraphPad Prism 9 for macOS (GraphPad Software, San Diego, CA, USA). The results were expressed as mean values ± standard deviation (SD) after carrying out the experiments in triplicate. ANOVA (One-way analysis of variance) using Dunnett's multiple comparison tests was used to examine sample differences. For FeMF and MF comparison unpaired Student's *t*-test was used. A *p*-value lower than 0.05 is considered as statistically significant.

3. Results

3.1. Content of Bioactive Compounds and Antioxidant Activities In Vitro

The bioactive compound content and the in vitro antioxidant activities, presented in Table 1, were determined in both fermented (FeMF) and unfermented millet flour (MF).

As demonstrated in Table 1, the levels of bioactive compounds exhibited a significant increase following millet fermentation. Specifically, total polyphenols were approximately 1.7-fold, and the flavonoid content was approximately 2.2-fold in the FeMF flour compared to the MF counterpart.

Table 1. Comparison of total polyphenols and flavonoids and in vitro antioxidant activities (DPPH, FRAP, and ORAC) between fermented (FeMF) and unfermented (MF) millet flour. Data represent mean values ± SD from three replicates. Unpaired Student *t*-test: significantly different from MF: * $p < 0.05$, ** $p < 0.01$, *** $p < 0.001$, ns: not statistically significant.

	TP (mg GAE/g dw)	TF (mg CE/g dw)	DPPH EC_{50} (mg mL^{-1})	FRAP (μM Fe^{2+})	ABTS (μmol TE/g dw)	ORAC (μmol TE/100g dw)
FeMF	3.25 ± 0.09 ***	2.84 ± 0.39 *	1.83 ± 0.23 ***	1476.3 ± 0.92 **	19.09 ± 2.64 ns	640.4 ± 30.97 *
MF	1.83 ± 0.02	1.29 ± 0.53	3.16 ± 0.08	1336.2 ± 30.33	15.65 ± 1.95	498.4 ± 78.48

TP: total polyphenols; GAE: gallic acid equivalents; dw: dry weight; TF: total flavonoids; CE: catechin equivalents; DPPH: 2,2-diphenyl-1-picrylhydrazyl; EC_{50}: half maximal effective concentration; FRAP: Ferric Reducing Antioxidant Power; TE: Trolox equivalent; ORAC: Oxygen Radical Absorbance Capacity.

In terms of in vitro antioxidant activity, various assays were employed, including DPPH, FRAP, ABTS, and ORAC. Similarly, fermentation substantially enhanced the antioxidant properties of millet flour. The FeMF sample displayed a slight increase in FRAP (1.1-fold), ABTS (1.2-fold), and ORAC (1.3-fold) values. Moreover, a 1.7-fold reduction of EC_{50} compared to the MF sample was observed for the DPPH assay.

3.2. HPLC-DAD Quantification of Phenolic Compounds

Table 2 presents the analysis of phenolic compounds in both MF and FeMF flours utilizing HPLC-DAD. Gallic acid, rutin, and quercetin were exclusively identified in the fermented millet flour. Instead, trans-*p*-coumaric acid was detected only in MF. The presence of 4-OH benzoic acid, vitexin, and vanillic acid was observed in both fractions, but their concentrations were significantly higher in fermented millet, exhibiting a fold increase of 1.4, 1.7, and 5.2, respectively.

Table 2. Phenolic compound, concentration (mg/kg dw), wavelengths (nm), class and retention time (Rt, min), in the fermented (FeMF) and unfermented (MF) millet flour. Unpaired Student *t*-test: significantly different from MF: *** $p < 0.001$.

Phenolic Compound	Concentration		Wavelength	Class	Rt
	FeMF	MF			
Gallic acid	106.26 ± 1.30	<LOD	320	Phenolic acid	2.669
4-OH Benzoic acid	141.15 ± 1.24 ***	98.80 ± 3.27	265	Carboxylic acid	4.830
Vanillic acid	39.58 ± 0.55 ***	23.87 ± 0.74	265	Phenolic acid	5.260
Rutin	4.49 ± 0.11	<LOD	265	Flavonol	5.955
Vitexin	296.88 ± 2.50 ***	56.88 ±1.80	320	Flavone	6.275
Iso-Quercitrin	<LOD	<LOD	372	Flavonol	6.987
trans-*p*-Coumaric acid	<LOD	6.48 ± 0.42	320	Hydroxycinnamic acid	7.444
trans-Ferulic acid	<LOD	<LOD	320	Hydroxycinnamic acid	8.219
Quercetin	0.96 ± 0.06	<LOD	372	Flavonol	17.363

LOD: limit of detection.

3.3. Hemolysis Test

The protective effect of growing concentrations of MF and FeMF flour extract toward AAPH-induced hemolysis stress in erythrocytes is illustrated in Figure 1.

As shown in Figure 1, both MF and FeMF, at 0.4 and 0.8 mg mL^{-1}, showed lysis protection values superimposable to the highest quercetin concentration (8 μM). Moreover, compared to control cells, MF and FeMF at the same doses significantly protected, with comparable effects, erythrocytes from oxidative hemolysis ($p < 0.001$ vs. control). Therefore, millet fermentation does not improve or worsen the ability of millet to protect erythrocytes from oxidative hemolysis.

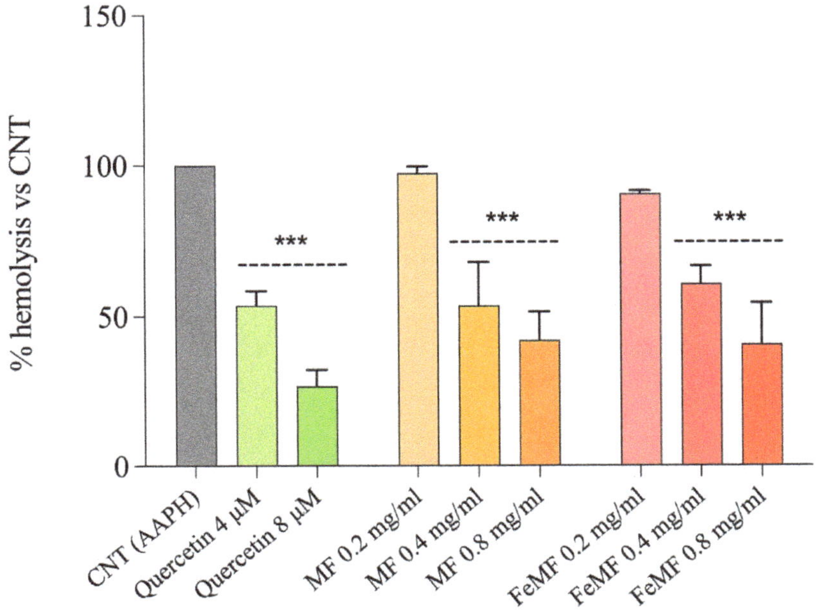

Figure 1. Protective effect of increasing concentrations (0.2, 0.4, and 0.8 mg mL^{-1}) of fermented (FeMF) and unfermented (MF) millet extracts on AAPH-induced hemolysis in erythrocytes. Quercetin (4 and 8 μM) was used as a standard. Results are the mean ± SD of at least three replicates. One-way ANOVA followed by Dunnet's Multiple Comparison test: significantly different from control, AAPH-treated cells (100% hemolysis), *** $p < 0.001$.

3.4. Protective Effect against TNFα-Induced Intestinal Alterations of FeMF and MF Extracts

The potential protective effect of MF and FeMF against TNF-α-induced alterations was investigated in a human colonic adenocarcinoma cell line. Cell viability was determined by MTT assay to ensure the absence of cytotoxic effects. Subsequently, the effect of MF and FeMF was evaluated under basal conditions and during TNF-α-induced inflammation by examining the gene expression levels of markers associated with inflammation (IL-8, COX-2, and ICAM-1), oxidative stress (HO-1), and apoptosis (BAX) (Figure 2).

Under the basal condition, pre-treatment with MF and FeMF did not alter the expression of IL-8, ICAM-1, HO-1, and BAX. Instead, pre-treatment with MF, at both concentrations, reduced considerably the basal level of COX-2 ($p < 0.05$ and $p < 0.01$ vs. control).

Conversely, TNF-α treatment led to an upregulation of all genes involved in inflammation (IL-8, COX-2, and ICAM-1; $p < 0.001$ for all) and apoptosis (BAX; $p < 0.01$) compared to the control cells. Under the inflammatory condition, pre-treatment with MF, at both concentrations, effectively mitigated the effects induced by TNF-α exposure. It significantly decreased the expression of IL-8 ($p < 0.001$), COX-2 ($p < 0.001$), and ICAM-1 ($p < 0.001$) while not affecting the expression levels of HO-1 and BAX. Similarly to MF, FeMF significantly reduced IL-8, COX-2, and ICAM-1 gene expression ($p < 0.001$ for all) in TNF-α-induced inflamed cells. Therefore, fermentation of millet with sourdough did not improve or worsen the anti-inflammatory potential of millet flour.

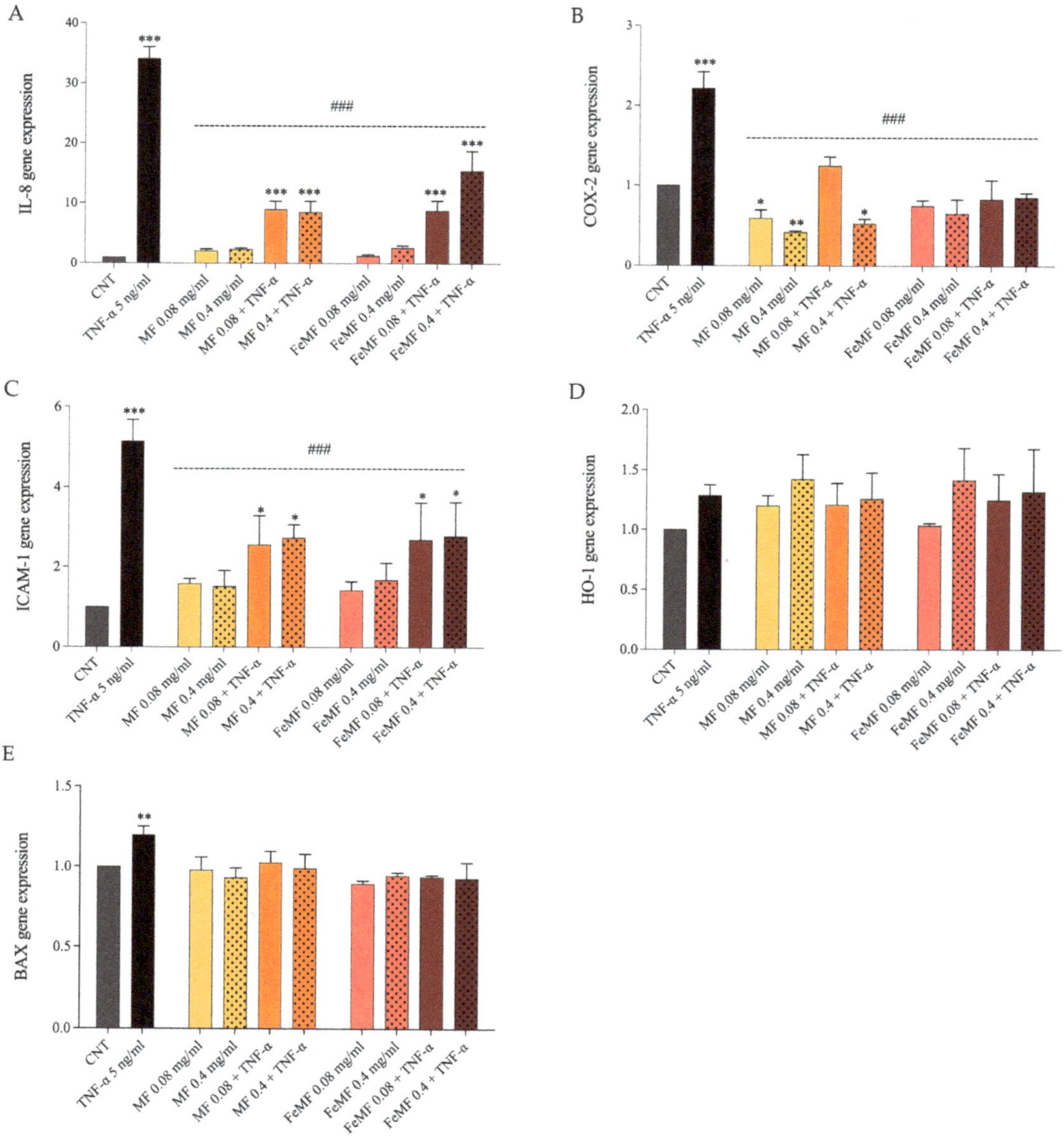

Figure 2. Evaluation of IL-8 (**A**), COX-2 (**B**), ICAM-1 (**C**), HO-1 (**D**), and BAX (**E**) expression by quantitative real-time RT-PCR in HT-29 cells. Cells were pre-treated for 1 h with either 0.08 or 0.4 mg mL^{-1} of fermented (FeMF) or unfermented (MF) millet extract, followed by a 24 h exposure to 5 ng mL^{-1} TNF-α. Experiments were carried out in triplicate and results are presented as the fold-change in expression levels relative to the control samples. One-way ANOVA with Dunnett's multiple comparison test: * different from control (CNT), * $p < 0.05$, ** $p < 0.01$, *** $p < 0.001$; different from TNF-α, ### $p < 0.001$.

4. Discussion

This work investigated the impact of sourdough fermentation on the bioactive composition and content of millet flour, together with its antioxidant and anti-inflammatory properties. Millet is an interesting gluten-free cereal, suitable for individuals with celiac disease, with similar or higher nutritional properties than conventional cereals, thanks to its high energy values, lipids, high-quality proteins, and a source of dietary fiber and micronutrients like calcium, iron, and zinc [31], besides climate-change resilient characteristics and low glycemic index suitable for diabetic patients [32].

Results showed that fermentation with sourdough increased the total polyphenol and flavonoid content of millet flour by approximately 1.7- and 2.2-fold, respectively. Moreover, following fermentation, we observed a change in the phenolic HPLC profile. The increased content of total polyphenols and flavonoids is in line with Gabaza et al. [33]. Moreover, fermentation enhances the release of flavonoids from cereals and induces the transformation of various polyphenolic compounds, improving their antioxidative activity [16]. Notably, the activity of enzymes produced and released by lactic bacteria during fermentation plays a crucial role in breaking down cereal cells, facilitating the extraction of flavonoids, and promoting the depolymerization of high molecular weight phenolic compounds [34,35]. Consequently, fermentation emerges as a valuable method for augmenting the availability of natural antioxidants.

The HPLC-DAD analysis revealed different phenolic compound concentrations and profiles in fermented and unfermented millet flour. The results indicated notable disparities in the phenolic compositions of the two types of flours. Among them, vitexin was the most abundant in fermented millet, followed by 4-OH benzoic acid, gallic acid, and vanillic acid. Moreover, gallic acid, quercetin, and rutin were exclusively identified in the fermented millet flour, indicating that fermentation might facilitate the release of these compounds, as supported by the literature [36]. These bioactive components enhance the body's redox defense, preventing oxidative stress and cellular injury caused by free radicals [36]. Contrary to the findings of Salar et al. [37], trans-*p*-coumaric acid was detected only in the unfermented millet flour.

The fermented millet flour showed an important increase in the in vitro antioxidant activities, compared to the unfermented one, in line with the increment in the phenolic and flavonoid content. Microbial enzymes, produced by fermentation can hydrolyze glucosides, and break down plant cell walls or starch, facilitating the extraction of bio-compounds, like quercetin and rutin, improving total phenolic content and their free radical scavenging capacity [16,38,39]. No significant increase was observed in the ABTS test values. This assay is based on the single-electron transfer (SET) mechanism, otherwise, the antioxidant tests DPPH, FRAP, and ORAC are based on a mechanism accompanied by the donation of a hydrogen atom (HAT) [40,41]. The latter correlates with the quantity of polyphenols and also with the quality of the phenolic compound, based on the number of hydroxyl groups in the molecule [42].

The hemolysis test was used to evaluate the antioxidant properties of bioactive compounds relying on the inhibition of lipid peroxidation, induced by AAPH radicals, on the erythrocyte membrane. The ability of the antioxidant to scavenge the peroxyl radicals correlates with the antihemolytic activity [43]. Our results indicate that both FeMF and MF, at the highest concentrations, provide significant protection against oxidative hemolysis compared to control cells, exposed only to AAPH, which show the highest hemolysis values. The level of protection provided by both MF and FeMF is comparable to the highest dose of quercetin, a well-known antioxidant compound [44]. These findings demonstrate that both MF and FeMF possess good anti-hemolytic properties, indicating their potential as protective agents against oxidative damage in erythrocytes; moreover, based on our results, the fermentation process did not improve or worse the antioxidant potential of millet flour against erythrocytes lysis.

Finally, the anti-inflammatory effect of pre-treatment with FeMF and MF extracts on the HT-29 inflamed cell model was investigated. The cells were exposed to TNF-α as an inflammatory stimulus for 24 h, and the expression of genes involved in inflammation (IL-8, COX-2, and ICAM-1), oxidative stress (HO-1), and apoptosis (BAX) was assessed. The results showed that pre-treatment with MF, at both concentrations (0.08 and 0.4 mg

mL^{-1}), was effective in mitigating the alterations induced by TNF-α exposure. This is evidenced by a significant decrease in the expression levels of IL-8, COX-2, and ICAM-1 genes, indicating the potential anti-inflammatory effects of MF pre-treatment. These findings are consistent with previous observations in lipopolysaccharide-inflamed HT-29 cells pre-treated with bound polyphenols from foxtail millet bran [45]. Instead, the expression levels of HO-1 and BAX genes were not affected by MF pre-treatment. Contrary to what was observed with fermented *T. dicoccum* (spelt) flour [13], our results indicate that the fermentation of millet flour with sourdough did not improve the anti-inflammatory properties of unfermented flour. Indeed, both MF and FeMF showed similar abilities in reducing IL-8, COX-2, and ICAM-1 gene expression in TNF-α-induced inflamed cells, although higher phenolic compounds were detected in FeMF than MF.

5. Conclusions

Our study points out that sourdough fermentation improves the phenolic profile and in vitro antioxidant capacity of millet flour while not affecting its anti-hemolytic and anti-inflammatory properties. Both fermented and unfermented millet flour provided significant dose-dependent protection to human erythrocyte cells against oxidative hemolysis. Similarly, pretreatment with fermented millet significantly reduced IL-8, ICAM-1, and COX-2 gene expression in inflamed HT-29, with comparable effects to those obtained with unfermented flour. The simplicity and affordability of the fermentation make it a practical choice for enhancing the overall quality of millet; in future studies, the selection of a single strand of lactic acid bacteria or a consortium of selected yeast and lactobacillus could be effective in improving the antioxidant and anti-inflammatory properties of millet on human cells, conferring an additional nutraceutical and nutritional value to millet flour.

Author Contributions: Conceptualization, M.G. and L.P.; validation, M.G.; formal analysis, M.G. and A.C.; investigation, M.G., A.C., E.T., N.A. and J.Á.; resources, M.G., J.Á., V.L. and L.P.; writing—original draft preparation, M.G., A.C. and E.T.; writing—review and editing, M.G. and L.P.; supervision, M.G.; project administration, L.P.; funding acquisition, M.G. and L.P. All authors have read and agreed to the published version of the manuscript.

Funding: This research received financial support from the CNR project FOE-2021 DBA.AD005.225.

Institutional Review Board Statement: Not applicable.

Data Availability Statement: The original contributions presented in the study are included in the article, further inquiries can be directed to the corresponding author.

Acknowledgments: The authors would like to thank Valter Lubrano and Fondazione G. Monasterio CNR-Regione Toscana (Pisa, Italy) for providing human blood samples.

Conflicts of Interest: The authors declare no conflict of interest. The funders had no role in the design of the study; in the collection, analyses, or interpretation of data; in the writing of the manuscript; or in the decision to publish the results.

Abbreviations

AAPH: 2,2′-azobis (2-amidinopropane) dihydrochloride; ABTS: 2,2-azinobis(3-ethylbenzthiazoline-6-sulphonic acid) diammonium salt; BAX: BCL2 associated X, apoptosis regulator; CE: catechin equivalents; COX-2: Cyclooxygenase-2; DMSO: dimethyl sulfoxide; DPPH: 1,1-diphenyl-2-picrylhydrazyl; DPPH: 2,2-diphenyl-1-picrylhydrazyl; dw: dry weight; EC_{50}: half maximal effective concentration; FBS: fetal bovine serum; FeMF: fermented millet flour; FRAP: Ferric Reducing Antioxidant Power; FRAP: Ferric Reducing Antioxidant Power; GAE: gallic acid equivalents; HO-1: heme oxygenase-1; HPLC-DAD: High-performance liquid chromatography with diode-array detection; ICAM-1: intracellular adhesion molecule; ICAM-1: Intracellular adhesion molecule-1; IL-8: C-X-C motif chemokine ligand 8; MF: unfermented millet flour; ORAC: Oxygen Radical Absorbance Capacity; TE: Trolox equivalent; TF: total flavonoids; TNF-α: tumor necrosis factor-alpha; TP: total polyphenols; Trolox: 6-hydroxy-2,5,7,8-tetramethylchromane-2-carboxylic acid.

References

1. Saleh, A.S.M.; Zhang, Q.; Chen, J.; Shen, Q. Millet Grains: Nutritional Quality, Processing, and Potential Health Benefits. *Compr. Rev. Food Sci. Food Saf.* **2013**, *12*, 281–295. [CrossRef]
2. International Year of Millets: Unleashing the Potential of Millets for the Well-Being of People and the Environment. Available online: https://www.fao.org/newsroom/detail/international-year-of-millets-unleashing-the-potential-of-millets-for-the-well-being-of-people-and-the-environment/en (accessed on 7 February 2024).
3. Taylor, J.R.N. Chapter 1—Sorghum and millets: Taxonomy, history, distribution, and production. In *Sorghum and Millets*, 2nd ed.; Taylor, J.R.N., Duodu, K.G., Eds.; AACC International Press: Concord, CA, USA, 2019; pp. 1–21, ISBN 978-0-12-811527-5.
4. Mahajan, P.; Bera, M.B.; Panesar, P.S.; Chauhan, A. Millet Starch: A Review. *Int. J. Biol. Macromol.* **2021**, *180*, 61–79. [CrossRef]
5. Serna-Saldivar, S.O.; Espinosa-Ramírez, J. Chapter 5—Grain structure and grain chemical composition. In *Sorghum and Millet*, 2nd ed.; Taylor, J.R.N., Duodu, K.G., Eds.; AACC International Press: Concord, CA, USA, 2019; pp. 85–129, ISBN 978-0-12-811527-5.
6. Abah, C.R.; Ishiwu, C.N.; Obiegbuna, J.E.; Oladejo, A.A. Nutritional Composition, Functional Properties and Food Applications of Millet Grains. *Asian Food Sci. J.* **2020**, *14*, 9–19. [CrossRef]
7. Sharma, R.; Sharma, S.; Dar, B.N.; Singh, B. Millets as Potential Nutri-Cereals: A Review of Nutrient Composition, Phytochemical Profile and Techno-Functionality. *Int. J. Food Sci. Technol.* **2021**, *56*, 3703–3718. [CrossRef]
8. Kam, J.; Puranik, S.; Yadav, R.; Manwaring, H.R.; Pierre, S.; Srivastava, R.K.; Yadav, R.S. Dietary Interventions for Type 2 Diabetes: How Millet Comes to Help. *Front. Plant Sci.* **2016**, *7*, 1454. [CrossRef]
9. Pei, J.; Umapathy, V.R.; Vengadassalapathy, S.; Hussain, S.F.J.; Rajagopal, P.; Jayaraman, S.; Veeraraghavan, V.P.; Palanisamy, C.P.; Gopinath, K. A Review of the Potential Consequences of Pearl Millet (*Pennisetum glaucum*) for Diabetes Mellitus and Other Biomedical Applications. *Nutrients* **2022**, *14*, 2932. [CrossRef]
10. Belton, P.S.; Taylor, J.R.N. Sorghum and Millets: Protein Sources for Africa. *Trends Food Sci. Technol.* **2004**, *15*, 94–98. [CrossRef]
11. Gabriele, M.; Pucci, L. Chapter 23—Fermentation and germination as a way to improve cereals antioxidant and antiinflammatory properties. In *Current Advances for Development of Functional Foods Modulating Inflammation and Oxidative Stress*; Hernández-Ledesma, B., Martínez-Villaluenga, C., Eds.; Academic Press: Cambridge, MA, USA, 2022; pp. 477–497, ISBN 978-0-12-823482-2.
12. Arora, K.; Ameur, H.; Polo, A.; Di Cagno, R.; Rizzello, C.G.; Gobbetti, M. Thirty Years of Knowledge on Sourdough Fermentation: A Systematic Review. *Trends Food Sci. Technol.* **2021**, *108*, 71–83. [CrossRef]
13. Gabriele, M.; Arouna, N.; Árvay, J.; Longo, V.; Pucci, L. Sourdough Fermentation Improves the Antioxidant, Antihypertensive, and Anti-Inflammatory Properties of *Triticum dicoccum*. *Int. J. Mol. Sci.* **2023**, *24*, 6283. [CrossRef]
14. Colosimo, R.; Gabriele, M.; Cifelli, M.; Longo, V.; Domenici, V.; Pucci, L. The Effect of Sourdough Fermentation on *Triticum dicoccum* from Garfagnana: 1H NMR Characterization and Analysis of the Antioxidant Activity. *Food Chem.* **2020**, *305*, 125510. [CrossRef]
15. Yang, T.; Ma, S.; Liu, J.; Sun, B.; Wang, X. Influences of Four Processing Methods on Main Nutritional Components of Foxtail Millet: A Review. *Grain Oil Sci. Technol.* **2022**, *5*, 156–165. [CrossRef]
16. Hur, S.J.; Lee, S.Y.; Kim, Y.-C.; Choi, I.; Kim, G.-B. Effect of Fermentation on the Antioxidant Activity in Plant-Based Foods. *Food Chem.* **2014**, *160*, 346–356. [CrossRef]
17. Shiferaw Terefe, N.; Augustin, M.A. Fermentation for Tailoring the Technological and Health Related Functionality of Food Products. *Crit. Rev. Food Sci. Nutr.* **2020**, *60*, 2887–2913. [CrossRef]
18. Yousaf, L.; Hou, D.; Liaqat, H.; Shen, Q. Millet: A Review of Its Nutritional and Functional Changes during Processing. *Food Res. Int.* **2021**, *142*, 110197. [CrossRef]
19. Martinez-Villaluenga, C.; Peñas, E.; Frias, J. Chapter 2—Bioactive peptides in fermented foods: Production and evidence for health effects. In *Fermented Foods in Health and Disease Prevention*; Frias, J., Martinez-Villaluenga, C., Peñas, E., Eds.; Academic Press: Cambridge, MA, USA, 2017; pp. 23–47, ISBN 978-0-12-802309-9.
20. Wu, H.; Rui, X.; Li, W.; Xiao, Y.; Zhou, J.; Dong, M. Whole-Grain Oats (*Avena sativa* L.) as a Carrier of Lactic Acid Bacteria and a Supplement Rich in Angiotensin I-Converting Enzyme Inhibitory Peptides through Solid-State Fermentation. *Food Funct.* **2018**, *9*, 2270–2281. [CrossRef]
21. Algonaiman, R.; Alharbi, H.F.; Barakat, H. Antidiabetic and Hypolipidemic Efficiency of *Lactobacillus plantarum* Fermented Oat (*Avena sativa*) Extract in Streptozotocin-Induced Diabetes in Rats. *Fermentation* **2022**, *8*, 267. [CrossRef]
22. Xiao, X.; Bai, J.; Zhang, J.; Wu, J.; Dong, Y. Inhibitory Effect of Fermented Selected Barley Extracts with *Lactobacillus plantarum* Dy-1 on the Proliferation of Human HT-29 Cells. *J. Food Biochem.* **2019**, *43*, e12989. [CrossRef]
23. Verni, M.; Verardo, V.; Rizzello, C.G. How Fermentation Affects the Antioxidant Properties of Cereals and Legumes. *Foods* **2019**, *8*, 362. [CrossRef]
24. Balli, D.; Bellumori, M.; Pucci, L.; Gabriele, M.; Longo, V.; Paoli, P.; Melani, F.; Mulinacci, N.; Innocenti, M. Does Fermentation Really Increase the Phenolic Content in Cereals? A Study on Millet. *Foods* **2020**, *9*, 303. [CrossRef]
25. Shahbazi, R.; Sharifzad, F.; Bagheri, R.; Alsadi, N.; Yasavoli-Sharahi, H.; Matar, C. Anti-Inflammatory and Immunomodulatory Properties of Fermented Plant Foods. *Nutrients* **2021**, *13*, 1516. [CrossRef]
26. Gabriele, M.; Pucci, L.; Árvay, J.; Longo, V. Anti-Inflammatory and Antioxidant Effect of Fermented Whole Wheat on TNFα-Stimulated HT-29 and NF-κB Signaling Pathway Activation. *J. Funct. Foods* **2018**, *45*, 392–400. [CrossRef]
27. Kim, D.O.; Chun, O.K.; Kim, Y.J.; Moon, H.-Y.; Lee, C.Y. Quantification of Polyphenolics and Their Antioxidant Capacity in Fresh Plums. *J Agric Food Chem* **2003**, *51*, 6509–6515. [CrossRef]

28. Guimarães, R.; Barros, L.; Barreira, J.C.M.; Sousa, M.J.; Carvalho, A.M.; Ferreira, I.C.F.R. Targeting Excessive Free Radicals with Peels and Juices of Citrus Fruits: Grapefruit, Lemon, Lime and Orange. *Food Chem. Toxicol.* **2010**, *48*, 99–106. [CrossRef]
29. Chelucci, E.; Chiellini, C.; Cavallero, A.; Gabriele, M. Bio-Functional Activities of Tuscan Bee Pollen. *Antioxidants* **2023**, *12*, 115. [CrossRef]
30. Mikstacka, R.; Rimando, A.M.; Ignatowicz, E. Antioxidant Effect of Trans-Resveratrol, Pterostilbene, Quercetin and Their Combinations in Human Erythrocytes In Vitro. *Plant Foods Hum. Nutr.* **2010**, *65*, 57–63. [CrossRef]
31. Hassan, Z.M.; Sebola, N.A.; Mabelebele, M. The Nutritional Use of Millet Grain for Food and Feed: A Review. *Agric. Food Secur.* **2021**, *10*, 16. [CrossRef]
32. Muthamilarasan, M.; Dhaka, A.; Yadav, R.; Prasad, M. Exploration of Millet Models for Developing Nutrient Rich Graminaceous Crops. *Plant Sci. Int. J. Exp. Plant Biol.* **2016**, *242*, 89–97. [CrossRef]
33. Gabaza, M.; Shumoy, H.; Muchuweti, M.; Vandamme, P.; Raes, K. Effect of Fermentation and Cooking on Soluble and Bound Phenolic Profiles of Finger Millet Sour Porridge. *J. Agric. Food Chem.* **2016**, *64*, 7615–7621. [CrossRef]
34. Đorđević, T.M.; Šiler-Marinković, S.S.; Dimitrijević-Branković, S.I. Effect of Fermentation on Antioxidant Properties of Some Cereals and Pseudo Cereals. *Food Chem.* **2010**, *119*, 957–963. [CrossRef]
35. Hubert, J.; Berger, M.; Nepveu, F.; Paul, F.; Daydé, J. Effects of Fermentation on the Phytochemical Composition and Antioxidant Properties of Soy Germ. *Food Chem.* **2008**, *109*, 709–721. [CrossRef]
36. Adebo, O.A.; Gabriela Medina-Meza, I. Impact of Fermentation on the Phenolic Compounds and Antioxidant Activity of Whole Cereal Grains: A Mini Review. *Molecules* **2020**, *25*, 927. [CrossRef]
37. Salar, R.K.; Purewal, S.S.; Sandhu, K.S. Fermented Pearl Millet (*Pennisetum glaucum*) with in Vitro DNA Damage Protection Activity, Bioactive Compounds and Antioxidant Potential. *Food Res. Int.* **2017**, *100*, 204–210. [CrossRef] [PubMed]
38. Chu, J.; Zhao, H.; Lu, Z.; Lu, F.; Bie, X.; Zhang, C. Improved Physicochemical and Functional Properties of Dietary Fiber from Millet Bran Fermented by *Bacillus natto*. *Food Chem.* **2019**, *294*, 79–86. [CrossRef] [PubMed]
39. Semwal, R.; Joshi, S.K.; Semwal, R.B.; Semwal, D.K. Health Benefits and Limitations of Rutin—A Natural Flavonoid with High Nutraceutical Value. *Phytochem. Lett.* **2021**, *46*, 119–128. [CrossRef]
40. Kotha, R.R.; Tareq, F.S.; Yildiz, E.; Luthria, D.L. Oxidative Stress and Antioxidants—A Critical Review on In Vitro Antioxidant Assays. *Antioxidants* **2022**, *11*, 2388. [CrossRef] [PubMed]
41. Moazzen, A.; Öztinen, N.; Ak-Sakalli, E.; Koşar, M. Structure-Antiradical Activity Relationships of 25 Natural Antioxidant Phenolic Compounds from Different Classes. *Heliyon* **2022**, *8*, e10467. [CrossRef] [PubMed]
42. Kang, J.; Li, Z.; Wu, T.; Jensen, G.S.; Schauss, A.G.; Wu, X. Anti-Oxidant Capacities of Flavonoid Compounds Isolated from Acai Pulp (*Euterpe oleracea* Mart.). *Food Chem.* **2010**, *122*, 610–617. [CrossRef]
43. Loganayaki, N.; Siddhuraju, P.; Manian, S. Antioxidant Activity and Free Radical Scavenging Capacity of Phenolic Extracts from *Helicteres isora* L. and *Ceiba pentandra* L. *J. Food Sci. Technol.* **2013**, *50*, 687–695. [CrossRef]
44. Singh, A.-K.; Patel, P.K.; Choudhary, K.; Joshi, J.; Yadav, D.; Jin, J.-O. Quercetin and Coumarin Inhibit Dipeptidyl Peptidase-IV and Exhibits Antioxidant Properties: In Silico, In Vitro, Ex Vivo. *Biomolecules* **2020**, *10*, 207. [CrossRef]
45. Shi, J.; Shan, S.; Li, H.; Song, G.; Li, Z. Anti-Inflammatory Effects of Millet Bran Derived-Bound Polyphenols in LPS-Induced HT-29 Cell via ROS/miR-149/Akt/NF-κB Signaling Pathway. *Oncotarget* **2017**, *8*, 74582–74594. [CrossRef]

foods

MDPI

Article

Fermented Tea as a Food with Functional Value—Its Microbiological Profile, Antioxidant Potential and Phytochemical Composition

Karolina Jakubczyk [1,*], Łukasz Łopusiewicz [2], Joanna Kika [1], Katarzyna Janda-Milczarek [1] and Karolina Skonieczna-Żydecka [3]

1 Department of Human Nutrition and Metabolomics, Pomeranian Medical University in Szczecin, 24 Broniewskiego Street, 71-460 Szczecin, Poland; joanna.kika@pum.edu.pl (J.K.); katarzyna.janda.milczarek@pum.edu.pl (K.J.-M.)
2 Center of Bioimmobilisation and Innovative Packaging Materials, Faculty of Food Sciences and Fisheries, West Pomeranian University of Technology, 35 K. Janickiego, 71-270 Szczecin, Poland; lukasz.lopusiewicz@zut.edu.pl
3 Department of Biochemical Science, Pomeranian Medical University in Szczecin, 24 Broniewskiego Street, 71-460 Szczecin, Poland; karolina.skonieczna-zydecka@pum.edu.pl
* Correspondence: karolina.jakubczyk@pum.edu.pl; Tel.: +48-91-441-48-11

Abstract: Kombucha is a fermented tea drink produced by a symbiotic culture of bacteria and yeast, known as SCOBY. Its base has traditionally been black tea, which has been recognized for its health-promoting properties, particularly its antioxidant activity based on its high content of pol-yphenolic compounds. A number of previous studies have demonstrated the equally favourable biochemical and phytochemical composition of green tea. The aim of this study was to analyse and compare the basic biochemical composition, microbiological composition and antioxidant properties of black and green tea-based Kombucha. The green tea-based Kombucha showed a quantitatively more abundant microbial composition (Lactic Acid Bacteria, *Acetobacter* sp., Yeast), a higher reducing potential (FRAP—4326.58 Fe(II)μM/L) and a higher content of total polyphenols (23.84 mg GAE/100 mL, reducing sugars (3212.00 mg/100 mL) as well as free amino acids (849.00 mg GLY/mL). Kombucha made from black tea, on the other hand, showed a higher anti-oxidant potential (1.17 Trolox (mM) TEAC), neutralising the DPPH radical at 94.33% and ABTS at 97.74%. It also had a higher level of acetic acid (0.08 g/100 mL). Green tea kombucha had a higher scavenging capacity of 90.6% for superoxide radical (O_2^-) and 69.28% for hydroxyl radical (·OH) than black tea kombucha. In the present study, both kombucha drinks tested were shown to be source of potent antioxidants. In addition, green tea, as a kombucha base, has proven to be as beneficial a raw material that will provide full nutritional and health-promoting values as traditional kombucha.

Keywords: kombucha; green tea; black tea; polyphenols; chemical composition; acetic acid; microbiological composition

1. Introduction

Kombucha is a fermented tea drink made with a symbiotic culture of bacteria and yeast, known as SCOBY, which generally consists of acetic acid bacteria, lactic acid bacteria and yeast (*Acetobacter xylinum*, *Gluconobacter* sp., *Saccharomyces cerevisiae*) [1,2]. Historically, kombucha originated in China, where it has been an established health-promoting drink for thousands of years. However, it owes its current name to a Korean doctor—Kombu—who used it for therapeutic use. Subsequently, kombucha became popular in Japan and expanded from Asia to Europe [3,4].

Kombucha is prepared by adding sugar (10%), inoculum from a previous fermentation (10%) and SCOBY to tea. The SCOBY added to the sweetened tea initiates fermentation, which results in the production of various new bioactive compounds. Fermentation is carried out at room temperature (20 to 25 °C) over a period of 7–14 days. Various types of tea can be used to make kombucha, whether green tea or fermented tea, e.g., red, black or yellow. However, black tea and white sugar (sucrose) are considered to be the traditional and at the same time the best ingredients for the correct composition of the finished beverage and its health-promoting effects. The taste of a tea drink is described as sour, slightly fruity and slightly effervescent [5–7].

Studies of kombucha have proven its antimicrobial, antioxidant, antidiabetic, cholesterol-lowering, immune system-supporting and liver detoxification-stimulating properties [8]. However, to date there are no clinical studies available to confirm the health-promoting properties of this beverage. Minerals mainly derived from tea (potassium, manganese, fluoride ions), vitamins (E, K, B), amino acids (especially theanine, a glutamine derivative), as well as other compounds formed by numerous reactions during tea fermentation have been found in kombucha drinks. When polyphenolic compounds are oxidised, they are converted into new molecules [8,9].

Various parameters influence the properties and composition of kombucha, including the type of tea used, the amount and type of carbohydrate source used, fermentation time, SCOBY composition or temperature [1,10]. Despite the rising popularity of consumption of this beverage, there is still incomplete information on the effect of different fermentation parameters or type of tea on properties or composition. One of the factors that can determine both composition and properties may be the type of tea.

Inappropriate lifestyles, intense exercise, stress and environmental pollutants are all factors that contribute to the excessive production of reactive oxygen species. Imbalance caused by free radicals leads to oxidative stress and damage to structures in the human body [11]. Diseases that may be underlain by free radical-related disorders are mainly atherosclerosis, neurodegenerative diseases such as Parkinson's and Alzheimer's disease or even obesity. In order to maintain a balance between the production and removal of reactive oxygen species, readily available sources of antioxidants are being sought [2,11]. The main and most widespread antioxidants are vitamins E, A and C and polyphenolic compounds [11]. Sources of antioxidants are primarily sought in natural plant materials. Antioxidants are present in many readily available sources such as tea, coffee, fruit, vegetables, spices and herbs. They supplement the daily diet, helping to maintain health.

There are limited works analyzing and comparing the detailed biochemical composition, phytochemical composition, antioxidant potential, and microbial composition in kombucha with different base raw materials. Therefore, the aim of the present study was to analyse and compare the basic biochemical composition, microbiological composition and antioxidant properties of black and green tea-based kombucha.

2. Materials and Methods

2.1. Material

The study material was Kombucha based on green and black tea available on the food market (Delikatna.bio, Gdynia, Poland). Kombucha from 3 different glass bottles, stored at 4 °C for each type, were included in the study. The entire experiment was performed in 9 replicates. The characteristics of the basic nutrients were taken from the product label provided by the manufacturer. (Table 1) The raw materials for kombucha preparation came from certified organic cultivation. Samples were kept refrigerated until analyses were performed. Prior to all biochemical analyses the samples were filtered through 0.22-μm nylon membrane filters (Sigma-Aldrich, Darmstadt, Germany). The obtained clear fluids were used for further analyses.

Table 1. Nutritional values in 100 mL of tested kombucha, claimed by the manufacturer.

Kombucha		Black	Green
ingredients	black tea *, cane sugar *, SCOBY culture		green tea *, concentrated apple juice *, SCOBY culture
Energy value	kJ/kcal	82/20	18/4
FAT	g	<0.1	<0.1
of which saturated fatty acids	g	<0.1	0
Carbohydrates	g	5	<0.5
of which sugars	g	5	<0.5
Protein	g	0.3	0
Salt	g	0.06	0
Ethanol	%	<1.2	<1.2

* from certified organic cultivation.

2.2. The Determination of Lactic Acid Bacteria, *Acetobacter sp.* and Yeast

The samples (1 mL) were collected and diluted with 9 mL of sterile buffered peptone water (Merck, Darmstad, Germany), and serial dilutions were prepared [12]. Lactic acid bacteria counts were determined on MRS (de Man, Rogosa and Sharpe) medium (Merck, Darmstad, Germany) after incubation at 37 °C under anaerobic conditions for 72 h, whereas *Acetobacter* bacteria were assayed on Acetobacter Agar (HiMedia, Mumbai, India), after incubation at 37 °C under aerobic conditions for 48 h. Yeast counts were determined on Rose Bengal Agar at 25 °C for 72 h [13]. The enumeration of microorganisms was performed in triplicate (by counting plates with 30–300 colonies) and the viable cell counts were expressed as CFU/mL of the samples.

2.3. The Determination of Reducing Sugars Content (RSC)

The RSC (Reducing Sugars Content) was determined by the DNS (3,5-dinitrosalicylic acid) method as described elsewhere [12]. Briefly, 1 mL of each sample was mixed with 1 mL of 0.05 M acetate buffer (pH 4.8) and 3 mL of DNS reagent, then shaken vigorously. The mixtures were incubated in boiling water for 5 min, and cooled at room temperature (20 to 25 °C). The absorbance value was measured at 540 nm using a microplate reader (Synergy LX, BioTek, Winooski, VT, USA) by placing the samples in a 96-well microplate. Glucose (0.01–10 mg/mL) in acetate buffer was used for the calibration curve.

2.4. The Determination of Total Free Amino Acids

Total free amino acids (TFAA) were analysed following previously described methodology using a Cd-ninhydrin reagent [13]. Exactly 1 mL of extracts were mixed with 2 mL of a Cd-ninhydrin reagent. The samples were shaken, heated at 84 °C for 5 min, cooled in ice water and the absorbance at 507 nm was determined by the microplate reader (Synergy LX, BioTek, Winooski, VT, USA). The results were expressed as mg glycine (Gly) per mL of sample with respect to the standard curve including the dilution factor. The standard curve was first prepared using glycine.

2.5. The Determination of Total Polyphenols Content (TPC)

Determination of polyphenols was performed according to method using the Folin-Ciocalteu reagent [14]. Absorbance at 765 nm was measured (8453UV, Agilent Technologies, Santa Clara, CA, USA). All assays were performed in triplicate. The content of polyphenols was determined from the calibration curve using gallic acid (GAE) as the reference standard (0–200 mg/L of gallic acid).

2.6. The Determination of the Total Flavonoids Content (TFC)

Determination of total flavonoids content was performed according to the Pękal and Pyrzynska [15] and Hu et al., methods [16]. Different concentrations of flavonoids were used in the plotting of the standard calibration curve. The content of flavonoids was determined from the calibration curve using rutin equivalent as the reference standard (0–120 mg/L of rutin equivalent). Absorbance at 510 nm was measured (8453UV, Agilent Technologies, Santa Clara, CA, USA). All assays were performed in triplicate.

2.7. The Determination of pH

The pH of beverages was determined by a pH meter (SCHOTT Instruments; SI Analytics Mainz, Germany).

2.8. The Determination of Acetic Acid

Acetic acid (AA) was analysed by High Performance Liquid Chromatography (HPLC) using a1200 series HPLC connected to a 1100 series RI detector (Agilent Technologies, Santa Clara, CA, USA) with a Rezex ROA-Organic Acid H+ (8%) column (Phenomenex, Torrance, CA, USA). The column was eluted with a degassed mobile phase containing 5 mM H_2SO_4, pH 2.25 at 60 °C with a flow rate of 0.5 mL/min for 30 min per sample [17,18]. The results are shown in g/100 mL.

2.9. Antioxidant Activity by the DPPH and TEAC Methods

The antioxidant activity of samples was measured with the spectrophotometric method using synthetic radical DPPH (2.2-diphenyl-1-picrylhydrazyl, Sigma, Poznań, Poland) according to Brand-Williams et al. [19] and Pekkarinen et al. [20]. The spectral absorbance was immediately measured at 518 nm (8453UV, Agilent Technologies, Santa Clara, CA, USA). All assays were performed in triplicate. The results are shown in % of DPPH radical inhibition and in Trolox equivalent as the reference standard. The range of Trolox concentrations used to run the calibration curve: 1, 0.6, 0.2, 0.12 μM Trolox.

Antioxidant potential (antioxidant activity, inhibition) of tested solutions has been expressed by the percent of DPPH inhibition, using the following formula:

$$\% \text{ inhibition} = \frac{A_0 - A_s}{A_0} \times 100 \quad (1)$$

where:

A_0—absorbance of DPPH solution at 518 nm without tested sample
A_s—absorbance of DPPH solution at 518 nm with tested sample
Total antioxidant potential by ABTS method

Antioxidants reduce $ABTS^{\bullet+}$ (oxidised form) to colourless ABTS (reduced form). The decrease in absorbance is a measure of the antioxidant content of the test material. Solution of ABTS in 96% ethanol, and 0.1 mL of the test sample were introduced into the vial. The prepared solution after mixing was placed for 6 min in a dark place. Ethanol was used as a reference solution and ABTS solution for A_0. Before the measurement, the vial contents were thoroughly mixed and poured into cuvettes. The spectral absorbance was immediately measured at 734 nm. All tests were performed in triplicate.

Antioxidant activity has been expressed by the percentage of ABTS inhibition, using the formula:

$$\% \text{ inhibition} = \frac{A_0 - A_s}{A_0} \times 100 \quad (2)$$

where:

A_0—absorbance of ABTS solution without the tested sample,
A_s—absorbance of ABTS solution with the tested sample.

2.10. The Determination of the Ferric Iron Reducing Antioxidant Power (FRAP) Method

The FRAP method, used to determine the total reduction potential, which also means the antioxidant properties of tested ingredient, is based on the ability of the test sample to reduce Fe^{3+} ions to Fe^{2+} ions. The FRAP unit determines the ability to reduce 1 micromole Fe^{3+} to Fe^{2+} according to Benzie and Strain [21,22]. Absorbance at 593 nm was measured (8453UV, Agilent Technologies, Santa Clara, CA, USA). All assays were performed in triplicate. The ferric ion reducing antioxidant power was determined from the calibration curve using Fe(II)/L as the reference standard (0–5000 µM Fe(II)/L).

2.11. The Determination of Superoxide (O_2^-) Scavenging Activity

Superoxide (O_2^-) scavenging activity was carried out based on pyrogallol oxidation inhibition following methodology described elsewhere [13]. Three mL of 50 mmol/L (pH 8.2) Tris-HCl buffer were mixed with 1 mL of the samples. These mixtures were mixed with a pyrogallol solution (0.3 mL, 7 mmol/L, preheated to 25 °C) and allowed to react for exactly 4 min, then 1 mL of 10 mmol/L of HCl was added to terminate the reaction, and absorbance was measured at 318 nm (microplate reader (Synergy LX, BioTek, Winooski, VT, USA).

2.12. The Determination of Hydroxyl (·OH) Scavenging Activity

The hydroxyl (·OH) scavenging assay was carried out based on method of Ye et al. [23]. One mL of the samples and 1.5 mL of orthophenanthroline solution (0.005 mmol/L) were mixed with 2 mL of phosphate buffer (pH 7.4, 0.05 mol/L). Then 1 mL of $FeSO_4$ solution (0.0075 mol/L) was added and then mixed with 1 mL of H_2O_2 (0.1%), and finally supplemented with distilled water to a total volume of 10 mL. The reaction solution was kept at 37 °C for 1 h in darkness, then the absorbance was measured at 536 nm (microplate reader (Synergy LX, BioTek, Winooski, VT, USA). The orthophenatroline solution without H_2O_2 addition (replaced by 1 mL of methanol) served as a blank.

2.13. Statistical Analysis

In all the experiments, three samples were analysed and all the assays were carried out at least in triplicate. The statistical analysis was performed using MedCalc® Statistical Software version 22.006 (MedCalc Software Ltd., Ostend, Belgium) and Microsoft Excel 2017. The results are expressed as mean values and standard deviation (SD) or median. To assess the differences between examined parameters, a Mann-Whitney test and a *t*-test was used. Correlations between parameters were determined using Spearman rank correlation coefficient. Differences were considered significant at $p \leq 0.05$.

3. Results

The kombucha drinks tested proved to be a valuable source of polyphenols, with concentrations of 20.7 and 23.95 mg GAE/100 mL in black tea kombucha and green tea kombucha, respectively. The total flavonoid content was significantly higher for traditional kombucha, at 138.12 mg RE/100 mL of beverage, compared to 98.2 mg RE/100 mL for green tea kombucha. The black tea kombucha exhibited a low pH (2.69) and high vitamin C content (27.62 mg/100 mL). The pH value of kombucha green tea was 2.99, significantly higher than that of traditional kombucha. The acetic acid content was non-significantly higher in the traditional kombucha and was 0.08 g/100 mL compared to the other type of kombucha tested, where acetic acid was determined at 0.03 g/100 mL. The green tea kombucha, on the other hand, was characterised by a significantly higher content of reducing sugars (3212 mg/100 mL) and total free amino acid content (849 mg GLY/mL). All results obtained relating to the biochemical composition of kombucha are presented in Tables 2 and 3.

Table 2. Parameters of tested kombucha beverages.

		TPC	TFC	Vitamin C
Kombucha		GAE [mg/100 mL]	RE [mg/100 mL]	[mg/100 mL]
Black	Median	20.697 *	138.120 *	27.132
	IQR	20.531 to 21.087	135.073 to 140.185	21.450 to 34.258
Green	Median	23.949 *	93.203 *	27.174
	IQR	23.699 to 24.068	86.600 to 104.219	24.606 to 27.690

* $p < 0.05$ statistically significant difference in column; IQR—interquartile range.

Table 3. Biochemical composition of tested kombucha beverages.

		pH	Acetic Acid	Reducing Sugars Content	Total Free Amino Acid
Kombucha			acetic acid [g/100 mL]	[mg/100 mL]	[mg GLY/mL]
Black	mean	2.69 *	0.08	2842.00 *	710.00 *
	SD	0.11	0.00	0.02	0.02
Green	mean	2.99 *	0.03	3212.00 *	849.00 *
	SD	0.01	0.00	0.02	0.08

* $p < 0.05$ statistically significant difference in column.

Given the proven antioxidant effects of kombucha and various types of tea, selected kombucha drinks made from green and black tea were tested to identify equally beneficial uses for other types of tea. The antioxidant potential depending on the method chosen for its determination was 3495 Fe(II)μM/L (FRAP) and 1.17 mM Trolox/L (TEAC) in black tea kombucha. The percentage of radical inhibition in this beverage was 94.33% (DPPH) and 97.36% (ABTS). The percentage superoxide radical scavenging activity (O_2^-) of the traditional kombucha was 90.03%. while the hydroxyl radical (·OH) scavenging activity was lower at 67.23%.

Green tea kombucha had a significantly higher scavenging capacity for superoxide radical (O_2^-)—90.6% and hydroxyl radical (–OH)—69.28% than black tea kombucha. Its antioxidant potential was 91.26% and 95.6% for the neutralisation of the DPPH and ABTS radicals. The reduction potential determined by the FRAP method was found to be significantly higher than that of black tea kombucha and was 4326.58 Fe(II)μM/L. In contrast, the antioxidant activity of green tea kombucha was significantly lower than that of black tea and was determined to be 1.14 mM Trolox/L.

The results relating to the antioxidant properties of kombucha are presented in Table 4.

Table 4. Antioxidant properties of the tested beverages.

		FRAP	TEAC	DPPH	ABTS	Superoxide Radical (O_2^-)	Hydroxyl Radical (·OH)
Kombucha		Fe(II)μM/L	Trolox (mM/L)	%	%	%	%
Black	Median	3495 *	1.168 *	94.33 *,a	97.358	90.03 *,a	67.23 *,a
	IQR	3389.500 to 3513.200	1.163 to 1.175	0.52 b	97.262 to 98.413	0.02 b	0.07 b
Green	Median	4350.6 *	1.140 *	91.26 *,a	95.604	90.60 *,a	69.28 *,a
	IQR	4304.400 to 4369.800	1.117 to 1.146	1.74 b	94.957 to 98.866	0.4 b	0.02 b

* $p < 0.05$ statistically significant difference in column; [a] Mean, [b] SD.

Depending on the SCOBY used the microbiological composition of the final product may vary. In order to identify the microorganisms present in the beverages tested, quantification of *Acetobacter* sp. Lactic Acid Bacteria and Yeast was carried out. Kombucha prepared on the basis of green tea, according to the results, contained a significantly more abundant quantitative microbial composition than kombucha based on black tea. The results are summarized in Table 5.

Table 5. Microbiological composition of the tested beverages.

Kombucha	Lactic Acid Bacteria	*Acetobacter* sp.	Yeast
Black	3.02×10^5 CFU/mL [a]	7.89×10^3 CFU/mL [a]	8.25×10^5 CFU/mL [a]
Green	1.06×10^6 CFU/mL [b]	1.06×10^4 CFU/mL [b]	8.50×10^5 CFU/mL [b]

[a,b] Means with different lowercase letters in the same column are significantly different at $p < 0.05$.

4. Discussion

Fermented beverages are becoming increasingly popular, as consumers perceive fermentation as a natural method of food preservation, and the products themselves are valued for their health-promoting properties, which are enhanced by the fermentation process [17]. Kombucha as a fermented tea drink is being consumed increasingly in Europe, despite being of Asian origin [24]. It is prepared primarily on the basis of black tea, however, other kombucha variants, such as those made from other types of tea, such as green, white or red tea and enriched with fruit or other flavours are increasingly popular on the food market [25,26]. Although, kombucha is well studied in terms of its microbial composition and its antimicrobial properties, there is a lack of research on the different base types of teas and their composition and health-promoting properties. For this reason, one of the most popular types, in addition to traditional black tea—green tea—was selected for our study, and the microbiological composition, phytochemical and antioxidant potential were checked depending on the type of tea chosen for Kombucha preparation.

Tea, which is one of the main ingredients in the kombucha drink, is abundant in polyphenols, which include theaflavin and thearubigin. The polyphenols present in tea are indirectly responsible for the final antioxidant activity of kombucha [6,9]. Chakravorty et al. [27] and de Noronha et al. [28] have noted, that during fermentation, the amount of polyphenols increased. The increase in polyphenolic compounds may be related to a number of reactions occurring during tea fermentation. e.g., during the oxidation of polyphenolic compounds by certain enzymes, catechins are formed, flavonoids and other compounds with health-promoting properties, including antioxidant properties [9]. Also microorganisms are capable of degrading polyphenols [29]. The catechins in tea can be broken down by bacterial and yeast activity to simpler molecules. thus increasing antioxidant power [9,30]. Based on our study, it can be concluded, that the total polyphenol content is dependent on the type of tea. Higher concentrations were recorded in kombucha made from green tea, whereas slightly lower for kombucha made from black, which is in the line with the results of other researchers [18,31]. The opposite results were obtained in the study by Cardoso et al. [32] where kombucha from black tea that showed higher levels of polyphenols (1.09 mg GAE/mL) than kombucha from green tea (0.7 mg GAE/mL). The total flavonoid content of black tea kombucha was 138.12 mg/L, which was the same as the results of other authors (126.7 mg/L). In contrast, in this work we obtained a 2-fold lower result for green tea kombucha (93.2 mg/L vs. 181.3 mg/L) [31].

Vitamin C has strong antioxidant activity due to its ability to effectively neutralize free radicals. During the fermentation process, the vitamin C content of beverages can increase [6,33]. Kaewkod et al. [18] observed higher vitamin C contents (0.61 g/L for green tea and 0.7 g/L for black tea) than in this study. However, it should be noted. that this vitamin is very unstable. It is estimated that 100 mL of kombucha based on black tea will provide 31% of the RDA for an adult male for this dietary component, while kombucha made from green tea will provide 30% of the RDA, so kombucha can be a good source of this component in the daily diet [34].

Kombucha, regardless of the type of tea used, is characterised by its high antioxidant potential [35,36]. In animal models, it was shown, that oral administration of kombucha tea to lead-exposed rats reduced lipid peroxidation and DNA damage, while increasing reduced glutathione levels and glutathione peroxidase (GPx) activity [37]. The determination of free radicals scavenging activity, which is hydroxyl (·OH) at 67.2–69.3% was higher in this study than observed by other researchers (44.8–52.3%) [9]. This work also determined the level of superoxide radical (O_2^-) neutralisation activity, which was 90% for traditional kombucha and 91% for green tea kombucha. This demonstrates the high antioxidant activity of the kombucha beverage tested, which, according to other reports, increases with the progress of fermentation [33]. In addition, the antioxidant potential measured by the DPPH method in the present study was determined at a significantly higher level for both teas (94.3% black tea and 91.3% green tea) than in the study by Malbaša et al. [33] (45% and 60%) and Jakubczyk et al. [31] (61.0% and 88.2%). Also in the above-mentioned works, the relationship of higher potential in kombucha from black tea was not preserved, as observed in the present study. A similar result (97.7% and 96.3%) was obtained for the ABTS radical scavenging assay. The reduction potential measured by the FRAP method was determined at a significantly higher level in our work than that of other authors. Jakubczyk et al. [31] observed a reduction potential in green tea kombucha of 3172.9 μM Fe(II)/L, where in this study its level was 4350.6 μM Fe(II)/L. Also, a significant difference was observed with respect to the results of kombucha made from traditional tea, with our result being 3495 μM Fe(II)/L, while the authors' comparison was 1573.9 μM Fe(II)/L. In the study by Chakravorty et al. [27] an increase in the antioxidant potential of Kombucha was observed compared to the base tea. It is therefore important to study the final product, not only the base raw materials. The present study showed, that the antioxidant properties of Kombucha depend on the type of base raw material.

An important parameter that changes during fermentation is the pH of the solution and the acidity i.e., the organic acid content. The microorganisms contained in the SCOBY process the sugar and substances contained in the tea, producing a variety of metabolites. The fermentation of kombucha is a combination of three types of fermentation, which are: lactic fermentation, alcoholic fermentation and acetic fermentation. Lactic fermentation is responsible for the breakdown of glucose, which occurs because of Lactic Acid Bacteria metabolic activity. Initially there is an increase in reducing sugar. which can be attributed to the hydrolysis of sucrose to glucose and fructose by yeast. As fermentation continues, the yeast anaerobically uses the sugar to produce ethanol [9]. Gaggìa et al. [17] checked the content of glucose, fructose and sucrose in kombucha prepared from black, green and red tea on days 7 and 14 of fermentation. The content of complex sugars. i.e., sucrose, decreased during fermentation. while simple sugars (such as glucose) increased. In our study, the content of reducing sugars was higher in kombucha based on green tea, which may be related to the potentially longer fermentation process. The carbohydrate source used may also have had an impact as well. Alcoholic fermentation is another type of fermentation. Yeast, which is part of the beverage's microflora, is responsible for metabolizing glucose into ethanol with the release of carbon dioxide. In the next stage, acetic acid bacteria (such as *Acetobacter* sp.) use ethanol as a substrate to produce acetic acid [8]. During fermentation, the alcohol content therefore decreases, while acidity increases and organic acids are produced. Acetic acid, which is the predominant organic acid present in the fermented solution. contributes to a decrease in pH from 5 even to 3 [9,38]. This is consistent with the results of the present study, as kombucha from black tea has a lower pH (2.69) than from green tea (2.99), while at the same time having a higher acetic acid content (0.08 g/100 mL vs. 0.03 g/100 mL). These values are within the consumer safety range (2.5–4.2 g/L) [32]. It should also be emphasised, that the pH of the solution and the presence of certain organic acids determine the growth of microorganisms and thus the chemical composition of the beverage [17]. Low pH and high acidity allow the growth of only those microbes, which are able to inhabit such extreme conditions and thus can provide some protection against undesirable microorganisms [39].

All of the microorganism species determined (Lactic Acid Bacteria, Yeast and *Acetobacter* sp.) were found in higher levels in kombucha green tea than in traditional kombucha tea, which may be related to both the different composition of the SCOBY starter (no manufacturer information available) and the type of base material. Thus, there is a need for further research to determine the optimal composition of SCOBY in terms of increasing the synthesis of compounds that exhibit health-promoting, particularly antioxidant, effects, while maintaining the flavor of the final product. In addition, it is worth extending the research on the impact of the choice of raw material from which the beverage is made on its final microbiological composition.

5. Conclusions

Kombucha (or fermented tea) has strong antioxidant properties. It exhibits high antioxidant potential, which can be attributed to its high content of polyphenols and flavonoids. The antioxidant activity, as well as the phytochemical composition of kombucha, varies depending on the type of tea used to prepare this beverage. Choosing a tea other than the traditionally used black tea and subjecting it to the fermentation process seems to be equally beneficial in terms of the health-promoting properties of the final product. However, there is a need for further research to possibly expand the range of health-promoting beverages.

Author Contributions: Conceptualization, K.J.; methodology, K.J. and Ł.Ł.; software, K.J. and K.S.-Ż.; validation, K.J., K.J.-M. and K.S.-Ż.; formal analysis, K.J., K.J.-M. and K.S.-Ż.; investigation, Ł.Ł., K.J. and J.K.; resources, K.J.; data curation, K.J.; writing—original draft preparation, J.K. and K.J.; writing—review and editing, K.J.; visualization, J.K.; supervision, K.J.; project administration, K.J.; funding acquisition, K.J. All authors have read and agreed to the published version of the manuscript.

Funding: This research was funded by Pomeranian Medical University in Szczecin.

Data Availability Statement: The data presented in this study are available on request from the corresponding author.

Acknowledgments: We would like to thank Emilia Drozłowska for her assistance in bioactivity determination.

Conflicts of Interest: The authors declare no conflict of interest.

References

1. de Miranda, J.F.; Ruiz, L.F.; Silva, C.B.; Uekane, T.M.; Silva, K.A.; Gonzalez, A.G.M.; Fernandes, F.F.; Lima, A.R. Kombucha: A Review of Substrates, Regulations, Composition, and Biological Properties. *J. Food Sci.* **2022**, *87*, 503–527. [CrossRef] [PubMed]
2. Anantachoke, N.; Duangrat, R.; Sutthiphatkul, T.; Ochaikul, D.; Mangmool, S. Kombucha Beverages Produced from Fruits, Vegetables, and Plants: A Review on Their Pharmacological Activities and Health Benefits. *Foods* **2023**, *12*, 1818. [CrossRef] [PubMed]
3. Da Júnior, J.C.S.; Meireles Mafaldo, Í.; de Lima Brito, I.; Tribuzy de Magalhães Cordeiro, A.M. Kombucha: Formulation, Chemical Composition, and Therapeutic Potentialities. *Curr. Res. Food Sci.* **2022**, *5*, 360–365. [CrossRef] [PubMed]
4. Chakravorty, S.; Bhattacharya, S.; Bhattacharya, D.; Sarkar, S.; Gachhui, R. Kombucha: A Promising Functional Beverage Prepared from Tea. In *Non-Alcoholic Beverages*; Elsevier: Amsterdam, The Netherlands, 2019; pp. 285–327, ISBN 978-0-12-815270-6.
5. Blanc, P.J. Characterization of the Tea Fungus Metabolites. *Biotechnol. Lett.* **1996**, *18*, 139–142. [CrossRef]
6. Antolak, H.; Piechota, D.; Kucharska, A. Kombucha Tea—A Double Power of Bioactive Compounds from Tea and Symbiotic Culture of Bacteria and Yeasts (SCOBY). *Antioxidants* **2021**, *10*, 1541. [CrossRef] [PubMed]
7. Tran, T.; Grandvalet, C.; Verdier, F.; Martin, A.; Alexandre, H.; Tourdot-Maréchal, R. Microbiological and Technological Parameters Impacting the Chemical Composition and Sensory Quality of Kombucha. *Compr. Rev. Food Sci. Food Saf.* **2020**, *19*, 2050–2070. [CrossRef] [PubMed]
8. Villarreal-Soto, S.A.; Beaufort, S.; Bouajila, J.; Souchard, J.-P.; Taillandier, P. Understanding Kombucha Tea Fermentation: A Review. *J. Food Sci.* **2018**, *83*, 580–588. [CrossRef] [PubMed]
9. Jayabalan, R.; Malini, K.; Sathishkumar, M.; Swaminathan, K.; Yun, S.-E. Biochemical Characteristics of Tea Fungus Produced during Kombucha Fermentation. *Food Sci. Biotechnol.* **2010**, *19*, 843–847. [CrossRef]
10. Su, J.; Tan, Q.; Tang, Q.; Tong, Z.; Yang, M. Research Progress on Alternative Kombucha Substrate Transformation and the Resulting Active Components. *Front. Microbiol.* **2023**, *14*, 1254014. [CrossRef]
11. Blokhina, O.; Virolainen, E.; Fagerstedt, K.V. Antioxidants, Oxidative Damage and Oxygen Deprivation Stress: A Review. *Ann. Bot.* **2003**, *91*, 179–194. [CrossRef]

12. Łopusiewicz, Ł.; Drozłowska, E.; Siedlecka, P.; Mężyńska, M.; Bartkowiak, A.; Sienkiewicz, M.; Zielińska-Bliźniewska, H.; Kwiatkowski, P. Development, Characterization, and Bioactivity of Non-Dairy Kefir-Like Fermented Beverage Based on Flaxseed Oil Cake. *Foods* **2019**, *8*, 544. [CrossRef] [PubMed]
13. Łopusiewicz, Ł.; Drozłowska, E.; Tarnowiecka-Kuca, A.; Bartkowiak, A.; Mazurkiewicz-Zapałowicz, K.; Salachna, P. Biotransformation of Flaxseed Oil Cake into Bioactive Camembert-Analogue Using Lactic Acid Bacteria, Penicillium Camemberti and Geotrichum Candidum. *Microorganisms* **2020**, *8*, 1266. [CrossRef] [PubMed]
14. Singleton, V.L.; Rossi, J.A. Colorimetry of Total Phenolics with Phosphomolybdic-Phosphotungstic Acid Reagents. *Am. J. Enol. Vitic.* **1965**, *16*, 144–158. [CrossRef]
15. Pękal, A.; Pyrzynska, K. Evaluation of Aluminium Complexation Reaction for Flavonoid Content Assay. *Food Anal. Methods* **2014**, *7*, 1776–1782. [CrossRef]
16. Hu, S.; Yuan, C.; Zhang, C.H.; Wang, P.; Li, Q.; Wan, J.; Chang, H.; Ye, J.; Guo, X. Comparative Study of Total Flavonoid Contents from the Different Tissues and Varieties of *Abelmoschus esculentus*. *Int. J. Med. Sci. Biotechnol.* **2013**, *1*, 26–30.
17. Gaggìa, F.; Baffoni, L.; Galiano, M.; Nielsen, D.S.; Jakobsen, R.R.; Castro-Mejía, J.L.; Bosi, S.; Truzzi, F.; Musumeci, F.; Dinelli, G.; et al. Kombucha Beverage from Green, Black and Rooibos Teas: A Comparative Study Looking at Microbiology, Chemistry and Antioxidant Activity. *Nutrients* **2018**, *11*, 1. [CrossRef] [PubMed]
18. Kaewkod, T.; Bovonsombut, S.; Tragoolpua, Y. Efficacy of Kombucha Obtained from Green, Oolong, and Black Teas on Inhibition of Pathogenic Bacteria, Antioxidation, and Toxicity on Colorectal Cancer Cell Line. *Microorganisms* **2019**, *7*, 700. [CrossRef]
19. Brand-Williams, W.; Cuvelier, M.E.; Berset, C. Use of a Free Radical Method to Evaluate Antioxidant Activity. *LWT–Food Sci. Technol.* **1995**, *28*, 25–30. [CrossRef]
20. Pekkarinen, S.S.; Stöckmann, H.; Schwarz, K.; Heinonen, I.M.; Hopia, A.I. Antioxidant Activity and Partitioning of Phenolic Acids in Bulk and Emulsified Methyl Linoleate. *J. Agric. Food Chem.* **1999**, *47*, 3036–3043. [CrossRef]
21. Benzie, I.F.; Strain, J.J. Ferric Reducing/Antioxidant Power Assay: Direct Measure of Total Antioxidant Activity of Biological Fluids and Modified Version for Simultaneous Measurement of Total Antioxidant Power and Ascorbic Acid Concentration. *Methods Enzymol.* **1999**, *299*, 15–27.
22. Benzie, I.F.; Strain, J.J. The Ferric Reducing Ability of Plasma (FRAP) as a Measure of "Antioxidant Power": The FRAP Assay. *Anal. Biochem.* **1996**, *239*, 70–76. [CrossRef] [PubMed]
23. Ye, M.; Wang, Y.; Guo, G.; He, Y.; Lu, Y.; Ye, Y.; Yang, Q.; Yang, P. Physicochemical Characteristics and Antioxidant Activity of Arginine-Modified Melanin from Lachnum YM-346. *Food Chem.* **2012**, *135*, 2490–2497. [CrossRef] [PubMed]
24. Batista, P.; Penas, M.R.; Pintado, M.; Oliveira-Silva, P. Kombucha: Perceptions and Future Prospects. *Foods* **2022**, *11*, 1977. [CrossRef] [PubMed]
25. Wang, B.; Rutherfurd-Markwick, K.; Zhang, X.-X.; Mutukumira, A.N. Kombucha: Production and Microbiological Research. *Foods* **2022**, *11*, 3456. [CrossRef] [PubMed]
26. Barakat, N.; Beaufort, S.; Rizk, Z.; Bouajila, J.; Taillandier, P.; El Rayess, Y. Kombucha Analogues around the World: A Review. *Crit. Rev. Food Sci. Nutr.* **2023**, *63*, 10105–10129. [CrossRef] [PubMed]
27. Chakravorty, S.; Bhattacharya, S.; Chatzinotas, A.; Chakraborty, W.; Bhattacharya, D.; Gachhui, R. Kombucha Tea Fermentation: Microbial and Biochemical Dynamics. *Int. J. Food Microbiol.* **2016**, *220*, 63–72. [CrossRef] [PubMed]
28. de Noronha, M.C.; Cardoso, R.R.; Dos Santos D'Almeida, C.T.; Vieira do Carmo, M.A.; Azevedo, L.; Maltarollo, V.G.; Júnior, J.I.R.; Eller, M.R.; Cameron, L.C.; Ferreira, M.S.L.; et al. Black Tea Kombucha: Physicochemical, Microbiological and Comprehensive Phenolic Profile Changes during Fermentation, and Antimalarial Activity. *Food Chem.* **2022**, *384*, 132515. [CrossRef]
29. Ettayebi, K.; Errachidi, F.; Jamai, L.; Tahri-Jouti, M.A.; Sendide, K.; Ettayebi, M. Biodegradation of Polyphenols with Immobilized Candida Tropicalis under Metabolic Induction. *FEMS Microbiol. Lett.* **2003**, *223*, 215–219. [CrossRef]
30. Tanaka, T.; Matsuo, Y.; Kouno, I. Chemistry of Secondary Polyphenols Produced during Processing of Tea and Selected Foods. *Int. J. Mol. Sci.* **2009**, *11*, 14–40. [CrossRef]
31. Jakubczyk, K.; Kałduńska, J.; Kochman, J.; Janda, K. Chemical Profile and Antioxidant Activity of the Kombucha Beverage Derived from White, Green, Black and Red Tea. *Antioxidants* **2020**, *9*, 447. [CrossRef]
32. Cardoso, R.R.; Neto, R.O.; dos Santos D'Almeida, C.T.; do Nascimento, T.P.; Pressete, C.G.; Azevedo, L.; Martino, H.S.D.; Cameron, L.C.; Ferreira, M.S.L.; Barros, F.A.R. de Kombuchas from Green and Black Teas Have Different Phenolic Profile, Which Impacts Their Antioxidant Capacities, Antibacterial and Antiproliferative Activities. *Food Res. Int.* **2020**, *128*, 108782. [CrossRef] [PubMed]
33. Malbaša, R.V.; Lončar, E.S.; Vitas, J.S.; Čanadanović-Brunet, J.M. Influence of Starter Cultures on the Antioxidant Activity of Kombucha Beverage. *Food Chem.* **2011**, *127*, 1727–1731. [CrossRef]
34. *Dietary Reference Intakes for Vitamin C, Vitamin E, Selenium, and Carotenoids*; National Academies Press: Washington, DC, USA, 2000; ISBN 978-0-309-06935-9.
35. Tefon Öztürk, B.E.; Eroğlu, B.; Delik, E.; Çiçek, M.; Çiçek, E. Comprehensive Evaluation of Three Important Herbs for Kombucha Fermentation. *Food Technol. Biotechnol.* **2023**, *61*, 127–137. [CrossRef] [PubMed]
36. Değirmencioğlu, N.; Yıldız, E.; Sahan, Y.; Güldas, M.; Gürbüz, O. Impact of Tea Leaves Types on Antioxidant Properties and Bioaccessibility of Kombucha. *J. Food Sci. Technol.* **2021**, *58*, 2304–2312. [CrossRef]
37. Dipti, P.; Yogesh, B.; Kain, A.K.; Pauline, T.; Anju, B.; Sairam, M.; Singh, B.; Mongia, S.S.; Kumar, G.I.D.; Selvamurthy, W. Lead Induced Oxidative Stress: Beneficial Effects of Kombucha Tea. *Biomed. Environ. Sci. BES* **2003**, *16*, 276–282.

38. Kluz, M.I.; Pietrzyk, K.; Pastuszczak, M.; Kacaniova, M.; Kita, A.; Kapusta, I.; Zaguła, G.; Zagrobelna, E.; Struś, K.; Marciniak-Lukasiak, K.; et al. Microbiological and Physicochemical Composition of Various Types of Homemade Kombucha Beverages Using Alternative Kinds of Sugars. *Foods* **2022**, *11*, 1523. [CrossRef]
39. Greenwalt, C.J.; Steinkraus, K.H.; Ledford, R.A. Kombucha, the Fermented Tea: Microbiology, Composition, and Claimed Health Effects. *J. Food Prot.* **2000**, *63*, 976–981. [CrossRef]

foods

MDPI

Review

Bioactive Peptides from Fermented Foods: Production Approaches, Sources, and Potential Health Benefits

Laryssa Peres Fabbri, Andrea Cavallero, Francesca Vidotto and Morena Gabriele *

Institute of Agricultural Biology and Biotechnology, National Research Council, 56124 Pisa, Italy; andrea.cavallero@ibba.cnr.it (A.C.); francesca.vidotto@ibba.cnr.it (F.V.)
* Correspondence: morena.gabriele@cnr.it

Abstract: Microbial fermentation is a well-known strategy for enhancing the nutraceutical attributes of foods. Among the fermentation outcomes, bioactive peptides (BAPs), short chains of amino acids resulting from proteolytic activity, are emerging as promising components thanks to their bioactivities. Indeed, BAPs offer numerous health benefits, including antimicrobial, antioxidant, antihypertensive, and anti-inflammatory properties. This review focuses on the production of bioactive peptides during the fermentation process, emphasizing how different microbial strains and fermentation conditions influence the quantity and quality of these peptides. Furthermore, it examines the health benefits of BAPs from fermented foods, highlighting their potential in disease prevention and overall health promotion. Additionally, this review addresses the challenges and future directions in this field. This comprehensive overview underscores the promise of fermented foods as sustainable and potent sources of bioactive peptides, with significant implications for developing functional foods and nutraceuticals.

Keywords: fermented foods; bioactive peptides; microbial fermentation; health benefits; functional foods; nutraceuticals; disease prevention

1. Introduction

As consumers are becoming increasingly knowledgeable about health and nutrition, the interest in fermented foods providing a less processed, naturally based, and health-promoting option is growing accordingly [1,2]. The fermentation process consists of a microbial-driven enzymatic transformation of substrates derived from various foods, including dairy products, grains, vegetables, meat, and many others. The microbial activity results in the generation of several products such as amino acids, organic acids, short-chain fatty acids (SCFAs), and bioactive peptides (BAPs), which endow fermented foods with a wide range of bioactivities [3].

BAPs are peptides consisting of 2 to 20 amino acids originating from precursor proteins following an activation process involving multiple reactions such as chemical hydrolysis, enzymatic hydrolysis operated by proteolytic enzymes, or microbial fermentation by proteolytic bacteria [4,5]. The health-promoting potential of BAPs, of either plant or animal origin, has been recognized and linked with several useful bioactivities including antihypertensive, antioxidant, antimicrobial, and anti-inflammatory activities [4,6–8]. The most investigated BAPs are those present in milk and dairy products, legumes, cereals, meat, and fish [8].

Owing to their promising properties, research has focused on overcoming the challenges linked with the production and delivery of BAPs. Indeed, fermentation often results in low yields of BAPs, which can be increased by modulating physicochemical parameters such as time, temperature, and pH and using emerging green technologies [9]. Moreover, the stability and bioavailability of BAPs in biological systems can be increased through encapsulation strategies and physicochemical processes [10].

Noteworthily, the attention toward fermentation-produced BAPs has experienced a surge in recent years. As shown in Figure 1, the input 'bioactive peptides from fermented foods' generated 375 hits in PubMed, spread between the years 1995 and 2024, and reached a peak of publications in 2021. Thus, an up-to-date review of this relatively new field is offered.

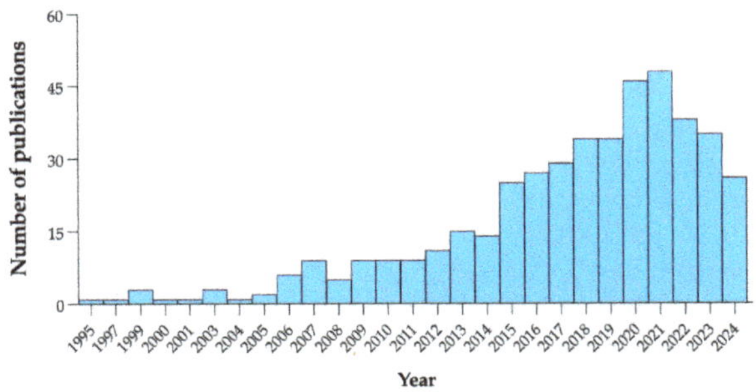

Figure 1. Bibliometric analysis of published scientific articles on 'bioactive peptides from fermented foods'. The search was carried out in PubMed® on articles from 1995 to 2024.

The present review summarizes the available knowledge on BAPs originated in fermented foods, focusing on their generation, sources, and physiological effects and challenges and strategies to enhance their content and delivery. A better understanding of these aspects could open a path toward innovative food products that may contribute to improving human health and nutrition.

2. Fermentation Process for the Synthesis of BAPs

Fermentation is an ancient practice used across several cultures worldwide, from Asia to Europe and the Middle East, to preserve and enhance the nutritional value of foods [11]. This process not only alters the taste, smell, and texture of food but also improves its digestibility and nutritional profile [12]. By breaking down complex carbohydrates and proteins, fermentation increases the intestinal absorption of nutrients, vitamins, and minerals, making them more bioavailable [8,12,13]. Additionally, fermentation lowers the glycemic index of foods thanks to the activity of microbial enzymes such as maltase and α-amylase, which convert starches into simple sugars that are readily absorbed by microorganisms, resulting in a low carbohydrate content in the final product [14].

One of the most exciting outcomes of fermentation is the production of BAPs. These peptides are formed through proteolytic enzymatic reactions, where proteins are hydrolyzed into smaller aminoacidic chains with beneficial health effects [15]. Lactic acid bacteria (LAB) and some fungi such as *Aspergillus oryzae* are central to this biosynthetic process. The most cited LABs for their effectiveness in producing BAPs are *Lactobacillus helveticus*, *Lactobacillus delbrueckii* ssp. *bulgaricus*, *Lactococcus lactis* ssp. *diacetylactis*, *L. lactis* ssp. *cremoris*, and *Streptococcus salivarius* ssp. *Thermophilus* [16]. In addition to LAB, other microorganisms such as *Bacillus* species and certain fungi play a significant role in BAP production. For example, *Bacillus subtilis* and *Bacillus licheniformis* are known for their robust proteolytic activity, allowing them to hydrolyze proteins into bioactive peptides [17,18]. Fungi such as *Aspergillus oryzae*, *Candida lipolytica*, *Saccharomyces cerevisiae*, and *Rhizopus oligosporus*, among others, are also extensively used in BAP production due to their variety of enzymes, including protease, capable of breaking down specific types of proteins derived from plants or animals. Fungal strains exhibit a remarkable metabolic diversity, enabling them to utilize diverse carbon and nitrogen sources for the new synthesis of proteins and generation of BAP [19].

LAB species have a highly developed proteolytic system composed of cell envelope proteinases (CEPs), peptide transporters, and intracellular peptidases. CEPs initiate hydrolysis by cleaving proteins into peptides, which are then transported into the cell and further degraded into amino acids by various peptidases [20].

Four types of CEP genes (PrtB, PrtP, PrtR, and PrtH) have been characterized in *Lactobacillus* species [20,21]. As an example, *L. helveticus* is highly proteolytic due to the presence of up to four PrtH paralogs with different specificities, making it particularly efficient at generating diverse BAPs [21–23]. Several factors, including variations in CEP gene expression, mutations, and specific hydrolysis conditions, influence the variability in BAP profiles among LAB strains [24].

Besides the microbial species, to obtain peptides with enhanced bioactivity, it is crucial to carefully manage the choice of substrate and maintain favorable environmental conditions, including pH and temperature, throughout the fermentation process [8,25]. Additionally, the diversity of animal- and plant-derived proteins plays a significant role in determining the variety of BAPs produced and their bioactivities. For instance, BAP-rich fermented products such as camel milk, Panxian and Spanish dry-cured ham, natto, and tempeh have shown notable health benefits, including antioxidant, antihypertensive, antibacterial, and anti-inflammatory activities [26–31].

Microbial fermentation, in general, has emerged as an efficient and cost-effective way of producing BAPs. This process is more convenient compared to conventional enzymatic hydrolysis, being cheaper and simplifying production by avoiding the multiple purification steps through fermentation [32]. However, challenges like low peptide yields and lack of specificity remain obstacles in industrial applications. Ongoing research aims to optimize fermentation conditions, microorganism selection, and substrate management to enhance peptide bioactivity and maximize health benefits [8].

3. Fermented Foods as Sources of BAPs

3.1. Milk and Dairy Products

BAPs in milk play a crucial role in human health [33]. Milk proteins, primarily caseins (α-casein, β-casein, and κ-casein) and whey proteins (β-lactoglobulin, α-lactalbumin, and lactoferrin), are rich in BAPs released during fermentation and digestion [34]. The use of LAB for proteolytic activation enriches milk-derived foods with bioactive compounds [35,36]. *Lactobacillus*, *Lactococcus*, and *Streptococcus* species, and some yeast strains, such as *Saccharomyces cerevisiae/paradoxus* and *Kluyveromyces marxianus*, have been identified in fermented dairy products and demonstrate the ability to release BAPs with antioxidant, antihypertensive, and antimicrobial properties [37–39].

Ripening and in vitro digestion processes in dairy products like Parmigiano Reggiano (PR) cheese enhance the release of peptides with various health benefits, including antihypertensive, antioxidant, immunomodulatory, anti-inflammatory effects [40]. Similarly, kefir, prepared by adding 4% kefir grains to pasteurized whole milk and fermenting at room temperature for 24 h, led to a 19% inhibition of angiotensin-converting enzyme (ACE) activity and a 37 mmHg reduction in systolic arterial pressure in hypertensive rats [41]. Several studies have shown that milk fermented with *L. helveticus* is rich in two ACE inhibitory tripeptides, which exert an antihypertensive effect by reducing angiotensin formation [42–44]. Peptides derived from lactoferrin, a protein found in the milk of all mammals, have been shown to possess antimicrobial and immunosuppressive properties [45]. Antioxidant and ACE inhibitory peptides have also been identified in donkey milk by multidimensional liquid chromatography and nano high-performance liquid chromatography (HPLC)–high-resolution mass spectrometry [46].

The characteristics of peptides derived from milk can change depending on various factors, such as the hydrolysis method used and the animal producing the milk. For instance, bioactive proteins and peptides with different bioactivities can be achieved from buffalo, camel, goat, sheep, mare, and yak milk [47].

3.2. Meat

Meat and its derivatives are regarded as rich sources of BAPs due to their high protein content. Numerous BAPs have been found in meat and fermented meat products. Mora et al. [48] reported that the fermentation of a dry Spanish sausage, made with 75% lean pork and 25% pork back fat and supplemented with various additives (NaCl, lactose, dextrin, sodium caseinate, glucose, sodium ascorbate, sodium nitrite, and potassium nitrate), inoculated with *Lactobacillus pentosus* and *Staphylococcus carnosus*, and ripened in two stages (22 h at 15–20 °C followed by 43 days at 9 °C), generated peptides with ACE-inhibitory activity, such as YQEPLV, YQEPVLGPVR, and YQEPVLGPVRGPFPI, as well as the peptide YQEPVVLGPVRGPFPIIV, which is known for its antimicrobial activity.

Additionally, it was shown that camel meat sausages treated with a range of bacterial strains and cured for up to 28 days exerted a higher peptide concentration and an enhanced antioxidant and antihypertensive capacity compared to non-inoculated sausages, with results being positively affected by the ripening time [49]. The choice of the bacterial species and strain is vital for guaranteeing the best recovery of BAPs from meat, as demonstrated by Takeda et al. [50]. When studying the fermentation of sausages, they found that specific strains of *Lactobacillus sakei* and *Lactobacillus curvatus* exerted better proteolysis as well as antioxidant and ACE-inhibitory activities than both the non-treated and the type-strain-treated sausages. Furthermore, the generation of BAPs is also influenced by the kind of meat. For instance, uncured fallow deer sausages had a higher nutritional value than beef sausages, thanks to their greater peptide content, L-carnitine concentration, and antioxidant property [51].

3.3. Plant-Based Foods

Cereals are widely utilized to create novel foods and ingredients due to their high protein content (10–15%), making them a suitable substrate for probiotic fermentation [52]. Cereal proteins can be broken down by microbial enzymes during fermentation, producing BAPs [53]. Recent studies have highlighted the potential of fermented cereal products in improving human health.

Peptides from fermented rice displayed promising results in preventing cognitive decline and promoting brain health in mice by upregulating the levels of brain-derived neurotrophic factor [54]. Fermented oats, when processed with specific microbes like *L. plantarum* and *Rhizopus oryzae*, yielded more BAPs and showed higher ACE-I inhibitory activity than unfermented oats, which could be beneficial for hypertension [55]. Additionally, the less studied pseudocereals also benefit from fermentation. For instance, a 72 h solid-state fermentation of quinoa with *L. plantarum* K779 demonstrated an enhancement in its antioxidant and antihypertensive properties [56]. Likewise, fermenting amaranth using 40 strains of *Lactobacillus* species as starters at 30 °C for 16 h could potentially lead to the production of the cancer-preventive peptide lunasin [57].

Moreover, fermentation improves the nutritional qualities of legumes such as soybeans and red beans. For example, a solid-state fermentation of soybeans with *Bacillus amyloliquefaciens* led to a reduction in antinutritional factors and an increase in antioxidant activity [58]. Similarly, solid-state fermentation of red beans with *Cordyceps militaris* enhanced both protein digestibility and bioactivity [59]. These findings highlight the health benefits of fermented legumes, including their potential in managing conditions like obesity and diabetes.

Overall, microbial fermentation of cereals, pseudocereals, and legumes can significantly enhance their nutritional and functional properties, offering promising health promotion and disease prevention applications.

3.4. Marine Organisms

Fish and shellfish are renowned for their high protein content and are traditionally preserved through salting and fermentation. Fermented fish and shellfish, often used as savory condiments, are processed with starter cultures such as *L. plantarum*, *Lactobacillus brevis*, and *Bacillus mojavensis* [60–62].

Recent research has highlighted the bioactive potential of peptides derived from fermented fish and shellfish. For instance, fermentation of *Ruditapes philippinarum* clams with *Bacillus natto* for 36 h at 37 °C and pH 7.2 resulted in a peptide with notable ACE-I-inhibitory activity, which remained stable during gastrointestinal digestion. The inhibitory peptide was purified sequentially using ultrafiltration, gel filtration chromatography, and reverse-phase HPLC (RP-HPLC). This peptide not only enhanced nitric oxide (NO) release but also inhibited endothelin-1 secretion and scavenged reactive oxygen species, demonstrating its potential as a pharmaceutical agent [18]. Budu, a Malaysian fermented fish sauce made from anchovy (*Ilisha Melastoma*) through over 120 days of fermentation, produced BAPs with strong antioxidant properties, such as LDDPVFIH and VAAGRTDAGVH, identified through LC-ESI-TOF analysis. These peptides, rich in hydrophobic and acidic amino acids, are believed to contribute to the high antioxidant activity observed in Budu [63]. Similarly, Pekasam, a traditional Malaysian fermented fish product, was fermented with *L. plantarum* for 15 days at 27 °C, with a pH range of 4.9 to 5.6. The fermentation process yielded two novel antioxidant peptides, identified by HPLC connected to tandem mass spectrometry (LC/MS/MS), which exhibited significant radical scavenging activity, attributed to their high content of hydrophobic amino acids [64]. Furthermore, commercial Thai fermented shrimp pastes, aged for six months, were found to contain peptides with antioxidant and ACE-I-inhibitory activity, isolated through sequential anion exchange chromatography and solid-phase extraction using a C18 matrix [65].

In another study, Zebra blenny, fermented with *Bacillus mojavensis* A21 at 37 °C and 200 rpm for 4 to 48 h and then fractionated using Sephadex G-25 gel filtration and RP-HPLC, produced antimicrobial peptides identified by nano ESI-LC–MS/MS. These peptides demonstrated significant antibacterial activity against various pathogens, including *Micrococcus luteus* and *Escherichia coli* [62].

Overall, these studies illustrate the significant health benefits of BAPs from fermented fish and shellfish, including antimicrobial, antioxidant, and antihypertensive properties, suggesting their potential use in both food supplements and functional foods.

4. Bioactivities of BAPs Derived from Fermented Foods

BAPs originated in fermented foods perform several biological functions as highlighted by extensive research. In the next sections, the antimicrobial, antihypertensive, antioxidant, and anti-inflammatory activities of different BAPs will be discussed and summarized in Table 1.

4.1. Antimicrobial Activity

Antimicrobial peptides (AMPs) are a kind of BAP present in almost all life forms as part of the innate immune system, being the first defense line against a wide range of pathogens [66]. Most AMPs are characterized by a dominance of β-sheet and α-helix structures, a net positive charge ranging from +2 to +9, and a high proportion of hydrophobic amino acid residues [67]. The positive charge in AMPs is a consequence of the highly represented lysine and arginine residues, which interact with the negatively charged outer membranes of Gram-negative bacteria, leading to membrane perturbation and bacterial cell death [68,69].

In addition to the endogenous ones, AMPs are also derived from fermented foods. A significant example of food-derived AMP production is provided by the bacterium *Lactiplantibacillus plantarum*, which improves the nutritional quality, organoleptic characteristics, and antioxidant and antimicrobial activities of food, extending its shelf life and reducing the presence of undesirable compounds [70,71]. For instance, camel milk fermented with

L. plantarum has demonstrated various health benefits, including antimicrobial, possibly due to the BAPs produced during fermentation [72]. A recent study identified a new low-molecular-weight AMP produced by *L. plantarum* FB-2, KMY15, which demonstrated a strong ability to inhibit the growth of *S. aureus* ATCC6538 and *E. coli* DH5. Furthermore, KMY15 maintained its efficacy against *S. aureus* ATCC6538, even in the presence of interfering substances such as proteins and lipids in contaminated milk [73]. In another study, antifungal peptides were released after the fermentation of kenaf seed proteins with *L. pentosus* RK3, identifying eight cationic peptides with molecular weights between 840 and 1876 Da. These peptides showed significant fungicidal activity against *Fusarium* sp. and *Aspergillus niger*, with a minimum inhibitory concentration (MIC) of 43 μg/mL and minimum fungicidal concentration (MFC) of 86 μg/mL, thus prolonging the latency phase of the fungi and increasing their membrane permeability [74]. All this evidence demonstrates the potential of AMPs as an alternative to traditional antibiotics due to their versatility, wide spectrum of activity, and the possibility of being modified to optimize their efficacy and stability under different conditions [75].

4.2. Antihypertensive Activity

ACE inhibitors work by blocking the enzyme ACE that converts angiotensin I into angiotensin II, a peptide that causes blood vessels to tighten. When angiotensin II is overproduced, it can lead to high blood pressure, which is a common problem in people with hypertension. Hypertension, characterized by persistently elevated blood pressure above 140/90 mmHg, is a prevalent risk factor associated with strokes and cardiovascular disease [76]. For this reason, ACE inhibitors are considered crucial for controlling blood pressure in hypertensive patients [77,78].

Recently, BAPs derived from food proteins have attracted significant interest as natural ACE inhibitors, offering a viable alternative to synthetic drugs, which often present relevant side effects [79]. Several ACE-inhibitory peptides (ACEIPs) have been isolated from fermented foods such as dairy products and vegetables [31,80–83]. In general, ACEIPs have similar primary structures, often featuring proline, tyrosine, cysteine, histidine, tryptophan, and methionine at the C-terminus, a factor that enhances their ACE inhibitory activity [84,85]. Moreover, the ACE active site contains a zinc coordination site and a hydrophobic pocket that allows interaction with aromatic residues, enhancing the binding of inhibitory peptides that contain them [86].

BAPs with all these features show strong ACE-inhibitory potential. They are likely absorbed by specific transporters across the intestinal epithelial wall, remain intact through gastrointestinal digestion, and maintain their ACE-inhibitory activities [87,88]. For instance, oral administration of the BAPs EAPLNPKANR and IVG, isolated from Cangkuk, a traditional Indonesian fermented beef, reduced blood pressure in spontaneously hypertensive rats 8 h after administration. The greatest effect was observed with IVG, due to its smaller size and the presence of isoleucine at the N-terminal, which facilitates its entry and binding to the active sites of ACE, thereby inhibiting the formation of angiotensin II and lowering blood pressure [89].

Among ACEIPs, the VPP (Valine–Proline–Proline) and IPP (Isoleucine–Proline–Proline) peptides have been extensively studied thanks to their potent ACE-inhibitory and pressure-lowering ability. These peptides were originally derived from the fermentation activity of bacteria like *L. helveticus* and yeasts like *Saccharomyces cerevisiae* in fermented milk and have since been demonstrated to yield marked antihypertensive activity in animal studies [90]. Additionally, clinical trials have shown that daily consumption of VPP/IPP tablets (10.2 mg/day) for 4 weeks decreased blood pressure levels and arterial stiffness in patients affected by metabolic syndrome [91]. Moreover, Moayedi et al. [92] used *Bacillus subtilis* to ferment tomato seeds and generate BAPs. Among the produced BAPs, the hexapeptide DGVVYY exhibited the strongest ACE-inhibitory activity, achieving an IC_{50} value of 2 μM.

4.3. Antioxidant Activity

BAPs derived from fermented foods have shown promising antioxidant activity by neutralizing reactive oxygen species (ROS) such as superoxide anions and hydroxyl radicals [93]. The body's enzymatic and non-enzymatic antioxidant systems effectively neutralize harmful free radicals. However, environmental stress, lifestyle, and disease can disrupt this balance, leading to oxidative stress. This imbalance is associated with cell apoptosis and various diseases, including diabetes, atherosclerosis, and cancer [94].

It has been shown that BAPs can scavenge free radicals, reduce lipid peroxidation, and chelate metal ions, which helps prevent oxidative damage [30,95]. Their antioxidant properties are closely linked to their aminoacidic composition. For instance, peptides rich in hydrophobic amino acids have been observed to interact more effectively with cell membranes, thereby enhancing their antioxidant effects [96]. An example is represented by the peptide VLPVPQK from cheddar cheese fermented with *L. helveticus*, which contains proline residues at the N-terminus, exhibiting an antioxidant capacity of 5.71 ± 0.59 mmol of Trolox equivalents/mg in a Trolox equivalent antioxidant capacity (TEAC) assay [97]. Furthermore, pre-treatment of rat fibroblast cells with this peptide, followed by induction of oxidative stress, suppressed ROS production and increased antioxidant enzymes [98]. Another example is the peptide SNLCRPCG, derived from chicken feathers fermented with *Bacillus subtilis*, which exhibited strong antioxidant activity due to its hydrophobic amino acids and cysteine's thiol (-SH) groups [99].

Furthermore, BAPs were found to modulate the antioxidant response by interacting with cellular signaling pathways. For instance, novel antioxidant peptides like LY-4, LP-5, and VL-9, identified from fermented broad bean paste, have been shown to activate the Kelch-like ECH-associated protein 1 (Keap1)-nuclear factor erythroid 2-related factor 2 (Nrf2) pathway, protecting liver cells (HepG2) from oxidative stress induced by 2,2′-Azobis(2-amidinopropane) dihydrochloride (AAPH), a radical generator [100]. Similarly, casein hydrolysate produced by *Lactobacillus reuteri* has been identified for its strong antioxidant potential, with the peptide VKEAMAPK showing significant effects by noncompetitively inhibiting Keap1, thereby activating the Nrf2 pathway, an essential protection mechanism against oxidative damage [101].

In another study, tomato seeds fermented with *Bacillus subtilis* produced antioxidant peptides such as GQVPP, which achieved a 97% 2,2-diphenyl-1-picrylhydrazyl (DPPH) radical scavenging activity at 0.4 mM, demonstrating its potential in reducing oxidative stress [92]. A peptide fraction with antioxidant activity was isolated and purified from whey protein metabolites fermented by *Lactobacillus rhamnosus B2-1*. The final purified peptide, B11, showed significant antioxidant activity, with a 2,2-azino-*bis*-3-ethylbenzothiazoline-6-sulphonic acid (ABTS) radical scavenging rate of 84.36%, a hydroxyl radical scavenging rate of 62.43%, and an oxygen radical absorbance capacity (ORAC) activity of 1726.44 μM Trolox equivalent/g. With an amino acid composition of 51.42%, dominated by glutamic and aspartic acids, B11 appears promising for food-based applications [102].

These findings highlight the diverse antioxidant mechanisms and health benefits of BAPs derived from fermented foods, making them promising natural alternatives to synthetic antioxidants.

4.4. Anti-Inflammatory Activity

The production of various immunomodulatory peptides with anti-inflammatory properties is facilitated by fermentation. They play a significant role in regulating immune response by modulating key interleukins such as IL-4 and IL-10 (anti-inflammatory), IL-1β and IL-2 (proinflammatory), IL-6 that can act as either, and cytokines such as tumor necrosis factor α (TNF-α) [103]. For instance, the hydrolysis of casein by the PrtB protease from *Lactobacillus. delbrueckii* subsp. *bulgaricus* 92,059 produced BAPs, which demonstrated immunomodulatory and anti-inflammatory effects in an in vitro assay of TNF-α-induced nuclear factor-κB (NF-κB) activation at concentrations of 5 mg/mL and 2.5 mg/mL, respectively [104]. Peptides from Xuanwei dry-cured ham have been found to reduce IL-6 and TNF-α levels in mice with dextran sodium sulfate-induced colitis, alleviating inflammatory bowel disease symptoms [105]. Moreover, soybean peptides have demonstrated beneficial effects in reducing LPS-induced intestinal inflammation in IEC-6 intestinal epithelial cells by decreasing NO production and downregulating the expression of inflammatory mediators, such as IL-1β, IL-6, and TNF-α [106].

Peptides with anti-inflammatory properties often feature positively charged and hydrophobic amino acids, especially at the C- and N-termini [107,108]. This structural composition facilitates their interaction with cell membranes and bacterial lipopolysaccharide (LPS), which is released from the wall of Gram-negative bacteria during infection and triggers inflammatory cytokines. However, the BAP's N-terminal region can block the release of LPS, thus stopping the inflammatory response. In a study where *L. plantarum* A3 and *L. rhamnosus* ATCC7469 were used to ferment broccoli, 17 novel anti-inflammatory peptides were identified, including GDRW, KASFAFAGL, FGDFNPGGRL, and ADLAHLPF, rich in hydrophobic amino acids, and HFKQPW, RFR, and KWR, with positively charged amino acid residues [109].

Table 1. Summary of BAP release under specific fermentation conditions and their bioactivity.

Fermented Food	Microorganisms	Fermentation Conditions	Peptide Sequence	Bioactivity	References
Milk	*L. plantarum* FB-2	37 °C for 20 h	KMYKKGRLWLVAGLS	Antimicrobial *S. aureus* and *L. monocytogenes* MIC = 256 μg/mL *E. coli* MIC = 128 μg/mL	[73]
Milk	*L. helveticus CP790* and *Saccharomyces cerevisiae*	37 °C for 24 h	VPP IPP	ACE-I IC50 = 9 μM IC50 = 5 μM	[110]
Milk	*Lacticaseibacillus rhamnosus NCDC24*	Not specified	AGWNIPM, ALPMHIR, VLPVPQKA YLGYLEQLLR	Antioxidant ABTS+ radical scavenging activity from 73.45 ± 0.57 (100 μg/mL) to 1.44 ± 0.22 (10 μg/mL)	[111]
Miso paste	*Aspergillus oryzae*	30 °C for 40 h	VPP, IPP	Antihypertensive	[112]
Milk	*Bifidobacterium bifidum* MF20/5	37 °C for 48 h	VLPVPQK LVYPFP	Antioxidant ACE-I IC_{50} = 132 μM	[113]
Sator bean (*Parkia speciosa*)	*L. fermentum ATCC9338*	37 °C for 8 days	EAKPSFYLK PVNNNAWAYATNFVPGK	Antioxidant DPPH activity = 78.48 ± 3.16% Antibacterial activity *S. typhi* (73.41 ± 0.08%) and *S. aureus* (64.70 ± 1.10%)	[114]
Skimmed milk	*Enterococcus faecalis* CECT 5727	30 °C for 48 h	LHLPLP	Antihypertensive IC_{50} = 59.6 μg/mL	[115]
Dry fermented sausage	*L. pentosus* and *Staphylococcus carnosus*	Two stages: 20 °C for 22 h; 9 °C for 43 days	YQEPVLGPVR, YQEPVLGPVRGPFPI, YQEPLV	ACE-I IC_{50} = 300 μM	[48]
Avena (*Avena sativa* L.)	*L. plantarum B1-6* and *Rhizopus oryzae*	30 °C for 72 h	Not specified	ACE-I IC_{50} = 0.42 mg protein/mL	[55]

Table 1. *Cont.*

Fermented Food	Microorganisms	Fermentation Conditions	Peptide Sequence	Bioactivity	References
Lupin, quinoa, and wheat	*L. reuteri* K777 and *L. plantarum* K779	35 °C for 72 h	Not specified	ACE-I from 25.3% to 58.9% Antioxidant DPPH radical scavenging activities from 25.0% to 65.0% Antiproliferative	[56]
Wheat, soybean, Barley, and amaranth	*L. curvatus SAL33* and *L. brevis AM7*	30 °C for 16 h	Lunasin (SKWQHQQD-SCRKQLQGVNLT-PCEKHIMEKIQ-GRGDDDDDDDDD)	Cancer preventive	[57]
Budu	Not specified	Over 120 days	VAAGRTDAGVH LDDPVFIH	Antioxidant DPPH radical scavenging activity $IC_{50} = 1.451 \pm 0.873$ (mg/mL) $IC_{50} = 0.844 \pm 0.203$ (mg/mL)	[63]
Zebra blenny (*Salaria basilisca*) muscle protein	*Bacillus mojavensis* A21	From 4 to 48 h at 37 °C	GLPPYPYAG, LVDGLDVGIL, ETPGGTPLAPEPD, LSYEEAITTY, HHPDDFNPSVH	Antibacterial *E. coli* MIC = 0.62 ± 0.01 mg/mL *K. pneumoniae* MIC = 1.23 ± 0.02 mg/mL ACE-I Antioxidant	[62]
Pekasan (Loma fish)	*L. plantarum* IFRPD P15	2 weeks at RT	AIPPHPYP IAEVFLITDPK	Antioxidant activity DPPH radical scavenging activity IC_{50} (mg/mL) = 1.38 ± 0.25 IC_{50} (mg/mL) = 0.897 ± 0.84	[64]
Manila clam (*Ruditapes philippinarum*)	*Bacillus natto*	37 °C for 36 h	VISDEDGVTH	ACE-I $IC_{50} = 8.16$ µM	[18]
Thai shrimp pastes	Not specified	Not specified	SV, IF, WP	ACE-I $IC_{50} = 60.68 \pm 1.06$ µM Antioxidant $ABTS^{+}$ $EC_{50} = 17.52 \pm 0.46$ µM	[65]
Kenaf seed	*L. casei*	37 °C for 72 h	AKVGLKPGGFFVLK, GSTIK, LLLSK, TAHDDYK	Antibacterial activity from 42.07% to 77.38%	[116]

Table 1. *Cont.*

Fermented Food	Microorganisms	Fermentation Conditions	Peptide Sequence	Bioactivity	References
Tomato waste proteins	*Bacillus subtilis*	37 °C for 24 h	DGVVYY GQVPP	ACE-I IC_{50} = 2 μM Antioxidant 97% DPPH scavenging activity at 0.4 mM	[92]
Cheddar cheese	*L. helveticus* and *Streptococcus thermophilus*	Not specified	EMPFPK, AVPYPQR, VLPVPQK, AMKPWIQPK	Antioxidant TEAC = 5.7 ± 0.6 mmol TE/mg	[97]
Feather hydrolysate	*Bacillus subtilis S1-4*	37 °C for 72 h	SNLCRPCG	Antioxidant DPPH IC_{50} = 0.39 mg/mL	[99]
Whey protein	*L. rhamnosus B2-1*	37 °C for 48 h	B11	Antioxidant $ABTS^{+}$ radical scavenging activities = 84.36%	[102]
Casein	*L. reuteri*	Not specified	VKEAMAPK	Antioxidant Decreased ROS activity by 45%	[101]
Broccoli	*L. plantarum A3* and *L. rhamnosus ATCC7469*	37 °C for 24 h	SIWYGPDRP	Anti-inflammatory Inhibits NO release from inflammatory cells at 25 μM, with an inhibition rate of 52.32 ± 1.48	[109]

RT: room temperature. ACE-I: angiotensin-converting enzyme-inhibitory activity. TEAC: Trolox equivalent antioxidant capacity. ROS: reactive oxygen species. Each letter in the peptide sequence corresponds to an amino acid: A = alanine, C = cysteine, D = aspartic acid, E = glutamic acid, F = phenylalanine, G = glycine, H = histidine, I = isoleucine, K = lysine, L = leucine, M = methionine, N = asparagine, P = proline, Q = glutamine, R = arginine, S = serine, T = threonine, V = valine, W = tryptophan, Y = tyrosine.

5. Challenges and Limitations in BAPs Production

The development of BAPs is a promising yet challenging area in biotechnology, with potential applications across the food, nutraceutical, and pharmaceutical industries. However, several significant obstacles impede their efficient development and commercialization. One of the foremost challenges in developing BAPs is the scalability of production processes. In microbial fermentation, issues like longer reaction times and low yield of peptide formation pose additional barriers. The complex proteolytic systems in different microorganisms result in variations in peptide structures, making it difficult to standardize production processes [24]. Even among the same species of bacteria, such as *Lactobacillus* or *Bacillus*, differences in proteolytic capabilities can result in different peptide profiles when exposed to the same substrates [117]. While enzymatic hydrolysis offers more controlled conditions, microbial fermentation involves a longer reaction time and greater unpredictability in the peptides generated. Factors such as the type of microorganism, the protein substrate used, fermentation time, and environmental conditions (e.g., pH, temperature, and oxygen levels) all impact the extent and specificity of hydrolysis [118].

Additionally, the purification of BAPs from fermented mixtures is technically demanding and costly. Microbial fermentation typically yields a complex mixture of peptides, alongside other microbial by-products such as exopolysaccharides, bacteriocins, and dead microbial cells. These additional compounds can interfere with the isolation of pure BAPs and complicate the identification of the specific bioactive properties of the peptides [119]. Purification processes are essential to isolate peptides with specific biological activities, yet they are time-consuming and costly. Ultrafiltration and chromatographic techniques, such as HPLC, are commonly used but are often not suitable for large-scale production due to issues like membrane fouling and poor reproducibility [120]. Once produced and purified, BAPs face challenges related to their stability and bioavailability, covered in depth in Section 6.

Furthermore, regulatory approval and clinical validation hinder the production of BAPs through microbial fermentation. Despite the promising bioactivities demonstrated in in vitro and animal studies, there is a lack of well-designed clinical trials to confirm the efficacy of food-derived peptides in humans [121]. Regulatory agencies, such as the European Food Safety Authority (EFSA), require rigorous evidence of the safety and efficacy of functional food ingredients before they can be marketed. Without sufficient clinical evidence, it is difficult for companies to commercialize BAPs or make health claims about their benefits, which hinders their marketability [5]. Therefore, addressing these issues is critical for the future success of BAPs in the health and wellness industries.

6. Approaches for Enhancing BAP Production, Stability, and Bioavailability

Since the fermentation process often results in low yields of BAPs, besides modulating physicochemical parameters such as temperature, pH, and time, emerging green technologies and encapsulation strategies have been employed to enhance BAP content, stability, and bioavailability. Figure 2 summarizes BAP production techniques, stability/bioavailability approaches, and the major mechanisms responsible for BAP biological activity.

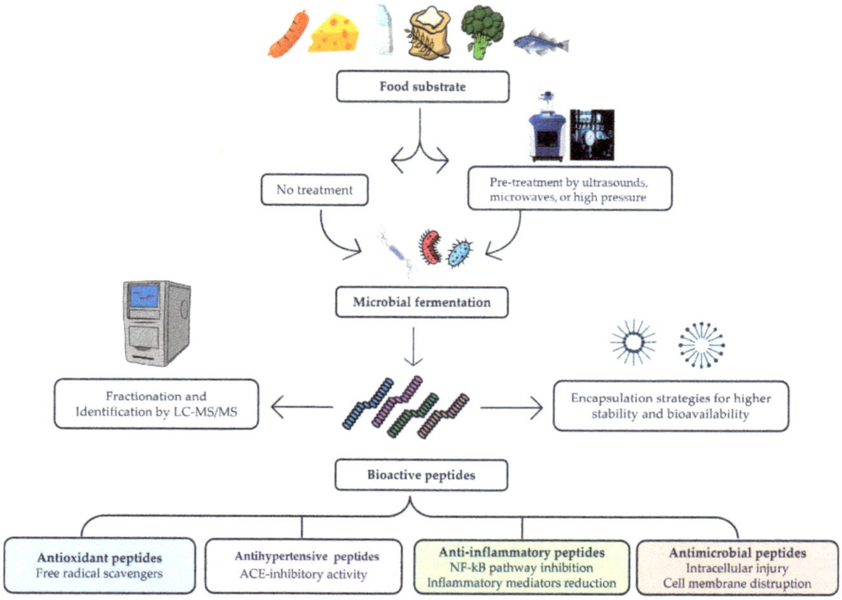

Figure 2. Schematic representation of BAP production and their associated bioactivities.

6.1. Strategies to Optimize the Production of BAPs in Fermented Foods

Microbial fermentation has been extensively used to improve the nutritional quality and health-promoting properties of various foodstuffs, for instance, by increasing the availability of BAPs. However, despite being economical and easy to perform, fermentation displays low yields of BAP production, a factor that limits its industrial escalation [9,122]. Thus, several strategies have been attempted to promote protein hydrolysis and increase the recovery of BAPs.

One such way consists of exposing the food matrix to ultrasounds, microwaves, or high pressure before the fermentation process [9,122]. In a study performed by Munir and colleagues [123], Cheddar cheese was produced with milk pre-treated by either ultrasonication, high-pressure processing, or microwave, and its antioxidant and ACE-inhibitory activities were followed during ripening.

It was revealed that the pre-processing steps increased the rate of proteolysis during cheese making and ripening, as well as the antioxidant and ACE-inhibitory activities of the resulting cheese, with the best results obtained with ultrasonication. Additionally, moderate ultrasound exposure during fermentation has been proven to affect the activity of microorganisms, as testified by the work of Xie et al. [124]. Indeed, they showed that ultrasound-assisted fermentation of okra improved its peptide and soluble protein contents, enhancing its antioxidant properties.

Finally, the food matrix can also be processed after fermentation to increase the yield of BAPs. As an example, milk fermented with *Lactobacillus delbrueckii* QS306 and subjected to ultrahigh-pressure treatment exerted a higher peptide concentration and variety, and an increased ACE-inhibitory activity compared to the non-processed sample [125].

Another strategy to enhance BAP yield is to optimize the fermentation parameters, including incubation time, temperature, pH, and inoculation rate to find the optimal conditions. Khakhariya et al. [126] explored this approach by studying the production of health-promoting BAPs in fermented buffalo and camel milk by *Limosilactobacillus fermentum* (KGL4) and *Saccharomyces cerevisiae* (WBS2A). They monitored the variation in ACE-inhibitory, antidiabetic, and proteolytic activity at 37 °C and at multiple time intervals (12, 24, 36, and 48 h) and inoculation rates (1.5%, 2.0%, and 2.5%), finding the optimal fermentation conditions to be a 2.5% inoculation rate and 48 h. One investigation on BAPs in flaxseed milk fermented with *L. plantarum* (NCDC 374) revealed that the optimal conditions to guarantee the best bioactivity (antioxidant, ACE-inhibition, and proteolysis) were a 4.20% inoculum size and 126 h of fermentation time [127]. Additionally, it was shown that pH is critical to optimizing BAP retrieval and bioactivity in fermented foods. For instance, the functional properties and BAP production in lentils fermented using *L. plantarum* combined with a commercial protease (Savinase® 16 L) benefited from a mild alkaline environment. A multivariate analysis highlighted a pH of 8.5 and 11.6 h of incubation time as the optimum [128]. Taken together, these studies highlight the possibility of improving the fermentation performance of foods, maximizing BAP release.

6.2. Strategies to Improve the Stability and Bioavailability of BAPs

Improving the stability and bioavailability of BAPs is essential for maximizing their health benefits in functional foods, nutraceuticals, and pharmaceuticals. Despite their promising biological activities, BAPs face several challenges that limit their application, particularly their susceptibility to enzymatic degradation, low bioavailability, and poor stability under various physiological and processing conditions [129]. Several advanced strategies have been developed to address these limitations, including encapsulation technologies, carrier matrices, and physicochemical modifications [10,130].

One major challenge in enhancing BAPs' stability and bioavailability is their susceptibility to gastrointestinal degradation. Peptidases in the stomach and intestinal lumen degrade BAPs, reducing their ability to reach the bloodstream in their active form [131]. Encapsulation is a commonly employed technique for maintaining bioactive compounds' functional and physicochemical properties. It involves surrounding solid or liquid particles with a coating or embedding them in a matrix [132].

Several materials can be used as carriers for peptide encapsulation, including proteins, polysaccharides, and lipids. Each type of carrier offers distinct advantages and faces specific challenges. Proteins, for instance, possess valuable functional properties, such as emulsification, water retention, and the ability to form gels, making them suitable candidates for encapsulation [133–135].

Polysaccharides are another promising class of carrier materials for encapsulation due to their structural stability, abundance, and low cost [136]. They have reactive functional groups that can form interactions with BAPs, helping to stabilize and protect them during processing and storage [132]. Polysaccharides such as starch, dextrin, gum arabic, pectin, and chitosan have been widely used in peptide encapsulation [132,137]. Their use is often combined with proteins to achieve better encapsulation efficiency and bioactive compound stabilization [131,138].

Lipid-based carriers, such as liposomes and nanoliposomes, have also gained significant attention as good systems for BAPs' encapsulation [132]. Liposomes consist of phospholipid bilayers, which can be useful to encapsulate both hydrophilic and lipophilic substances [131,139]. This versatility makes liposomes an attractive alternative for delivering a wide range of bioactive compounds. Nanoliposomes, which are smaller in size, provide even better protection and improved bioavailability by facilitating the controlled release of encapsulated peptides [140,141]. For instance, BAPs derived from shrimp waste have been stabilized through chitosan-coated nanoliposome formation [142].

Among the methods for peptide encapsulation, spray drying is a popular method for encapsulating bioactive compounds, including peptides, because it is economical, flexible, and capable of producing stable powders that can be easily incorporated into food products. During the spray drying process, peptides are mixed with carrier materials that form a protective film around the peptide [132]. This approach prevents peptide degradation during processing and storage but also effectively masks undesirable flavors, such as bitterness, as demonstrated by Sarabandi et al. [134] through spray drying encapsulation to stabilize peptides from oleaster seeds.

Another method is nanoencapsulation, which represents a promising approach for improving peptide stability and bioavailability. Nanoparticles can be engineered to protect peptides from degradation, enhance their absorption in the intestine, and ensure their targeted delivery to specific tissues [143]. Nanoparticles made from materials such as dextran, chitosan, or amphiphilic molecules can improve the bioaccessibility and controlled release of peptides. These nanoscale systems offer several advantages over traditional encapsulation methods, including increased surface area, enhanced cellular uptake, and the ability to penetrate physiological barriers [144].

Physicochemical modifications of peptides such as the Maillard reaction (MR) also hold the potential for improving their stability and bioavailability. The MR involves the interaction between amino acids and reducing sugars, which leads to the formation of peptide–sugar conjugates [145]. These conjugates have been shown to enhance the stability of peptides during processing and storage and improve their taste and overall sensory properties. Additionally, MR-conjugated peptides may exhibit enhanced biological activities, such as antioxidant or antimicrobial effects, further increasing their potential as functional ingredients in foods and nutraceuticals [130].

Therefore, improving the stability and bioavailability of BAPs is crucial for their successful incorporation into functional foods, nutraceuticals, and pharmaceuticals. Encapsulation and physicochemical modifications are among the most promising strategies for achieving these goals.

7. Conclusions and Future Perspectives

By reviewing a substantial slice of the literature, we have confirmed microbial fermentation as a valuable natural technology extensively used to enrich several animal- and plant-derived foods with BAPs. This approach offers structurally variable peptides due to the diversity of microbial proteases but also supports a wide range of beneficial biological activities, namely antimicrobial, anti-inflammatory, antioxidant, and antihypertensive. Adopting innovative technologies such as ultrasound, microwave, and high pressure, combined with optimizing fermentation parameters, represents a promising avenue for scaling up this process. Additionally, finding appropriate delivery strategies is crucial for guaranteeing the stability and efficacy of BAPs. As research and fermentation technologies continue to advance, new perspectives are emerging for developing functional products with high added value. This will also be achieved by establishing interdisciplinary collaborations, involving food technologists, microbiologists, biochemists, and medical professionals, ensuring the scalability of BAPs' production processes.

Author Contributions: Conceptualization, L.P.F. and M.G.; writing—original draft preparation, L.P.F., A.C., and F.V.; writing—review and editing, L.P.F., A.C., F.V., and M.G.; visualization, L.P.F. and M.G.; supervision, M.G. All authors have read and agreed to the published version of the manuscript.

Funding: This research received no external funding.

Institutional Review Board Statement: Not applicable.

Informed Consent Statement: Not applicable.

Data Availability Statement: The original contributions presented in the study are included in the article, further inquiries can be directed to the corresponding author.

Conflicts of Interest: The authors declare no conflict of interest.

References

1. Marco, M.L.; Sanders, M.E.; Gänzle, M.; Arrieta, M.C.; Cotter, P.D.; De Vuyst, L.; Hill, C.; Holzapfel, W.; Lebeer, S.; Merenstein, D.; et al. The International Scientific Association for Probiotics and Prebiotics (ISAPP) Consensus Statement on Fermented Foods. *Nat. Rev. Gastroenterol. Hepatol.* **2021**, *18*, 196–208. [CrossRef] [PubMed]
2. García-Burgos, M.; Moreno-Fernández, J.; Alférez, M.J.M.; Díaz-Castro, J.; López-Aliaga, I. New Perspectives in Fermented Dairy Products and Their Health Relevance. *J. Funct. Foods* **2020**, *72*, 104059. [CrossRef]

3. Diez-Ozaeta, I.; Astiazaran, O.J. Fermented Foods: An Update on Evidence-Based Health Benefits and Future Perspectives. *Food Res. Int.* **2022**, *156*, 111133. [CrossRef]
4. Korhonen, H.; Pihlanto, A. Bioactive Peptides: Production and Functionality. *Int. Dairy. J.* **2006**, *16*, 945–960. [CrossRef]
5. Chalamaiah, M.; Keskin Ulug, S.; Hong, H.; Wu, J. Regulatory Requirements of Bioactive Peptides (Protein Hydrolysates) from Food Proteins. *J. Funct. Foods* **2019**, *58*, 123–129. [CrossRef]
6. Mirzaei, M.; Shavandi, A.; Mirdamadi, S.; Soleymanzadeh, N.; Motahari, P.; Mirdamadi, N.; Moser, M.; Subra, G.; Alimoradi, H.; Goriely, S. Bioactive Peptides from Yeast: A Comparative Review on Production Methods, Bioactivity, Structure-Function Relationship, and Stability. *Trends Food Sci. Technol.* **2021**, *118*, 297–315. [CrossRef]
7. Manzoor, M.; Singh, J.; Gani, A. Exploration of Bioactive Peptides from Various Origin as Promising Nutraceutical Treasures: In Vitro, in Silico and in Vivo Studies. *Food Chem.* **2022**, *373*, 131395. [CrossRef]
8. Martinez-Villaluenga, C.; Peñas, E.; Frias, J. Bioactive Peptides in Fermented Foods. In *Fermented Foods in Health and Disease Prevention*; Elsevier: Amsterdam, The Netherlands, 2017; pp. 23–47.
9. Tadesse, S.A.; Emire, S.A. Production and Processing of Antioxidant Bioactive Peptides: A Driving Force for the Functional Food Market. *Heliyon* **2020**, *6*, e04765. [CrossRef]
10. Jiang, Y.; Sun, J.; Chandrapala, J.; Majzoobi, M.; Brennan, C.; Zeng, X.; Sun, B. Recent Progress of Food-derived Bioactive Peptides: Extraction, Purification, Function, and Encapsulation. *Food Front.* **2024**, *5*, 1240–1264. [CrossRef]
11. Mannaa, M.; Han, G.; Seo, Y.-S.; Park, I. Evolution of Food Fermentation Processes and the Use of Multi-Omics in Deciphering the Roles of the Microbiota. *Foods* **2021**, *10*, 2861. [CrossRef]
12. Wang, Z.; Wang, L. Impact of Sourdough Fermentation on Nutrient Transformations in Cereal-Based Foods: Mechanisms, Practical Applications, and Health Implications. *Grain Oil Sci. Technol.* **2024**, *7*, 124–132. [CrossRef]
13. Day, C.N.; Morawicki, R.O. Effects of Fermentation by Yeast and Amylolytic Lactic Acid Bacteria on Grain Sorghum Protein Content and Digestibility. *J. Food Qual.* **2018**, *2018*, 1–8. [CrossRef]
14. Nkhata, S.G.; Ayua, E.; Kamau, E.H.; Shingiro, J. Fermentation and Germination Improve Nutritional Value of Cereals and Legumes through Activation of Endogenous Enzymes. *Food Sci. Nutr.* **2018**, *6*, 2446–2458. [CrossRef] [PubMed]
15. Ulug, S.K.; Jahandideh, F.; Wu, J. Novel Technologies for the Production of Bioactive Peptides. *Trends Food Sci. Technol.* **2021**, *108*, 27–39. [CrossRef]
16. Hernández-Ledesma, B.; del Mar Contreras, M.; Recio, I. Antihypertensive Peptides: Production, Bioavailability and Incorporation into Foods. *Adv. Colloid. Interface Sci.* **2011**, *165*, 23–35. [CrossRef]
17. Kumari, R.; Sharma, N.; Sharma, S.; Samurailatpam, S.; Padhi, S.; Singh, S.P.; Kumar Rai, A. Production and Characterization of Bioactive Peptides in Fermented Soybean Meal Produced Using Proteolytic Bacillus Species Isolated from Kinema. *Food Chem.* **2023**, *421*, 136130. [CrossRef]
18. Chen, Y.; Gao, X.; Wei, Y.; Liu, Q.; Jiang, Y.; Zhao, L.; Ulaah, S. Isolation, Purification and the Anti-Hypertensive Effect of a Novel Angiotensin I-Converting Enzyme (ACE) Inhibitory Peptide from *Ruditapes philippinarum* Fermented with *Bacillus natto*. *Food Funct.* **2018**, *9*, 5230–5237. [CrossRef]
19. Ahmed, T.; Juhász, A.; Bose, U.; Shiferaw Terefe, N.; Colgrave, M.L. Research Trends in Production, Separation, and Identification of Bioactive Peptides from Fungi—A Critical Review. *J. Funct. Foods* **2024**, *119*, 106343. [CrossRef]
20. Ji, D.; Ma, J.; Xu, M.; Agyei, D. Cell-envelope Proteinases from Lactic Acid Bacteria: Biochemical Features and Biotechnological Applications. *Compr. Rev. Food Sci. Food Saf.* **2021**, *20*, 369–400. [CrossRef]
21. Raveschot, C.; Cudennec, B.; Coutte, F.; Flahaut, C.; Fremont, M.; Drider, D.; Dhulster, P. Production of Bioactive Peptides by Lactobacillus Species: From Gene to Application. *Front. Microbiol.* **2018**, *9*, 2354. [CrossRef]
22. Sadat-Mekmene, L.; Jardin, J.; Corre, C.; Mollé, D.; Richoux, R.; Delage, M.-M.; Lortal, S.; Gagnaire, V. Simultaneous Presence of PrtH and PrtH2 Proteinases in *Lactobacillus helveticus* Strains Improves Breakdown of the Pure α_{s1}-Casein. *Appl. Environ. Microbiol.* **2011**, *77*, 179–186. [CrossRef] [PubMed]
23. Miyamoto, M.; Ueno, H.M.; Watanabe, M.; Tatsuma, Y.; Seto, Y.; Miyamoto, T.; Nakajima, H. Distinctive Proteolytic Activity of Cell Envelope Proteinase of *Lactobacillus helveticus* Isolated from Airag, a Traditional Mongolian Fermented Mare's Milk. *Int. J. Food Microbiol.* **2015**, *197*, 65–71. [CrossRef] [PubMed]
24. Cruz-Casas, D.E.; Aguilar, C.N.; Ascacio-Valdés, J.A.; Rodríguez-Herrera, R.; Chávez-González, M.L.; Flores-Gallegos, A.C. Enzymatic Hydrolysis and Microbial Fermentation: The Most Favorable Biotechnological Methods for the Release of Bioactive Peptides. *Food Chem. Mol. Sci.* **2021**, *3*, 100047. [CrossRef] [PubMed]
25. Melini, F.; Melini, V.; Luziatelli, F.; Ficca, A.G.; Ruzzi, M. Health-Promoting Components in Fermented Foods: An Up-to-Date Systematic Review. *Nutrients* **2019**, *11*, 1189. [CrossRef]
26. Dharmisthaben, P.; Basaiawmoit, B.; Sakure, A.; Das, S.; Maurya, R.; Bishnoi, M.; Kondepudi, K.K.; Hati, S. Exploring Potentials of Antioxidative, Anti-Inflammatory Activities and Production of Bioactive Peptides in Lactic Fermented Camel Milk. *Food Biosci.* **2021**, *44*, 101404. [CrossRef]
27. Li, H.; Wu, J.; Wan, J.; Zhou, Y.; Zhu, Q. Extraction and Identification of Bioactive Peptides from Panxian Dry-Cured Ham with Multifunctional Activities. *LWT* **2022**, *160*, 113326. [CrossRef]
28. Folliero, V.; Lama, S.; Franci, G.; Giugliano, R.; D'Auria, G.; Ferranti, P.; Pourjula, M.; Galdiero, M.; Stiuso, P. Casein-Derived Peptides from the Dairy Product Kashk Exhibit Wound Healing Properties and Antibacterial Activity against *Staphylococcus aureus*: Structural and Functional Characterization. *Food Res. Int.* **2022**, *153*, 110949. [CrossRef]

29. Taniguchi, M.; Aida, R.; Saito, K.; Ochiai, A.; Takesono, S.; Saitoh, E.; Tanaka, T. Identification and Characterization of Multifunctional Cationic Peptides from Traditional Japanese Fermented Soybean *Natto* Extracts. *J. Biosci. Bioeng.* **2019**, *127*, 472–478. [CrossRef]
30. Gallego, M.; Mora, L.; Toldrá, F. Characterisation of the Antioxidant Peptide AEEEYPDL and Its Quantification in Spanish Dry-Cured Ham. *Food Chem.* **2018**, *258*, 8–15. [CrossRef]
31. Gibbs, B.F.; Zougman, A.; Masse, R.; Mulligan, C. Production and Characterization of Bioactive Peptides from Soy Hydrolysate and Soy-Fermented Food. *Food Res. Int.* **2004**, *37*, 123–131. [CrossRef]
32. Nasri, R.; Abdelhedi, O.; Nasri, M.; Jridi, M. Fermented Protein Hydrolysates: Biological Activities and Applications. *Curr. Opin. Food Sci.* **2022**, *43*, 120–127. [CrossRef]
33. Sánchez, A.; Vázquez, A. Bioactive Peptides: A Review. *Food Qual. Saf.* **2017**, *1*, 29–46. [CrossRef]
34. Tsermoula, P.; Khakimov, B.; Nielsen, J.H.; Engelsen, S.B. WHEY—The Waste-Stream That Became More Valuable than the Food Product. *Trends Food Sci. Technol.* **2021**, *118*, 230–241. [CrossRef]
35. Mann, B.; Athira, S.; Sharma, R.; Bajaj, R. Bioactive Peptides in Yogurt. In *Yogurt in Health and Disease Prevention*; Elsevier: Amsterdam, The Netherlands, 2017; pp. 411–426.
36. Chourasia, R.; Abedin, M.M.; Chiring Phukon, L.; Sahoo, D.; Singh, S.P.; Rai, A.K. Biotechnological Approaches for the Production of Designer Cheese with Improved Functionality. *Compr. Rev. Food Sci. Food Saf.* **2021**, *20*, 960–979. [CrossRef]
37. Maïworé, J.; Tatsadjieu Ngoune, L.; Piro-Metayer, I.; Montet, D. Identification of Yeasts Present in Artisanal Yoghurt and Traditionally Fermented Milks Consumed in the Northern Part of Cameroon. *Sci. Afr.* **2019**, *6*, e00159. [CrossRef]
38. Dullius, A.; Rama, G.R.; Giroldi, M.; Goettert, M.I.; Lehn, D.N.; Volken de Souza, C.F. Bioactive Peptide Production in Fermented Foods. In *Current Developments in Biotechnology and Bioengineering*; Elsevier: Amsterdam, The Netherlands, 2022; pp. 47–72.
39. Padhi, S.; Sharma, S.; Sahoo, D.; Montet, D.; Rai, A.K. Potential of Lactic Acid Bacteria as Starter Cultures for Food Fermentation and as Producers of Biochemicals for Value Addition. In *Lactic Acid Bacteria in Food Biotechnology*; Elsevier: Amsterdam, The Netherlands, 2022; pp. 281–304.
40. Martini, S.; Conte, A.; Tagliazucchi, D. Effect of Ripening and in Vitro Digestion on the Evolution and Fate of Bioactive Peptides in Parmigiano-Reggiano Cheese. *Int. Dairy. J.* **2020**, *105*, 104668. [CrossRef]
41. Amorim, F.G.; Coitinho, L.B.; Dias, A.T.; Friques, A.G.F.; Monteiro, B.L.; de Rezende, L.C.D.; Pereira, T.d.M.C.; Campagnaro, B.P.; De Pauw, E.; Vasquez, E.C.; et al. Identification of New Bioactive Peptides from Kefir Milk through Proteopeptidomics: Bioprospection of Antihypertensive Molecules. *Food Chem.* **2019**, *282*, 109–119. [CrossRef]
42. Chen, Y.F.; Zhao, W.J.; Wu, R.N.; Sun, Z.H.; Zhang, W.Y.; Wang, J.C.; Bilige, M.; Zhang, H.P. Proteome Analysis of *Lactobacillus helveticus* H9 during Growth in Skim Milk. *J. Dairy. Sci.* **2014**, *97*, 7413–7425. [CrossRef]
43. Chen, Y.; Li, C.; Xue, J.; Kwok, L.; Yang, J.; Zhang, H.; Menghe, B. Characterization of Angiotensin-Converting Enzyme Inhibitory Activity of Fermented Milk Produced by *Lactobacillus helveticus*. *J. Dairy. Sci.* **2015**, *98*, 5113–5124. [CrossRef]
44. Baptista, D.P.; Galli, B.D.; Cavalheiro, F.G.; Negrão, F.; Eberlin, M.N.; Gigante, M.L. *Lactobacillus helveticus* LH-B02 Favours the Release of Bioactive Peptide during Prato Cheese Ripening. *Int. Dairy. J.* **2018**, *87*, 75–83. [CrossRef]
45. Bielecka, M.; Cichosz, G.; Czeczot, H. Antioxidant, Antimicrobial and Anticarcinogenic Activities of Bovine Milk Proteins and Their Hydrolysates—A Review. *Int. Dairy. J.* **2022**, *127*, 105208. [CrossRef]
46. Zenezini Chiozzi, R.; Capriotti, A.L.; Cavaliere, C.; La Barbera, G.; Piovesana, S.; Samperi, R.; Laganà, A. Purification and Identification of Endogenous Antioxidant and ACE-Inhibitory Peptides from Donkey Milk by Multidimensional Liquid Chromatography and NanoHPLC-High Resolution Mass Spectrometry. *Anal. Bioanal. Chem.* **2016**, *408*, 5657–5666. [CrossRef] [PubMed]
47. El-Salam, M.H.A.; El-Shibiny, S. Bioactive Peptides of Buffalo, Camel, Goat, Sheep, Mare, and Yak Milks and Milk Products. *Food Rev. Int.* **2013**, *29*, 1–23. [CrossRef]
48. Mora, L.; Escudero, E.; Aristoy, M.-C.; Toldrá, F. A Peptidomic Approach to Study the Contribution of Added Casein Proteins to the Peptide Profile in Spanish Dry-Fermented Sausages. *Int. J. Food Microbiol.* **2015**, *212*, 41–48. [CrossRef]
49. Mejri, L.; Vásquez-Villanueva, R.; Hassouna, M.; Marina, M.L.; García, M.C. Identification of Peptides with Antioxidant and Antihypertensive Capacities by RP-HPLC-Q-TOF-MS in Dry Fermented Camel Sausages Inoculated with Different Starter Cultures and Ripening Times. *Food Res. Int.* **2017**, *100*, 708–716. [CrossRef]
50. Takeda, S.; Matsufuji, H.; Nakade, K.; Takenoyama, S.; Ahhmed, A.; Sakata, R.; Kawahara, S.; Muguruma, M. Investigation of Lactic Acid Bacterial Strains for Meat Fermentation and the Product's Antioxidant and Angiotensin-I-converting-enzyme Inhibitory Activities. *Anim. Sci. J.* **2017**, *88*, 507–516. [CrossRef]
51. Kononiuk, A.D.; Karwowska, M. Bioactive Compounds in Fermented Sausages Prepared from Beef and Fallow Deer Meat with Acid Whey Addition. *Molecules* **2020**, *25*, 2429. [CrossRef]
52. Chai, K.F.; Voo, A.Y.H.; Chen, W.N. Bioactive Peptides from Food Fermentation: A Comprehensive Review of Their Sources, Bioactivities, Applications, and Future Development. *Compr. Rev. Food Sci. Food Saf.* **2020**, *19*, 3825–3885. [CrossRef]
53. Gabriele, M.; Pucci, L. Fermentation and Germination as a Way to Improve Cereals Antioxidant and Antiinflammatory Properties. In *Current Advances for Development of Functional Foods Modulating Inflammation and Oxidative Stress*; Elsevier: Amsterdam, The Netherlands, 2022; pp. 477–497.
54. Corpuz, H.M.; Fujii, H.; Nakamura, S.; Katayama, S. Fermented Rice Peptides Attenuate Scopolamine-Induced Memory Impairment in Mice by Regulating Neurotrophic Signaling Pathways in the Hippocampus. *Brain Res.* **2019**, *1720*, 146322. [CrossRef]

55. Wu, H.; Rui, X.; Li, W.; Xiao, Y.; Zhou, J.; Dong, M. Whole-Grain Oats (*Avena sativa* L.) as a Carrier of Lactic Acid Bacteria and a Supplement Rich in Angiotensin I-Converting Enzyme Inhibitory Peptides through Solid-State Fermentation. *Food Funct.* **2018**, *9*, 2270–2281. [CrossRef]
56. Ayyash, M.; Johnson, S.K.; Liu, S.-Q.; Mesmari, N.; Dahmani, S.; Al Dhaheri, A.S.; Kizhakkayil, J. In Vitro Investigation of Bioactivities of Solid-State Fermented Lupin, Quinoa and Wheat Using *Lactobacillus* spp. *Food Chem.* **2019**, *275*, 50–58. [CrossRef] [PubMed]
57. Rizzello, C.G.; Nionelli, L.; Coda, R.; Gobbetti, M. Synthesis of the Cancer Preventive Peptide Lunasin by Lactic Acid Bacteria During Sourdough Fermentation. *Nutr. Cancer* **2012**, *64*, 111–120. [CrossRef] [PubMed]
58. Chi, C.-H.; Cho, S.-J. Improvement of Bioactivity of Soybean Meal by Solid-State Fermentation with *Bacillus amyloliquefaciens* versus *Lactobacillus* Spp. and *Saccharomyces cerevisiae*. *LWT—Food Sci. Technol.* **2016**, *68*, 619–625. [CrossRef]
59. Xiao, Y.; Sun, M.; Zhang, Q.; Chen, Y.; Miao, J.; Rui, X.; Dong, M. Effects of *Cordyceps militaris* (L.) Fr. Fermentation on the Nutritional, Physicochemical, Functional Properties and Angiotensin I Converting Enzyme Inhibitory Activity of Red Bean (*Phaseolus angularis* [Willd.] W.F. Wight.) Flour. *J. Food Sci. Technol.* **2018**, *55*, 1244–1255. [CrossRef]
60. Ihn, H.J.; Kim, J.A.; Lim, S.; Nam, S.-H.; Hwang, S.H.; Lim, J.; Kim, G.-Y.; Choi, Y.H.; Jeon, Y.-J.; Lee, B.-J.; et al. Fermented Oyster Extract Prevents Ovariectomy-Induced Bone Loss and Suppresses Osteoclastogenesis. *Nutrients* **2019**, *11*, 1392. [CrossRef]
61. Kim, J.A.; Yao, Z.; Kim, H.-J.; Kim, J.H. Properties of Gul Jeotgal (Oyster Jeotgal) Prepared with Different Types of Salt and *Bacillus subtilis* JS2 as Starter. *Microbiol. Biotechnol. Lett.* **2018**, *46*, 1–8. [CrossRef]
62. Jemil, I.; Abdelhedi, O.; Mora, L.; Nasri, R.; Aristoy, M.-C.; Jridi, M.; Hajji, M.; Toldrá, F.; Nasri, M. Peptidomic Analysis of Bioactive Peptides in Zebra Blenny (*Salaria basilisca*) Muscle Protein Hydrolysate Exhibiting Antimicrobial Activity Obtained by Fermentation with Bacillus Mojavensis A21. *Process Biochem.* **2016**, *51*, 2186–2197. [CrossRef]
63. Najafian, L.; Babji, A.S. Purification and Identification of Antioxidant Peptides from Fermented Fish Sauce (*Budu*). *J. Aquat. Food Product. Technol.* **2019**, *28*, 14–24. [CrossRef]
64. Najafian, L.; Babji, A.S. Fractionation and Identification of Novel Antioxidant Peptides from Fermented Fish (Pekasam). *J. Food Meas. Charact.* **2018**, *12*, 2174–2183. [CrossRef]
65. Kleekayai, T.; Harnedy, P.A.; O'Keeffe, M.B.; Poyarkov, A.A.; CunhaNeves, A.; Suntornsuk, W.; FitzGerald, R.J. Extraction of Antioxidant and ACE Inhibitory Peptides from Thai Traditional Fermented Shrimp Pastes. *Food Chem.* **2015**, *176*, 441–447. [CrossRef]
66. Duarte-Mata, D.I.; Salinas-Carmona, M.C. Antimicrobial Peptides' Immune Modulation Role in Intracellular Bacterial Infection. *Front. Immunol.* **2023**, *14*, 1119574. [CrossRef] [PubMed]
67. Zhang, Q.-Y.; Yan, Z.-B.; Meng, Y.-M.; Hong, X.-Y.; Shao, G.; Ma, J.-J.; Cheng, X.-R.; Liu, J.; Kang, J.; Fu, C.-Y. Antimicrobial Peptides: Mechanism of Action, Activity and Clinical Potential. *Mil. Med. Res.* **2021**, *8*, 48. [CrossRef] [PubMed]
68. Powers, J.-P.S.; Hancock, R.E.W. The Relationship between Peptide Structure and Antibacterial Activity. *Peptides* **2003**, *24*, 1681–1691. [CrossRef]
69. Ciulla, M.G.; Gelain, F. Structure–Activity Relationships of Antibacterial Peptides. *Microb. Biotechnol.* **2023**, *16*, 757–777. [CrossRef]
70. Echegaray, N.; Yilmaz, B.; Sharma, H.; Kumar, M.; Pateiro, M.; Ozogul, F.; Lorenzo, J.M. A Novel Approach to *Lactiplantibacillus plantarum*: From Probiotic Properties to the Omics Insights. *Microbiol. Res.* **2023**, *268*, 127289. [CrossRef]
71. Yilmaz, B.; Bangar, S.P.; Echegaray, N.; Suri, S.; Tomasevic, I.; Manuel Lorenzo, J.; Melekoglu, E.; Rocha, J.M.; Ozogul, F. The Impacts of *Lactiplantibacillus plantarum* on the Functional Properties of Fermented Foods: A Review of Current Knowledge. *Microorganisms* **2022**, *10*, 826. [CrossRef]
72. Algboory, H.L.; Muhialdin, B.J. Novel Peptides Contribute to the Antimicrobial Activity of Camel Milk Fermented with *Lactobacillus plantarum* IS10. *Food Control* **2021**, *126*, 108057. [CrossRef]
73. Yu, S.; Qian, Y.; Gao, Q.; Yan, Y.; Huang, Y.; Wu, Z.; Luo, X.; Shen, J.; Liu, Y. Discovery, Characterization, and Application of a Novel Antimicrobial Peptide Produced by *Lactiplantibacillus plantarum* FB-2. *Food Biosci.* **2024**, *58*, 103663. [CrossRef]
74. Arulrajah, B.; Qoms, M.S.; Muhialdin, B.J.; Hasan, H.; Zarei, M.; Meor Hussin, A.S.; Chau, D.-M.; Saari, N. Antibacterial and Antifungal Activity of Kenaf Seed Peptides and Their Effect on Microbiological Safety and Physicochemical Properties of Some Food Models. *Food Control* **2022**, *140*, 109119. [CrossRef]
75. Liu, H.; Zhang, L.; Yu, J.; Shao, S. Advances in the Application and Mechanism of Bioactive Peptides in the Treatment of Inflammation. *Front. Immunol.* **2024**, *15*, 1413179. [CrossRef]
76. Li, E.C.; Heran, B.S.; Wright, J.M. Angiotensin Converting Enzyme (ACE) Inhibitors versus Angiotensin Receptor Blockers for Primary Hypertension. In *Cochrane Database of Systematic Reviews*; Li, E.C., Ed.; John Wiley & Sons, Ltd.: Chichester, UK, 2011.
77. Balti, R.; Bougatef, A.; Sila, A.; Guillochon, D.; Dhulster, P.; Nedjar-Arroume, N. Nine Novel Angiotensin I-Converting Enzyme (ACE) Inhibitory Peptides from Cuttlefish (*Sepia officinalis*) Muscle Protein Hydrolysates and Antihypertensive Effect of the Potent Active Peptide in Spontaneously Hypertensive Rats. *Food Chem.* **2015**, *170*, 519–525. [CrossRef] [PubMed]
78. Majumder, K.; Chakrabarti, S.; Morton, J.S.; Panahi, S.; Kaufman, S.; Davidge, S.T.; Wu, J. Egg-Derived ACE-Inhibitory Peptides IQW and LKP Reduce Blood Pressure in Spontaneously Hypertensive Rats. *J. Funct. Foods* **2015**, *13*, 50–60. [CrossRef]
79. Daskaya-Dikmen, C.; Yucetepe, A.; Karbancioglu-Guler, F.; Daskaya, H.; Ozcelik, B. Angiotensin-I-Converting Enzyme (ACE)-Inhibitory Peptides from Plants. *Nutrients* **2017**, *9*, 316. [CrossRef] [PubMed]
80. Chen, L.; Zhang, Q.; Ji, Z.; Shu, G.; Chen, H. Production and Fermentation Characteristics of Angiotensin-I-Converting Enzyme Inhibitory Peptides of Goat Milk Fermented by a Novel Wild *Lactobacillus plantarum* 69. *LWT* **2018**, *91*, 532–540. [CrossRef]

81. Wu, N.; Xu, W.; Liu, K.; Xia, Y. Shuangquan Angiotensin-Converting Enzyme Inhibitory Peptides from *Lactobacillus delbrueckii* QS306 Fermented Milk. *J. Dairy. Sci.* **2019**, *102*, 5913–5921. [CrossRef]
82. Gallego, M.; Mora, L.; Escudero, E.; Toldrá, F. Bioactive Peptides and Free Amino Acids Profiles in Different Types of European Dry-Fermented Sausages. *Int. J. Food Microbiol.* **2018**, *276*, 71–78. [CrossRef]
83. Jakubczyk, A.; Karaś, M.; Złotek, U.; Szymanowska, U.; Baraniak, B.; Bochnak, J. Peptides Obtained from Fermented Faba Bean Seeds (*Vicia faba*) as Potential Inhibitors of an Enzyme Involved in the Pathogenesis of Metabolic Syndrome. *LWT* **2019**, *105*, 306–313. [CrossRef]
84. Wu, J.; Aluko, R.E.; Nakai, S. Structural Requirements of Angiotensin I-Converting Enzyme Inhibitory Peptides: Quantitative Structure-Activity Relationship Modeling of Peptides Containing 4-10 Amino Acid Residues. *QSAR Comb. Sci.* **2006**, *25*, 873–880. [CrossRef]
85. Wu, J.; Aluko, R.E.; Nakai, S. Structural Requirements of Angiotensin I-Converting Enzyme Inhibitory Peptides: Quantitative Structure−Activity Relationship Study of Di- and Tripeptides. *J. Agric. Food Chem.* **2006**, *54*, 732–738. [CrossRef]
86. Bhuyan, B.J.; Mugesh, G. Antioxidant Activity of Peptide-Based Angiotensin Converting Enzyme Inhibitors. *Org. Biomol. Chem.* **2012**, *10*, 2237. [CrossRef]
87. Roberts, P.R.; Burney, J.D.; Black, K.W.; Zaloga, G.P. Effect of Chain Length on Absorption of Biologically Active Peptides from the Gastrointestinal Tract. *Digestion* **1999**, *60*, 332–337. [CrossRef] [PubMed]
88. Shen, W.; Matsui, T. Intestinal Absorption of Small Peptides: A Review. *Int. J. Food Sci. Technol.* **2019**, *54*, 1942–1948. [CrossRef]
89. Mirdhayati, I.; Zain, W.N.H.; Fatah, A.; Yokoyama, I.; Arihara, K. Purification of Angiotensin Converting Enzyme Inhibitory Peptides and Antihypertensive Effect Generated from Indonesian Traditional Fermented Beef (Cangkuk). *Anim. Biosci.* **2024**, *37*, 1799–1808. [CrossRef] [PubMed]
90. Miyazaki, H.; Nakamura, T.; Ohki, K.; Nagai, K. Effects of the Bioactive Peptides Ile-Pro-Pro and Val-Pro-Pro upon Autonomic Neurotransmission and Blood Pressure in Spontaneously Hypertensive Rats. *Auton. Neurosci.* **2017**, *208*, 88–92. [CrossRef]
91. Cicero, A.F.G.; Colletti, A.; Rosticci, M.; Cagnati, M.; Urso, R.; Giovannini, M.; Borghi, C.; D'Addato, S. Effect of Lactotripeptides (Isoleucine–Proline–Proline/Valine–Proline–Proline) on Blood Pressure and Arterial Stiffness Changes in Subjects with Suboptimal Blood Pressure Control and Metabolic Syndrome: A Double-Blind, Randomized, Crossover Clinical Trial. *Metab. Syndr. Relat. Disord.* **2016**, *14*, 161–166. [CrossRef]
92. Moayedi, A.; Mora, L.; Aristoy, M.C.; Safari, M.; Hashemi, M.; Toldrá, F. Peptidomic Analysis of Antioxidant and ACE-Inhibitory Peptides Obtained from Tomato Waste Proteins Fermented Using *Bacillus subtilis*. *Food Chem.* **2018**, *250*, 180–187. [CrossRef]
93. Theansungnoen, T.; Yaraksa, N.; Daduang, S.; Dhiravisit, A.; Thammasirirak, S. Purification and Characterization of Antioxidant Peptides from Leukocyte Extract of *Crocodylus siamensis*. *Protein J.* **2014**, *33*, 24–31. [CrossRef]
94. Puchalska, P.; Marina, M.L.; García, M.C. Isolation and Identification of Antioxidant Peptides from Commercial Soybean-Based Infant Formulas. *Food Chem.* **2014**, *148*, 147–154. [CrossRef]
95. Gaspar-Pintiliescu, A.; Oancea, A.; Cotarlet, M.; Vasile, A.M.; Bahrim, G.E.; Shaposhnikov, S.; Craciunescu, O.; Oprita, E.I. Angiotensin-converting Enzyme Inhibition, Antioxidant Activity and Cytotoxicity of Bioactive Peptides from Fermented Bovine Colostrum. *Int. J. Dairy. Technol.* **2020**, *73*, 108–116. [CrossRef]
96. Zou, T.-B.; He, T.-P.; Li, H.-B.; Tang, H.-W.; Xia, E.-Q. The Structure-Activity Relationship of the Antioxidant Peptides from Natural Proteins. *Molecules* **2016**, *21*, 72. [CrossRef]
97. Yang, W. Evaluation of the Antioxidant Activity and Identification of Potential Antioxidant Peptides in Commercially Available Probiotic Cheddar Cheese. *LWT* **2024**, *205*, 116486. [CrossRef]
98. Devi, S.; Kumar, N.; Kapila, S.; Mada, S.B.; Reddi, S.; Vij, R.; Kapila, R. Buffalo Casein Derived Peptide Can Alleviates H_2O_2 Induced Cellular Damage and Necrosis in Fibroblast Cells. *Exp. Toxicol. Pathol.* **2017**, *69*, 485–495. [CrossRef] [PubMed]
99. Wan, M.-Y.; Dong, G.; Yang, B.-Q.; Feng, H. Identification and Characterization of a Novel Antioxidant Peptide from Feather Keratin Hydrolysate. *Biotechnol. Lett.* **2016**, *38*, 643–649. [CrossRef]
100. Lin, H.; Zhao, J.; Xie, Y.; Tang, J.; Wang, Q.; Zhao, J.; Xu, M.; Liu, P. Identification and Molecular Mechanisms of Novel Antioxidant Peptides from Fermented Broad Bean Paste: A Combined in Silico and in Vitro Study. *Food Chem.* **2024**, *450*, 139297. [CrossRef]
101. Zhang, A.; Cui, L.; Tu, X.; Liang, Y.; Wang, L.; Sun, Y.; Kang, X.; Wu, Z. Peptides Derived from Casein Hydrolyzed by Lactobacillus: Screening and Antioxidant Properties in H_2O_2-Induced HepG2 Cells Model. *J. Funct. Foods* **2024**, *117*, 106221. [CrossRef]
102. Guo, H.; Fan, L.; Ding, L.; Yang, W.; Zang, C.; Guan, H. Separation and Purification of Antioxidant Peptide from Fermented Whey Protein by *Lactobacillus rhamnosus* B2-1. *Food Sci. Anim. Resour.* **2023**, *43*, 10–24. [CrossRef]
103. Pavlicevic, M.; Marmiroli, N.; Maestri, E. Immunomodulatory Peptides—A Promising Source for Novel Functional Food Production and Drug Discovery. *Peptides* **2022**, *148*, 170696. [CrossRef]
104. Li, B.; Habermann, D.; Kliche, T.; Klempt, M.; Wutkowski, A.; Clawin-Rädecker, I.; Koberg, S.; Brinks, E.; Koudelka, T.; Tholey, A.; et al. Soluble *Lactobacillus delbrueckii* Subsp. *Bulgaricus* 92059 PrtB Proteinase Derivatives for Production of Bioactive Peptide Hydrolysates from Casein. *Appl. Microbiol. Biotechnol.* **2019**, *103*, 2731–2743. [CrossRef]
105. Xing, L.; Fu, L.; Hao, Y.; Miao, Y.; Zhang, W. Xuanwei Ham Derived Peptides Exert the Anti-Inflammatory Effect in the Dextran Sulfate Sodium-Induced C57BL/6 Mice Model. *Food Biosci.* **2022**, *48*, 101800. [CrossRef]
106. Wen, L.; Bi, H.; Zhou, X.; Jiang, Y.; Zhu, H.; Fu, X.; Yang, B. Structure Characterization of Soybean Peptides and Their Protective Activity against Intestinal Inflammation. *Food Chem.* **2022**, *387*, 132868. [CrossRef]

107. Vogel, H.J.; Schibli, D.J.; Jing, W.; Lohmeier-Vogel, E.M.; Epand, R.F.; Epand, R.M. Towards a Structure-Function Analysis of Bovine Lactoferricin and Related Tryptophan- and Arginine-Containing Peptides. *Biochem. Cell Biol.* **2002**, *80*, 49–63. [CrossRef] [PubMed]
108. Guha, S.; Majumder, K. Structural-Features of Food-Derived Bioactive Peptides with Anti-Inflammatory Activity: A Brief Review. *J. Food Biochem.* **2019**, *43*, e12531. [CrossRef]
109. Li, Y.; Gao, X.; Pan, D.; Liu, Z.; Xiao, C.; Xiong, Y.; Du, L.; Cai, Z.; Lu, W.; Dang, Y.; et al. Identification and Virtual Screening of Novel Anti-Inflammatory Peptides from Broccoli Fermented by *Lactobacillus strains*. *Front. Nutr.* **2023**, *9*, 1118900. [CrossRef]
110. Nakamura, Y.; Yamamoto, N.; Sakai, K.; Okubo, A.; Yamazaki, S.; Takano, T. Purification and Characterization of Angiotensin I-Converting Enzyme Inhibitors from Sour Milk. *J. Dairy. Sci.* **1995**, *78*, 777–783. [CrossRef]
111. Srivastava, U.; Nataraj, B.H.; Kumari, M.; Kadyan, S.; Puniya, A.K.; Behare, P.V.; Nagpal, R. Antioxidant and Immunomodulatory Potency of *Lacticaseibacillus rhamnosus* NCDC24 Fermented Milk-Derived Peptides: A Computationally Guided in-Vitro and Ex-Vivo Investigation. *Peptides* **2022**, *155*, 170843. [CrossRef]
112. Inoue, K.; Gotou, T.; Kitajima, H.; Mizuno, S.; Nakazawa, T.; Yamamoto, N. Release of Antihypertensive Peptides in Miso Paste during Its Fermentation, by the Addition of Casein. *J. Biosci. Bioeng.* **2009**, *108*, 111–115. [CrossRef]
113. Gonzalez-Gonzalez, C.; Gibson, T.; Jauregi, P. Novel Probiotic-Fermented Milk with Angiotensin I-Converting Enzyme Inhibitory Peptides Produced by Bifidobacterium Bifidum MF 20/5. *Int. J. Food Microbiol.* **2013**, *167*, 131–137. [CrossRef]
114. Muhialdin, B.J.; Abdul Rani, N.F.; Meor Hussin, A.S. Identification of Antioxidant and Antibacterial Activities for the Bioactive Peptides Generated from Bitter Beans (*Parkia speciosa*) via Boiling and Fermentation Processes. *LWT* **2020**, *131*, 109776. [CrossRef]
115. Quirós, A.; Ramos, M.; Muguerza, B.; Delgado, M.A.; Miguel, M.; Aleixandre, A.; Recio, I. Identification of Novel Antihypertensive Peptides in Milk Fermented with *Enterococcus faecalis*. *Int. Dairy. J.* **2007**, *17*, 33–41. [CrossRef]
116. Arulrajah, B.; Muhialdin, B.J.; Zarei, M.; Hasan, H.; Saari, N. Lacto-Fermented Kenaf (*Hibiscus cannabinus* L.) Seed Protein as a Source of Bioactive Peptides and Their Applications as Natural Preservatives. *Food Control* **2020**, *110*, 106969. [CrossRef]
117. Sanjukta, S.; Rai, A.K.; Muhammed, A.; Jeyaram, K.; Talukdar, N.C. Enhancement of Antioxidant Properties of Two Soybean Varieties of Sikkim Himalayan Region by Proteolytic *Bacillus subtilis* Fermentation. *J. Funct. Foods* **2015**, *14*, 650–658. [CrossRef]
118. Purohit, K.; Reddy, N.; Sunna, A. Exploring the Potential of Bioactive Peptides: From Natural Sources to Therapeutics. *Int. J. Mol. Sci.* **2024**, *25*, 1391. [CrossRef] [PubMed]
119. Martínez-Augustin, O.; Rivero-Gutiérrez, B.; Mascaraque, C.; Sánchez de Medina, F. Food Derived Bioactive Peptides and Intestinal Barrier Function. *Int. J. Mol. Sci.* **2014**, *15*, 22857–22873. [CrossRef] [PubMed]
120. Hayes, M.; Tiwari, B. Bioactive Carbohydrates and Peptides in Foods: An Overview of Sources, Downstream Processing Steps and Associated Bioactivities. *Int. J. Mol. Sci.* **2015**, *16*, 22485–22508. [CrossRef] [PubMed]
121. Ye, H.; Tao, X.; Zhang, W.; Chen, Y.; Yu, Q.; Xie, J. Food-Derived Bioactive Peptides: Production, Biological Activities, Opportunities and Challenges. *J. Future Foods* **2022**, *2*, 294–306. [CrossRef]
122. Murtaza, M.A.; Irfan, S.; Hafiz, I.; Ranjha, M.M.A.N.; Rahaman, A.; Murtaza, M.S.; Ibrahim, S.A.; Siddiqui, S.A. Conventional and Novel Technologies in the Production of Dairy Bioactive Peptides. *Front. Nutr.* **2022**, *9*, 780151. [CrossRef]
123. Munir, M.; Nadeem, M.; Mahmood Qureshi, T.; Gamlath, C.J.; Martin, G.J.O.; Hemar, Y.; Ashokkumar, M. Effect of Sonication, Microwaves and High-Pressure Processing on ACE-Inhibitory Activity and Antioxidant Potential of Cheddar Cheese during Ripening. *Ultrason. Sonochem* **2020**, *67*, 105140. [CrossRef]
124. Xie, M.; Ma, Y.; An, F.; Yu, M.; Zhang, L.; Tao, X.; Pan, G.; Liu, Q.; Wu, J.; Wu, R. Ultrasound-Assisted Fermentation for Antioxidant Peptides Preparation from Okara: Optimization, Stability, and Functional Analyses. *Food Chem.* **2024**, *439*, 138078. [CrossRef]
125. Wu, N.; Zhang, F.; Shuang, Q. Peptidomic Analysis of the Angiotensin-Converting-Enzyme Inhibitory Peptides in Milk Fermented with *Lactobacillus delbrueckii* QS306 after Ultrahigh Pressure Treatment. *Food Res. Int.* **2023**, *164*, 112406. [CrossRef]
126. Khakhariya, R.; Basaiawmoit, B.; Sakure, A.A.; Maurya, R.; Bishnoi, M.; Kondepudi, K.K.; Padhi, S.; Rai, A.K.; Liu, Z.; Hati, S. Production and Characterization of ACE Inhibitory and Anti-Diabetic Peptides from Buffalo and Camel Milk Fermented with Lactobacillus and Yeast: A Comparative Analysis with in Vitro, In Silico, and Molecular Interaction Study. *Foods* **2023**, *12*, 2006. [CrossRef]
127. Sharma, P.; Sharma, D.; Kaur, S.; Borah, A. Optimization of Flaxseed Milk Fermentation for the Production of Functional Peptides and Estimation of Their Bioactivities. *Food Sci. Technol. Int.* **2021**, *27*, 585–597. [CrossRef] [PubMed]
128. Bautista-Expósito, S.; Peñas, E.; Silván, J.M.; Frias, J.; Martínez-Villaluenga, C. PH-Controlled Fermentation in Mild Alkaline Conditions Enhances Bioactive Compounds and Functional Features of Lentil to Ameliorate Metabolic Disturbances. *Food Chem.* **2018**, *248*, 262–271. [CrossRef] [PubMed]
129. Akbarbaglu, Z.; Ayaseh, A.; Ghanbarzadeh, B.; Sarabandi, K. Biological Stabilization of Arthrospira Bioactive-Peptides within Biopolymers: Functional Food Formulation; Bitterness-Masking and Nutritional Aspects. *LWT* **2024**, *191*, 115653. [CrossRef]
130. Abdo, A.A.A.; Al-Dalali, S.; Hou, Y.; Aleryani, H.; Shehzad, Q.; Asawmahi, O.; AL-Farga, A.; Mohammed, B.; Liu, X.; Sang, Y. Modification of Marine Bioactive Peptides: Strategy to Improve the Biological Activity, Stability, and Taste Properties. *Food Bioproc Tech.* **2024**, *17*, 1412–1433. [CrossRef]
131. Atma, Y.; Murray, B.S.; Sadeghpour, A.; Goycoolea, F.M. Encapsulation of Short-Chain Bioactive Peptides (BAPs) for Gastrointestinal Delivery: A Review. *Food Funct.* **2024**, *15*, 3959–3979. [CrossRef]

132. Pérez-Pérez, V.; Jiménez-Martínez, C.; González-Escobar, J.L.; Corzo-Ríos, L.J. Exploring the Impact of Encapsulation on the Stability and Bioactivity of Peptides Extracted from Botanical Sources: Trends and Opportunities. *Front. Chem.* **2024**, *12*, 1423500. [CrossRef]
133. Lekshmi, R.G.K.; Tejpal, C.S.; Anas, K.K.; Chatterjee, N.S.; Mathew, S.; Ravishankar, C.N. Binary Blend of Maltodextrin and Whey Protein Outperforms Gum Arabic as Superior Wall Material for Squalene Encapsulation. *Food Hydrocoll.* **2021**, *121*, 106976. [CrossRef]
134. Sarabandi, K.; Karami, Z.; Akbarbaglu, Z.; Duangmal, K.; Jafari, S.M. Spray-Drying Stabilization of Oleaster-Seed Bioactive Peptides within Biopolymers: Pan-Bread Formulation and Bitterness-Masking. *Food Biosci.* **2024**, *58*, 103837. [CrossRef]
135. Wang, Z.; Ju, X.; He, R.; Yuan, J.; Wang, L. The Effect of Rapeseed Protein Structural Modification on Microstructural Properties of Peptide Microcapsules. *Food Bioproc. Tech.* **2015**, *8*, 1305–1318. [CrossRef]
136. Thongcumsuk, B.; Woraprayote, W.; Janyaphisan, T.; Cheunkar, S.; Oaew, S. Microencapsulation and Peptide Identification of Purified Bioactive Fraction from Spirulina Protein Hydrolysates with Dipeptidyl Peptidase IV (DPP-IV) Inhibitory Activity. *Food Biosci.* **2023**, *56*, 103438. [CrossRef]
137. Gharehbeglou, P.; Homayouni-Rad, A.; Jafari, S.M.; Sarabandi, K.; Akbarbaglu, Z. Stabilization of Chlorella Bioactive Hydrolysates Within Biopolymeric Carriers: Techno-Functional, Structural, and Biological Properties. *J. Polym. Environ.* **2024**, *32*, 763–779. [CrossRef]
138. Devi, N.; Sarmah, M.; Khatun, B.; Maji, T.K. Encapsulation of Active Ingredients in Polysaccharide–Protein Complex Coacervates. *Adv. Colloid. Interface Sci.* **2017**, *239*, 136–145. [CrossRef] [PubMed]
139. Thapa, R.K.; Kim, J.O. Nanomedicine-Based Commercial Formulations: Current Developments and Future Prospects. *J. Pharm. Investig.* **2023**, *53*, 19–33. [CrossRef] [PubMed]
140. Javadi, B.; Farahmand, A.; Soltani-Gorde-Faramarzi, S.; Hesarinejad, M.A. Chitosan-Coated Nanoliposome: An Approach for Simultaneous Encapsulation of Caffeine and Roselle-Anthocyanin in Beverages. *Int. J. Biol. Macromol.* **2024**, *275*, 133469. [CrossRef]
141. Mazloomi, S.N.; Mahoonak, A.S.; Ghorbani, M.; Houshmand, G. Physicochemical Properties of Chitosan-Coated Nanoliposome Loaded with Orange Seed Protein Hydrolysate. *J. Food Eng.* **2020**, *280*, 109976. [CrossRef]
142. Khalatbari, S.; Hasani, M.; Khoshvaght-Aliabadi, M. Investigating the Characteristics of Nanoliposomes Carrying Bioactive Peptides Obtained from Shrimp Waste. *Int. J. Pept. Res. Ther.* **2024**, *30*, 10. [CrossRef]
143. Pateiro, M.; Gómez, B.; Munekata, P.E.S.; Barba, F.J.; Putnik, P.; Kovačević, D.B.; Lorenzo, J.M. Nanoencapsulation of Promising Bioactive Compounds to Improve Their Absorption, Stability, Functionality and the Appearance of the Final Food Products. *Molecules* **2021**, *26*, 1547. [CrossRef]
144. Intiquilla, A.; Jiménez-Aliaga, K.; Iris Zavaleta, A.; Gamboa, A.; Caro, N.; Diaz, M.; Gotteland, M.; Abugoch, L.; Tapia, C. Nanoencapsulation of Antioxidant Peptides from *Lupinus mutabilis* in Chitosan Nanoparticles Obtained by Ionic Gelling and Spray Freeze Drying Intended for Colonic Delivery. *Food Biosci.* **2022**, *50*, 102055. [CrossRef]
145. Chen, X.; Fang, F.; Wang, S. Physicochemical Properties and Hepatoprotective Effects of Glycated Snapper Fish Scale Peptides Conjugated with Xylose via Maillard Reaction. *Food Chem. Toxicol.* **2020**, *137*, 111115. [CrossRef]

foods

MDPI

Review

Plant-Based Fermented Beverages: Nutritional Composition, Sensory Properties, and Health Benefits

Belén Hidalgo-Fuentes [1], Edgar de Jesús-José [1], Anselmo de J. Cabrera-Hidalgo [2], Ofelia Sandoval-Castilla [3], Teodoro Espinosa-Solares [3,4], Ricardo. M. González-Reza [5], María L. Zambrano-Zaragoza [5], Andrea M. Liceaga [6,*,†] and José E. Aguilar-Toalá [7,*,†]

[1] Licenciatura en Ciencia y Tecnología de Alimentos, División de Ciencias Biológicas y de la Salud, Universidad Autónoma Metropolitana, Unidad Lerma, Av. de las Garzas 10. Col. El Panteón, Lerma de Villada 52005, Estado de Mexico, Mexico; 2193039504@correo.ler.uam.mx (E.d.J.-J.)

[2] TecNM-Instituto Tecnológico Superior de Tlatlauquitepec (ITSTL), Carretera Federal Amozoc-Nautla km 122+600 Almoloni, Tlatlauquitepec 73907, Puebla, Mexico; anselmo.cabrera@tlatlauquitepec.tecnm.mx

[3] Departamento de Ingeniería Agroindustrial, Universidad Autónoma Chapingo, Texcoco 56230, Estado de Mexico, Mexico

[4] Agricultural Research and Extension Center, Southern University, Baton Rouge, LA 70813, USA

[5] Laboratorio de Procesos de Transformación y Tecnologías Emergentes de Alimentos-UIM, FES-Cuautitlán, Universidad Nacional Autónoma de México, Cuautitlán Izcalli 54714, Estado de Mexico, Mexico; luz.zambrano@unam.mx (M.L.Z.-Z.)

[6] Protein Chemistry and Bioactive Peptides Laboratory, Department of Food Science, Purdue University, 745 Agriculture Mall, West Lafayette, IN 47907, USA

[7] Departamento de Ciencias de la Alimentación, División de Ciencias Biológicas y de la Salud, Universidad Autónoma Metropolitana, Unidad Lerma, Av. de las Garzas 10. Col. El Panteón, Lerma de Villada 52005, Estado de Mexico, Mexico

* Correspondence: aliceaga@purdue.edu (A.M.L.); j.aguilar@correo.ler.uam.mx (J.E.A.-T.)

† These authors contributed equally to this work.

Abstract: Plant-based beverages have gained consumers' attention for being the main substitutes for dairy milk, especially for people with lactose intolerance, milk allergies, and a prevalence of hypercholesterolemia. Moreover, there is a growing demand for a more sustainable diet and plant-based lifestyle due to concerns related to animal wellbeing, environmental impacts linked to dairy production, and the rising cost of animal-derived foods. However, there are some factors that restrict plant-based beverage consumption, including their nutritional quality and poor sensory profile. In this context, fermentation processes can contribute to the improvement of their sensory properties, nutritional composition, and functional/bioactive profile. In particular, the fermentation process can enhance flavor compounds (e.g., acetoin and acetic acid) while decreasing off-flavor components (e.g., hexanal and hexanol) in the substrate. Furthermore, it enhances the digestibility and bioavailability of nutrients, leading to increased levels of vitamins (e.g., ascorbic acid and B complex), amino acids (e.g., methionine and tryptophan), and proteins, while simultaneously decreasing the presence of anti-nutritional factors (e.g., phytic acid and saponins). In contrast, plant-based fermented beverages have been demonstrated to possess diverse bioactive compounds (e.g., polyphenols and peptides) with different biological properties (e.g., antioxidant, anti-inflammatory, and antihypertensive). Therefore, this article provides an overview of plant-based fermented beverages including their production, technological aspects, and health benefits.

Keywords: dairy milk alternatives; plant-based beverages; functional foods; fermentation technology; health benefits

1. Introduction

Nowadays, plant-based beverages have gained interest from the scientific community, food industry, and consumers as milk alternatives. The main reasons include that they are a good alternative for people with lactose intolerance, milk allergies, and a prevalence of hypercholesterolemia [1]. In addition, a wide sector of the population is seeking a more sustainable diet, moving towards a more plant-based lifestyle (i.e., vegetarianism, veganism, and flexitarianism); also relevant are the growing ethical concerns related to animal welfare, the negative environmental impacts associated with dairy production, and the rising prices of animal-derived foods [1–3]. In particular, plant-based beverages can be considered to be sustainable food systems because they potentially require less resources (e.g., less water and land use) and generate a lower carbon footprint compared to dairy milk production. For example, it is estimated that the carbon footprint of milk production is around 3.2 kg of CO_2 equivalent per liter of milk produced, whereas the carbon footprint of different plant-based beverages can range from 0.7 to 1.2 kg of CO_2 equivalent per liter of beverage produced [4–8]. The considerations stated above are supported by the fact that the worldwide market of plant-based beverages is expected to grow by 15% (Compound Annual Growth Rate) annually from 2023 to 2028 [9]. In contrast, some regional markets have experienced a decline in fluid dairy milk sales. In this context, the estimated US fluid milk sales of total conventional fluid milk products had decreased 6.5% by April 2023 from a year earlier [10]. This information suggests that plant-based beverages have strong market potential with expected full market growth and a clear consumer target.

Plant-based beverages are defined as fluids that result from water extraction processes of plant material through their breakdown (size reduction) and/or homogenization [1,11,12]. Generally, the resulting particles have a size distribution ranging from 5 to 20 µm, allowing these water-soluble extracts of plant material to resemble bovine milk in appearance [1,12]. From a physicochemical perspective, plant-based beverages are colloidal suspensions or emulsions of dispersed plant material [13]. Normally, plant-based beverages are characterized as oil-in-water (O/W) emulsions, where oil is the dispersed phase and water is the aqueous/continuous phase [14,15]. This type of emulsion allows the imitation of some characteristics found in dairy milk, which is also an O/W emulsion, such as appearance, consistency, stability, mouthfeel, and taste [14,16,17]. In dairy milk, the emulsifier agents are phospholipids and milk proteins [18], while for plant-based beverages these can be biosurfactants, phospholipids, proteins, and polysaccharides that are naturally present in the plant matrix or added as additives during beverage production [15,19]. It is important to note that while these types of beverages are also referred to as "milk", such as plant milks, vegetable milks, vegetable/plant-based milks, or non-dairy milks, this seems to not be permitted according to the legislation in many countries. For example, in the United States, the Food and Drug Administration (FDA) had defined milk as the lacteal secretion obtained from healthy cows [20]. However, in 2023, the FDA provided a new guidance for plant-based beverages to be called "milk". According to this guidance, companies are also encouraged to voluntarily include extra nutrition labels that note when the plant-based beverages have lower levels of nutrients than dairy milk (e.g., vitamin D, calcium, and magnesium) [21]. In 2020, European Union regulations stated that the term "milk" could only be used to market and advertise products derived from animal milk, thus banning dairy-related terms for plant-based alternatives. Similarly, in Mexico, the Mexican Official Standard (NOM-155-SCFI-2012 [22]) mentions that milk is derived exclusively from the mammary secretion obtained from cows. Thus, in Mexico, plant-based beverages cannot be called milks and are referred to as "beverages" or "drinks".

The growing interest in plant-based beverages is highlighted in Figure 1, which shows the literature accumulating ca. 4100 publications, available from 2018 to 2023, related to plant-based beverages from any of the five categories: legume, nut, cereal, pseudo-cereal, and seeds. Of these publications, a total of 247 are directly related to fermented beverages. Soy-, rice-, corn-, coconut-, and peanut-based beverages have the highest number of publications (>190), with soy-based beverages being noticeably the most studied (ca. 1300 publications). These beverages have been mainly evaluated for their nutritional characterization, off-odor compounds, and antinutrient profile, as well as for starch, lipid, and protein extraction [23–25], and little attention has been paid to their fermentation [26–28]. Nut- and cereal-based beverages have also gained popularity in the last 5 years as they are considered rich in nutrients and bioactive compounds [29,30].

Initial research on plant-based beverages focused attention towards soy-based beverages as the main alternative to dairy milk, due to their nutritive value, as well as higher protein content and lower price [1,31]. Nevertheless, in recent years, the use of other plant materials such as legumes, nuts, seeds, cereals, and pseudo-cereals have been explored in the formulation of plant-based beverages [32]. In addition, mixing two or more plant materials to produce plant-based beverages could add or complement the nutritional value and at the same time produce a novel product [12]. One way to innovate and meet the needs of consumers is the elaboration of plant-based fermented beverages [33], which not only improves their sensory properties and nutritional composition but also offers the opportunity to develop functional foods with health benefits. This gives a whole new set of possibilities for novel beverages with different tastes, aromas, flavors, textures, mouthfeels, and nutritive characteristics. Therefore, this review aims to provide an in-depth exploration of the main types and characteristics of plant-based beverages, while emphasizing the significant influence of the fermentation process on their sensory properties, nutritional compositions, and the biological activity associated with their health benefits.

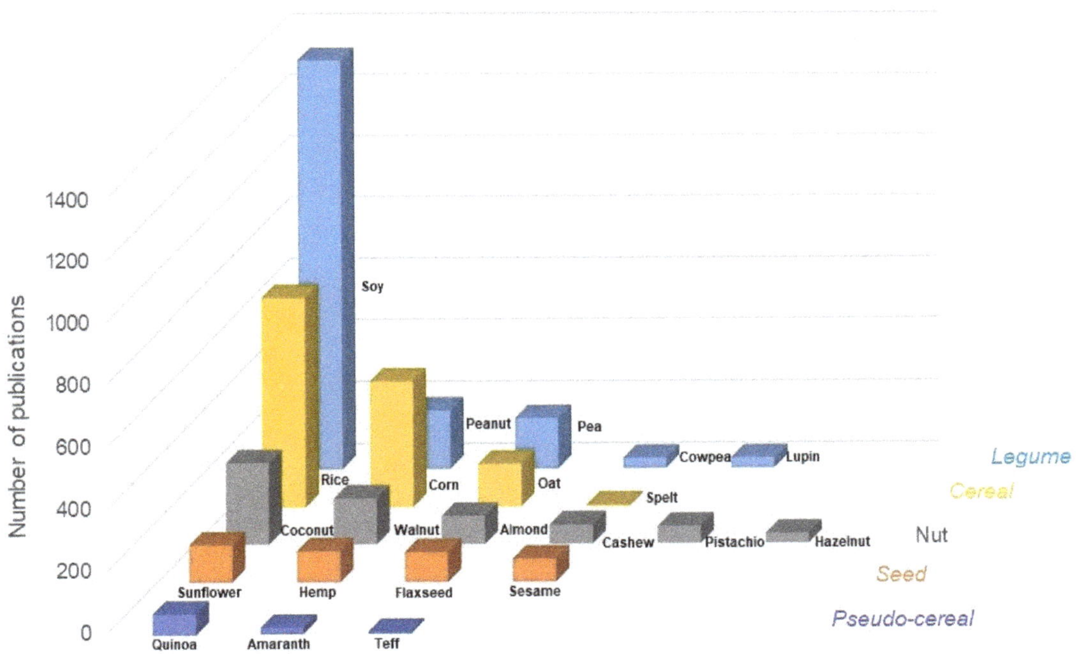

Figure 1. Comparative chart of recent publications related to plant-based beverages. The search was conducted on the Web of Science Core Collection database (Clarivate analytics, USA) on December 2023. Original scientific studies dating from January 2018 to November 2023 were included. The main search terms used were the "name" of the plant-based beverage AND "milk" (e.g., soy milk OR soymilk) OR "-based beverage" OR "extract", consulting scientific studies published in English.

2. Plant-Based Beverages: Types and Characteristics

Plant-based beverages are growing in interest and acceptance from consumers, while also being visualized as promising and viable alternatives to dairy-based products [34]. Although the soy-based beverages were the first produced, at the present time, there are a wide variety of these beverages from various plant sources [1,31]. In accordance with the above, plant-based beverages can be classified into five main categories based on their source [1,32]:

1. Legume beverages: soy, peanut, pea, lupin, and cowpea;
2. Nut beverages: almond, coconut, hazelnut, pistachio, walnut, and cashew;
3. Cereal beverages: oat, rice, corn, and spelt;
4. Pseudo-cereal beverages: quinoa, teff, and amaranth;
5. Seed beverages: sesame, flaxseed, hemp, sunflower.

In search of dairy alternatives, both the scientific community as well as the food and pharmaceutical industries have investigated various plants matrices such as legumes, cereals, nuts, pseudo-cereals, and seeds for their functional properties because of their content of nutritional and health compounds [16]. In this context, legume-based beverages are characterized as having a balanced composition, since these beverages have a protein content (ca. 3–4%) similar to cow´s milk (ca. 3.5%) [35]. In this group of beverages, soy is the most representative and most consumed legume-based beverage. Around 2000 years ago, this plant was the initial source utilized in China to produce a substitute for milk [16]. Among these legume-based beverages, the peanut-based beverage introduces a roasted and nutty taste, while the pea-based beverage offers a slightly sweet flavor [36]. In contrast, nut-based beverages encompass a variety of plant-based drinks derived from different dry fruits composed of a hard inedible husk and a seed, which are characterized by their high lipid (ca. 3–5%) and protein (ca. 0.8–1.3%) content [30,37,38]. Almond milk, one of the most widely consumed nut-based beverages, has a mild taste and versatility of use. Coconut milk gives off a characteristic tropical flavor and creamy texture, making it a favorite in various culinary applications [39]; hazelnut milk introduces a rich and slightly sweet profile; and pistachio milk offers a unique and nutty taste [40].

On the other hand, cereal-based beverages are characterized as having a protein and lipid content ranging from 0.3 to 3% and 2.3 to 7%, respectively, in addition to their naturally subtle sweetness [29,37]. In particular, in this group, oat- and rice-based beverages are praised for their versatility in coffee, cooking, and baking applications [41]. For example, oat-based beverages are characterized by their noticeable dietary fiber (3–4%) content [1,38], while rice-based beverages are rich in carbohydrates (ca. 9%) [42]. In contrast, beverages made from corn and spelt are valued for their sweet–sour taste and thin yet gritty–smooth consistency [43]. Similarly, pseudo-cereal beverages include quinoa, teff, and amaranth, which can have a variety of flavors from nutty and earthy to slightly sweet or bitter [44–46]. Principally, these beverages possess a range of protein and lipid contents between 0.5 and 3.5% and 0.6 and 3%, respectively. Meanwhile, seed-based beverages including sesame, flaxseed, hemp, and sunflower offer delicate nutty and earthy undertones [45,46]. In particular, seed-based beverages are characterized for their variable lipid (ca. 1–7%) and protein (ca. 1–5%) content [37,38].

In general, plant-based beverages have several advantages and disadvantages compared with dairy milk. For example, these beverages do not contain lactose, milk protein, or cholesterol, which is beneficial for people with lactose intolerance, milk allergies, and a prevalence of hypercholesterolemia, respectively [47,48]. Regarding this last matter, plant-based beverages are low in saturated fat, while containing more polyunsaturated fatty acids, which are associated with a lower incidence of chronic diseases [49]. Additionally, in contrast to dairy milk, plant-based beverages contain fiber [33,49].

Some disadvantages of plant-based beverages are associated mainly with their low micronutrient content as well as lower protein quality [38,50]. In this context, most plant-based beverages tend to have lowers levels of essential micronutrients such as vitamins (i.e., B complex, B_6, K, E, and folate), magnesium, potassium, and zinc. In addition, these beverages are characterized by their low protein content, except for specific cases such as soy protein, as well as a lack of all the essential amino acids in the optimal ratios required by the human body in comparison with dairy milk [37,38,50]. For example, the protein quality reported as the Digestible Indispensable Amino Acid Score (DIAAS) is low for rice protein isolate (ca. 37%) and relatively high for soy protein (ca. 90%), while dairy milk possess a DIAAS of 100% [50]. In addition, they can contain antinutrient factors such as inositol phosphates, phytic acid, and trypsin inhibitors, as well as protein allergens [13,50]. Another concern with plant-based beverages is that they can have a negative impact on the sensory characteristics of these beverages, including unpleasant aromas and flavors and poor texture [51]. For example, soy beverages tend to have a beany flavor and seed-based beverages can present bitterness [1,26]. In particular, in legume-based beverages, the compounds hexanal and hexanol have been identified as responsible for the beany flavors [23,26]. Similarly, Pérez-González et al. (2015) [52] identified several aldehydes (e.g., pentanal and hexanal) and alcohols (e.g., heptanol) in an almond beverage. These compounds are formed via lipid oxidation processes and are responsible for the off-flavor components in plant-based beverages. Some compounds such as alkaloids, phenols, saponins, cyanogenic glycoside, flavonoids, terpenes, and glucosinolates are known to impart bitterness and astringency to plant-based beverages [26,53]. The content and diversity of these compounds depended on the plant source, which results in a difference in acceptability for these beverages. In this context, Jeske et al. [54] conducted a sensory acceptance test of commercial plant-based beverages (i.e., oat, rice, hemp, almond, soy, and lentil) and found that the most accepted beverages were oat and rice, while the least accepted was hemp.

In this sense, for the case of plant-based beverages, the fermentation processes could be a useful technology that contributes to the improvement of sensory properties and nutritional composition, similar to what occurs with other fermented foods such as cheese, yogurt, and bread [26,55].

3. Impact of Fermentation on Nutritional Composition, Sensory Properties, and Bioactive Profile of Plant-Based Beverages

Fermentation has been used since ancient times for preserving or preparing new foods and beverages [56] and improving their flavor and texture [57]. It is an enzyme-driven process facilitated by microorganisms, breaking complex macromolecules into simpler ones [28,55,56]. This process can produce various compounds such as proteins, amino acids, fatty acids, vitamins, acetic acid, and volatile compounds, imparting different flavors, aromas, and textures to foods such as cheese, yogurt, kefir, and bread [55,57,58]. Moreover, it is cost-effective and operationally easy, making it a feasible option for food processing [59]. In the case of plant-based beverages, the fermentation process can vary depending on the composition of the specific plant matrix being fermented and the microorganisms involved, for example, lactic acid and alcoholic fermentations, or in some cases, a combination of both. Other types involve, to a lesser extent, acetic acid fermentation [2,60]. Typically, lactic acid bacteria such as *Lactobacillus* and *Streptococcus* participate in lactic acid fermentation, where carbohydrates are converted into lactic acid that imparts sensory properties and also helps preserve the fermented product. In contrast, during alcoholic fermentation, yeast such as *Saccharomyces cerevisiae* participates in the conversion of carbohydrates into alcohol (e.g., ethanol) and carbon dioxide [17,32,57,60].

The typical production process for plant-based beverages starts with water extraction of the raw material either via soaked/wet-milling or dry-milling and extraction of powder/flour. This is followed by the separation of solids via filtration, product formulation using different additives and ingredients, homogenization, heat treatment to inactivate the endogenous enzymes and microbial load followed by the fermentation process, and finally product packaging (Figure 2). In the case of plant-based beverages, it has been reported that fermentation contributes to the improvement of their sensory properties, safety quality, nutritional composition, and bioactive profile [26–28]. The above is attributed mainly to microorganisms that utilize nutrients during their growth in the matrix of the plant-based beverages, which lead to changes in the components and their concentrations [61].

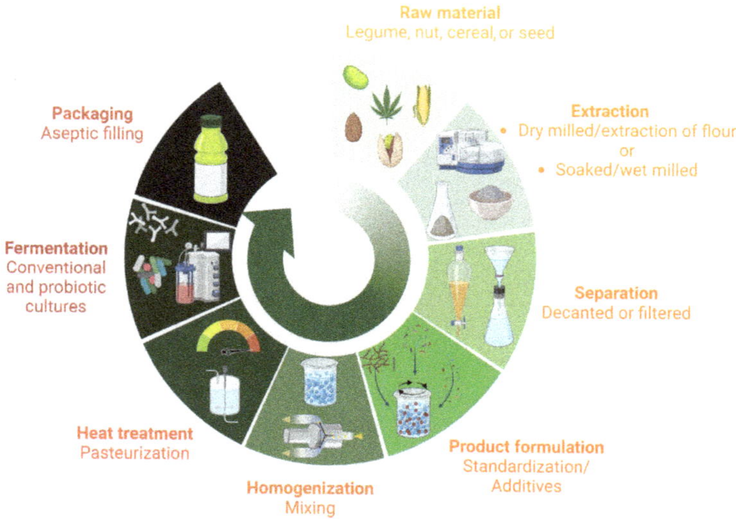

Figure 2. General diagram of the production process for plant-based fermented beverages. Common procedures were adapted from Alexandre et al. [62] and Gil-Serna et al. [63]. Created using Biorender.com (accessed on 11 December 2023).

3.1. Impact on Nutritional Composition

The nutritional composition of plant-based beverages is improved by the fermentation processes by enhancing the digestibility and availability of nutrients, as well as producing nutritional factors such as vitamins. In addition, this biochemical process can help in the elimination of anti-nutritional factors, which is beneficial for enhanced nutrient bioavailability. During the fermentation process, enzymes and microorganisms break down complex macromolecules (i.e., proteins and carbohydrates) present in the plant matrix into simpler forms. In the case of proteins, this process can lead to improved protein digestibility as proteins are partially hydrolyzed into polypeptides, oligopeptides, and free amino acids, making them more accessible and easier for the body to absorb [64]. In addition, the amino acid composition of the plant-based beverages is also modified during fermentation because some amino acids are synthesized or released from peptides by microorganisms, leading to changes in the overall amino acid profile. The resulting amino acid profile can vary depending on the initial protein composition of the specific plant matrix being fermented and the microbial enzymes (e.g., cell-envelope proteinases and intracellular peptidases) involved [65]. For example, lactic acid bacteria contain cell-envelope proteinases that start the proteolysis of plant proteins, breaking them down into oligopeptides that are then used by cells through specific peptide-transport systems or degraded into small peptides and free amino acids via the collective action of various intracellular peptidases [66]. Fermentation can also enhance mineral bioavailability by decreasing the presence of anti-nutritional factors such as phytic acid, saponins, condensed tannins, and trypsin inhibitors, among others, which can bind to minerals and hinder their absorption [67,68]. During fermentation, some microorganisms can produce enzymes that degrade anti-nutritional factors or they can modify them as part of their microbial metabolism. Some changes in pH can also affect the stability of these factors [67,69]. For example, microbial phytases and tannases, derived from lactic acid bacteria or yeast, can degrade phytates and tannins, respectively [69–71]. Similarly, it has been reported that some lactic acid bacteria can metabolize oxalates, reducing their concentration in the final fermented product [72].

In this sense, Pontonio et al. [73] reported that a fermented beverage prepared using a mix of rice, lentil, and chickpea flours showed a higher biological value, protein efficiency ratio, essential amino acid index, and total of free amino acids compared to the unfermented beverage. In particular, the authors found that the in vitro protein digestibility increased from 67 to 79% after the fermentation process. In addition, lower contents of the antinutritive factors (i.e., phytic acid, condensed tannins, saponins, and raffinose) were found in the fermented beverage. Likewise, Vila-Real et al. [74] reported that a fermented millet-based, yoghurt-like beverage showed a higher content of threonine, arginine, glutamine, and gamma-aminobutyric acid (GABA) compared to the unfermented beverage. The authors reported that protein digestibility increased from 25 to 64% after fermentation. In another study, Sanni et al. [75] reported that pea-based fermented beverages showed higher protein, total ash, calcium, potassium, phosphorus, lysine, methionine, tryptophan, riboflavin, niacin, and thiamine than their unfermented counterparts. The antinutritive factors (i.e., phytic acid and trypsin inhibitor) in this fermented beverage decreased after the fermentation process. Likewise, Rekha and Vijayalakshmi [76] reported that the mineral bioavailability (i.e., calcium, magnesium, and zinc) and vitamin B complex (i.e., thiamin, riboflavin, and niacin) increased in soy-based fermented beverages compared with the unfermented versions. In addition, the authors observed a decrease in phytic acid in the final fermented beverage. Other studies [77] found that a soy-based beverage fermented by *Streptococcus thermophilus* 14085 and *Bifidobacterium infantis* 14603 showed a lower content of saponins and phytates. Hou et al. [78] reported that soy-based beverages fermented by *Bifidobacterium longum* B6 and *B. infantis* CCRC 14633 showed a significant increase in protein, thiamin, and riboflavin after 48 h of fermentation. On the other hand, using a Caco-2 cell model, Bernat et al. [79] described that the fermentation of an almond-based beverage using different potentially probiotic bacteria increased the bioavailability of dietary iron. Santos et al. [80] reported that a peanut–soy beverage fermented using six different lactic acid bacteria (LAB) strains, including probiotic strains, and yeasts, increased the amount of total amino acids compared with the unfermented beverage. The authors also reported that, when co-cultured, the LAB strains were more efficient at utilizing carbohydrates and releasing metabolites compared to the single culture fermentations. In a related study, using a liquid formulation of the serine–endoprotease subtilisin (derived from *Bacillus subtilis*) on soy pulp (okara) water extract (a soy by-product beverage) resulted in higher extracted proteins [81]. Likewise, Karovičová et al. [82] reported an increase in protein content by ca. 40% after fermentation of a quinoa-based beverage. Similarly, Jeske et al. [83] also reported that a quinoa-based beverage treated using different enzymes increased the protein content by 1.5-fold compared to the control (no enzyme added) beverage.

3.2. Impact on Sensory Properties

With regard to the sensory characteristics of plant-based beverages, the fermentation processes seem to increase the production of flavor compounds such as organic acids and volatile metabolites. For instance, Zheng et al. [84] found that the sensory properties of a soy-based beverage fermented by *Lactobacillus harbinensis* M1 were improved by the production of 2,3-butanedione and acetoin, which impart a buttery aroma. The authors reported that, after fermentation, the hexanal content decreased to undetectable levels. Similarly, soy-based beverages supplemented with okara flour and fermented by *Lactobacillus acidophilus* LA3 increased significantly ($p < 0.05$) in their content of organic acids (e.g., lactic and acetic acids) and also showed enhanced sensory acceptability [85]. Likewise, Nissen et al. [86] reported a shift in the aromatic profile of hemp-based beverages fermented by probiotic bacteria (*Lactobacillus fermentum*, *Lb. plantarum*, and *Bifidobacterium bifidum*), generating 2,3-butanedione and acetoin, which conferred a pleasant buttery taste to the drinks and improved the overall mouthfeel. In contrast, Menezes et al. [87] reported that maize-based beverages fermented by probiotic yeast and lactic acid bacteria (e.g., *Lactobacillus paracasei* LBC-81, *Saccharomyces cerevisiae* CCMA 0731, *S. cerevisiae* CCMA 0732, and *Pichia kluyveri* CCMA 0615) produced 70 volatile compounds, including acids, alcohols, aldehydes, esters, and ketones, with lactic and acetic acids being the main metabolites detected. In the same way, Horáčková et al. [88] reported that a soy-based beverage fermented by a yoghurt culture (YC-381) in combination with two probiotic bacteria (*Bifidobacterium animalis* subsp. *lactis* BB 12 and *Bifidobacterium bifidum* CCDM 94) produced considerable amounts of lactic acid, acetic acid, and acetaldehyde, which provide the typical aroma and flavor of yoghurt. Similarly, Demarinis et al. [89] determined the sensory characteristics of various plant-based beverages (i.e., lupin and pea) fermented by specific lactic acid bacteria strains. The lupin- and pea-based beverages underwent an in-depth sensory evaluation using a simplified check-all-that-apply (CATA) method, displaying sensory notes such as cooked ham, almonds, and sandalwood. Also, diverse cereal-based beverages fermented using different lactobacilli strains exhibited different sensory characteristics, with fermented beverages with a higher acetaldehyde content having better sensory acceptability [90].

On the other hand, other strategies to improve the sensory properties of this type of plant-based fermented beverage include the blending of different plant sources or the addition of fruit pulps. For instance, studies show that mixing two plant-based beverages (i.e., soy and almond, 50:50% v/v) significantly improved ($p < 0.05$) the sensory properties (e.g., color, flavor, mouthfeel, and overall acceptability) of the final beverage compared with the 100% fermented soy-based beverage (nine-point hedonic scale, $n = 10$) [91]. Similar results were obtained by Antoine et al. [92] by mixing 20% v/v of cashew nut (*Anacardium occidental*) and 80% v/v of soy (*Glycine max* L.) beverages, which increased their sensory properties (e.g., sweetness and aroma) according to the descriptive analysis (N = 10) of those beverages fermented using probiotic potential strains compared with those fermented using conventional yoghurt strains. The addition of fruit pulps or flavoring agents can help improve the acceptability of the fermented beverages. Karovičová, Kohajdová, Minarovičová, Lauková, Greifová, Greif, and Greif [82] reported a significant ($p < 0.05$) improvement in the acceptability of a quinoa-based fermented beverage when a raspberry syrup (10% v/v) was added (ca. 91% acceptability compared to the non-supplemented beverage, N = 11). Similarly, a rice-based fermented beverage enriched with 10% v/v of cactus pear (*Opuntia* spp.) and 10% v/v of physalis (*Physalis peruviana*) fruit pulps significantly enhanced ($p < 0.05$) the overall acceptability (on a seven-point hedonic scale, N = 10) compared with the non-enriched rice-based fermented beverage [93]. Likewise, Sanni, Onilude, and Adeleke [75] reported significantly ($p < 0.05$) more overall acceptability (on a nine-point hedonic scale, N = 9) of pea-based fermented beverages with added banana and strawberry flavors than the control (unflavored) fermented beverage. The addition of fruit pulps can enhance the sensory acceptability of the final fermented product because they provide natural sweetness. In addition, the fermentation of the simple sugars present in these fruit pulps can generate new sweet taste profiles [94,95]. Fruits pulps also contain aromatic compounds (e.g., aldehydes), and their consistency can improve the mouthfeel of the plant-based fermented beverages [95,96].

3.3. Impact on Bioactive Profile

During the fermentation of plant-based beverages, a diverse group of bioactive compounds and metabolites are generated. In particular, microorganisms used for the fermentation process hydrolyze oligosaccharides and proteins present in plants that are not commonly digested in the human gut [28,97]. According to Table 1, the main bioactive compounds and metabolites generated and/or liberated from the plant matrix of the beverages include polyphenols (e.g., flavonoids), vitamins (e.g., ascorbic acid and tocopherol), isoflavone aglycones (e.g., genistein, daidzein, and glycitein), bioactive peptides, amino acids (e.g., gamma-aminobutyric acid, alanine, and arginine), and organic acids (e.g., lactic acid). In addition, in the majority of cases, plant-based fermented beverages are produced from soy and, to a lesser extent, from almond, oat, flaxseed, rice, cashew, coconut, and hemp, among others (Table 1). This may be due to the fact that, among the above-mentioned plant-based beverages, soy has a higher and more balanced nutrient content comparable to that of cow´s milk (e.g., protein and fat) [98] and is one of the most accepted plant-based beverages [54]. Various other sources have also been used to produce these fermented beverages, but to a relatively minor degree, such as lupin, lentil, chickpea, camelia, and apricot, to name a few.

Research has determined the presence and/or content of peptides, total polyphenols, flavonoids, and tocopherol in soy-based fermented beverages, which increase during the fermentation time [82,99–101], and their production/profile depends on the metabolic activity of microorganisms involved [61]. Some of these studies reported a relationship between the increase in specific bioactive compounds and the bioactivity tested. For example, Tonolo, Moretto, Folda, Scalcon, Bindoli, Bellamio, Feller, and Rigobello [99] reported that, after fermentation, the content of peptides and polyphenols as well as the antioxidant activity of a soy-based fermented beverage increased during storage time (weeks). Similarly, Undhad Trupti, Das, Solanki, Kinariwala, and Hati [100] reported that during the fermentation of a soy-based beverage, bioactive peptides were released with antihypertensive activity towards the inhibition of angiotensin-converting enzyme (ACE). Likewise, de Queirós, de Ávila, Botaro, Chirotto, Macedo, and Macedo [101] and Karovičová, Kohajdová, Minarovičová, Lauková, Greifová, Greif, and Greif [82] found that soy-based and quinoa-based beverages, respectively, increased in their total polyphenolic content and antioxidant activity. In particular, in soy-based fermented beverages, isoflavone aglycones (daidzein, genistein, and glycitein) have been identified. To obtain these aglycone forms, soy isoflavones should be converted from their glycosylated (e.g., daidzin, genistin, and glycitin) to aglycones forms through the action of β-glycosidase from microorganisms used to fermented the soy-based beverages [102,103]. There are some reasons for the interest in transforming glycosylated isoflavones into their aglycone forms because this increases their bioavailability, their bioactivity/health benefits, and facilitates their transformation into other bioactive secondary metabolites (e.g., equol) [102]. The above-mentioned isoflavones are known to exhibit antioxidant, chemoprotective, osteogenetic, and anti-inflammatory properties [61,103].

In other studies, the presence of polyphenols, flavonoids, and ascorbic acids in flaxseed- [104] and lupin-based [105] fermented beverages showed antioxidant properties. Similarly, fermented beverages derived from an almond–soy blend [106] and hemp [107] demonstrated antioxidant activity as well as a noticeable content of total polyphenols and flavonoids. In addition, other bioactive compounds such as GABA were found in fermented beverages derived from a rice–chickpea blend [73], rice [108], apricot seed [109], and coconut [110]. These fermented beverages showed antioxidant and antimicrobial activities, as well as angiotensin-converting enzyme (ACE), lipase, α-amylase, and α-glucosidase inhibition properties. These properties were attributed to the presence of GABA and other compounds such as polyphenols and lactic acid [73,108–110]. In contrast, other studies have demonstrated the release of bioactive peptides after fermentation from proteins in soy [100,111], flaxseed [112], coconut [113], and oat [114] beverages, which presented antioxidant and antimicrobial properties as well as ACE inhibition [91,100,113,114].

Table 1. Bioactive compounds and their bioactivities present in fermented plant-based beverages.

Fermented Plant-Based Beverage	Microorganisms	Bioactive Compounds	Bioactivity	References
Soy (*Glycine max*)	*Lactobacillus delbrueckii* subs. *Bulgaricus* and *Streptococcus thermophilus*	Polyphenols, tocopherol, and peptides	Antioxidant activity	[99]
Soy (*Glycine max*)	*Lactobacillus rhamnosus* CRL981	Isoflavone aglycon	Antioxidant activity	[115]
Soy (*Glycine max*)	*Bacillus subtilis* MTCC5480 and *Bacillus subtilis* MTCC1747	Peptides and polyphenols	Antioxidant activity	[111]
Soy (*Glycine max*)	*Bacillus subtilis* 10160	Polyphenols and flavonoids	Antioxidant activity and ACE inhibition	[116]

Table 1. *Cont.*

Fermented Plant-Based Beverage	Microorganisms	Bioactive Compounds	Bioactivity	References
Soy (*Glycine max*)	*Lactobacillus* sp. FTDC 2113, *Lactobacillus acidophilus* FTDC 8033, *Lactobacillus acidophilus* ATCC 4356, *Lactobacillus casei* ATCC 393, *Bifidobacterium* FTDC 8943, and *Bifidobacterium longum* FTDC 8643	Isoflavone aglycones (genistein)	ACE inhibition	[117]
Soy (*Glycine max*)	*Lactobacillus fermentum* M2 and *Lactobacillus casei* NK9	Peptides	ACE inhibition	[100]
Soy (*Glycine max*)	*Bifidobacterium animalis* ssp. *lactis* BLC 1, *Lactobacillus acidophilus* LA 3, *Streptococcus thermophilus* ST 066, *Lactobacillus casei* MB151 (ATCC 334), *Lactobacillus delbrueckii* ssp. *bulgaricus* MB153 (ATCC 9649), *Lactobacillus rhamnosus* MB154 (ATCC 7469), and *Lactobacillus kefiri* CBMAI21	Total polyphenols as well as isoflavone aglycones (daidzein, genistein, and glycitein)	Antioxidant activity	[101]
Soy (*Glycine max*)	*Lactobacillus curieae* CCTCC M2011381	Flavonoids	Antioxidant activity, ACE inhibition, and HMGR inhibition	[118]
Oat (*Avena sativa*)	*Lactobacillus plantarum* 22158, *Lactobacillus acidophilus* 6089, *Lactobacillus casei* 6117, *Lactobacillus delbrueckii* subsp. *bulgaricus* 57004, and *Streptococcus thermophilus* 58013	Polyphenols and flavonoids	N.D.	[119]
Flaxseed (*Linum usitatissimum*)	Commercial Kefir grains (Yoghurt-Tek®, Lactoferm Kefir Series, Kefir-31) and Biochem S.R.L. (Rome, Italy)	Polyphenols, flavonoids, and ascorbic acid	Antioxidant activity	[104]
Flaxseed (*Linum usitatissimum*)	*Lactobacillus plantarum* (NCDC374)	Peptides	Antioxidant activity, ACE inhibition	[112]
Sweet blue lupin (*Lupinus angustifolius* L. cv. "Boregine")	Commercial Kefir grains (Yoghurt-Tek®, Lactoferm Kefir Series, Kefir-31) and Biochem S.R.L. (Rome, Italy)	Polyphenols, flavonoids, and ascorbic acid	Antioxidant activity	[105]
Rice (*Oryza sativa*)–Lentil (Lens culinaris)–Chickpea (*Cicer arietinum* L.)	*Lactobacillus plantarum* DSM33326, *Lactobacillus brevis* DSM33325, and *Lactobacillus rhamnosus* SP1	GABA	Antioxidant activity	[73]
Camelina (*Camelina sativa* L.) seed	Commercial yogurt starter culture YO 122	Polyphenols	Antioxidant activity	[120]
Baru (*Dipteryx alata* Vog.) almond	Commercial yogurt starter culture YF-L811, Christian Hansen® Probiotic culture (*Lactobacillus casei* 01, Christian Hansen®	Polyphenols	Antioxidant activity, α-amylase inhibition, and α-glucosidase inhibition	[121]
Almond (*Prunus dulcis*)	Kefir grains	N.D.	Antimicrobial activity	[122]
Soy (*Glycine max*)–Almond (*Prunus dulcis*)	*Bifidobacterium longum* DSM 20219 and *Bifidobacterium animalis* subsp. *lactis* DSM 10140	Polyphenols and flavonoids	Antioxidant activity	[106]
Cashew nut (*Anacardium occidentale*)–Soy (*Glycine max* L.)	*Weissella paramesenteroides* TC6 and *Enterococcus faecalis* A4	N.D.	Antioxidant and anti-inflammatory activities	[92]

Table 1. *Cont.*

Fermented Plant-Based Beverage	Microorganisms	Bioactive Compounds	Bioactivity	References
Coconut (*Cocos nucifera* L.)	Kefir grains	Peptides	Antimicrobial and antioxidant activities	[113]
Coconut (*Cocos nucifera* L.)	*Lactiplantibacillus plantarum* ngue16	GABA, lactic acid, alanine, and arginine	Antimicrobial and antioxidant activities	[110]
Hemp (*Cannabis sativa*)	Commercial Kefir grains (Yoghurt-Tek®, Lactoferm Kefir Series, Kefir-31) and Biochem S.R.L. (Rome, Italy)Commercial yogurt starter culture YO 122	Polyphenols and flavonoids	Antioxidant activity	[107]
Hemp (*Cannabis sativa*)	*Bifidobacterium longum* B 379M	Flavonoids and curcumin (added)	Antioxidant activity	[123]
Apricot seed (*Prunus armeniaca*)	Commercial Kefir culture (KF2 100 MU), Maysa Company (Istanbul, Turkey)	GABA	Antioxidant activity and ACE inhibition	[109]
Cashew (*Anacardium occidentale*)	*Lactobacillus acidophilus* TISTR 1338, *Lactobacillus casei* TISTR 390, and *Lactobacillus plantarum* TISTR 543	Polyphenols and ascorbic acid	Antioxidant activity	[124]
Oat (*Avena sativa*)	Commercial yogurt starter culture (CSL, Italy)Probiotic culture (*Lactobacillus casei* 01, Christian Hansen®	Peptides	ACE inhibition	[114]
Brown rice (*Oryza sativa*)	*Lactobacillus pentosus* 9D3	Polyphenols and GABA	Antioxidant activity, lipase inhibition, α-amylase inhibition, and α-glucosidase inhibition	[108]

ACE: Angiotensin I-converting enzyme; HMGR: 3-hydroxy-3-methylglutaryl-coenzyme A reductase (its inhibition suggested a cholesterol-lowering effects); GABA: gamma-aminobutyric acid; N.D.: Not determined or not reported.

As previously discussed, fermentation processes can improve the overall nutritional value, sensory characteristics and acceptability, and bioactive properties of plant-based beverages via the modification of their components and their bioavailability (Figure 3). In most cases, an increase in different nutritional components (e.g., proteins, amino acids, and vitamins) as well as an improvement in the digestibility of proteins and minerals bioavailability was observed. Additionally, a general decrease in anti-nutritional factors (e.g., phytic acid, condensed tannins, and saponins), which contribute to enhanced nutrient bioavailability, as found. Both observations are related to the improvement of the nutritional value of plant-based beverages as a result of the fermentation process. Similarly, the sensory properties of fermented plant-based beverages are further favored by the generation of flavor compounds during fermentation, which improve the sensory acceptability of these beverages. Likewise, the impact of fermentation on the bioactive profile is reflected because fermentation generates a diverse group of bioactive compounds/metabolites with unique biological properties that are beneficial for the health of consumers; this can contribute to the growing preference of consumers for these fermented beverages. Some mechanisms of modification of the nutritional, sensory, and bioactive profiles that occur during the fermentation process of plant-based beverages include chemical modifications (e.g., decarboxylation and hydrolysis) or the biosynthesis/release of novel compounds/metabolites during the fermentation process due to the use of nutrients as substrates and the metabolic activity of microorganisms leading to changes in the level of some constituents.

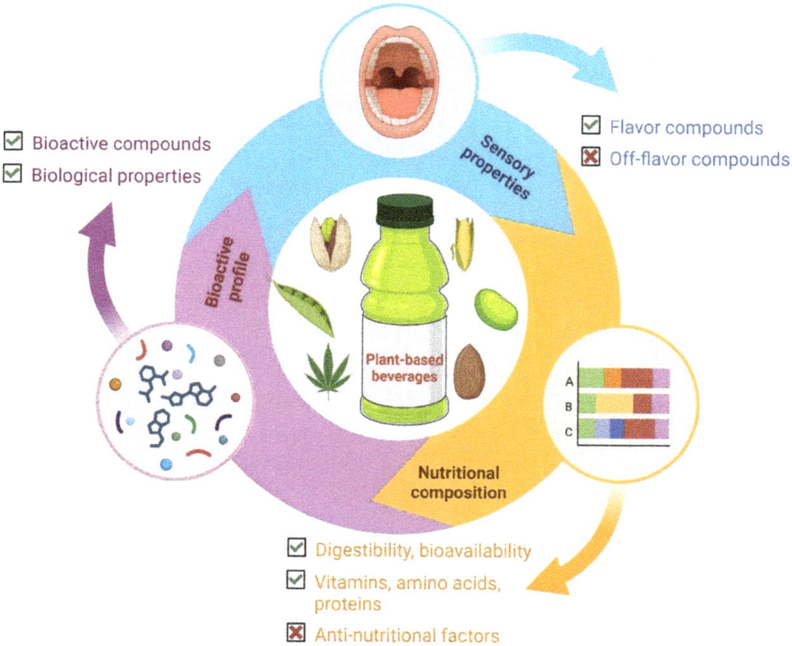

Figure 3. General impact of the fermentation processes on main plant-based beverage aspects. ☑ indicates increased; ☒ indicates decreased. Created using Biorender.com (accessed on 11 December 2023).

4. Health Benefits of Plant-Based Fermented Beverages

As mentioned in the previous section, fermented plant-based beverages contain a variety of bioactive compounds, primarily phytochemicals, that have demonstrated diverse biological properties. These properties are associated with different health benefits when these fermented products are consumed. The health effects of plant-based fermented beverages have been tested using in vivo (animal) models and clinical (human) trials, which are summarized in Table 2.

Table 2. Health benefits of plant-based fermented beverages associated with bioactive components.

Plant-Based Fermented Beverage	Identification of Bioactive Compounds	In Vivo Model	Dose and Duration	Health Benefits	Reference
Soy (*Glycine max*)	Aglycone isoflavones	Rats fed a high-fat diet	10 mL per kg body weight/daily/6 weeks	Antihyperlipidemic	[125]
Soy (*Glycine max*)	N.D.	Type 2 diabetic kidney patients	200 mL per patient/daily/8 weeks	Antidiabetic	[126]
Soy (*Glycine max*)	Aglycone isoflavones and vitamins	Rats fed a hypercaloric high-fat, high-fructose diet	10 mL per kg body weight/daily/90 days	Antidiabetic and anti-obesity	[127]
Soy (*Glycine max*)	Aglycone isoflavones	Murine model of aging induced by D-galactose	0.2 mL per animal/daily/8 weeks	Antioxidant and anti-aging	[128]
Oat (*Avena sativa*)	N.D.	Celiac patients	200 mL per patient/daily/6 months	Antihyperlipidemic	[129]

Table 2. *Cont.*

Plant-Based Fermented Beverage	Identification of Bioactive Compounds	In Vivo Model	Dose and Duration	Health Benefits	Reference
Oat (*Avena sativa*)	Polyphenols and GABA	Rat model with diabetes induced by streptozotocin	7 mL per animal/daily/6 weeks	Antidiabetic and hypolipidemic	[130]
Rice (*Oryza sativa*)	GABA, α-tocopherol, and polyphenols	Rat colitis model	150 mg dissolved in PBS per kg body weight/daily/10 days	Antioxidant	[131]
Rice (*Oryza sativa*)	Polyphenols	Stroke-prone, spontaneously hypertensive rats	40 mg beverage per kg body weight/16 h fasting	Antihypertensive	[132]
Sweet cucumber (*Solanum muricatum*)	GABA	Spontaneously hypertensive rats	2.5 mL per animal/daily/8 weeks	Antihypertensive	[133]
Aksu apple (*Rosaceae*, Malus, *Fuji*)	N.D.	C57BL/6J mice	10 mL beverage per kg body weight/daily/4 weeks	Antidiabetic	[134]

GABA: gamma-aminobutyric acid; N.D.: not determined.

For example, Chen, Wu, Yang, Xu, and Meng [125] reported that the administration (10 mL per day for 6 weeks) of a soy-based beverage fermented by *Lactococcus acidophilus* decreased total cholesterol, triglycerides, and low-density lipoprotein cholesterol levels in the serum and liver of animals in a hyperlipidemic rat model. The authors reported an improvement in the antioxidant activities of superoxide dismutase and glutathione peroxidase, as well as increased total antioxidant capacity in blood serum. Similarly, Miraghajani, Zaghian, Mirlohi, Feizi, and Ghiasvand [126] found that the administration (200 mL per day for 8 weeks) of a probiotic-fermented soy beverage improved some oxidative stress factors, including antioxidant enzymes such as glutathione peroxidase, and glutathione reductase. Another study reported the potential antidiabetic and anti-obesity properties of a soy-based beverage (10 mL per day for 90 days) fermented using kefir grains using an animal model where rats were fed a hypercaloric high-fat, high-fructose diet. This fermented beverage was able to inhibit α-amylase and lipase enzymes in the pancreas and small intestine, while decreasing total cholesterol, LDL-cholesterol, and glucose levels in blood serum (Tiss, Souiy, Abdeljelil, Njima, Achour, and Hamden) [127].

Hu, Chen, Qian, Ye, Long, Park, and Zhao [128] reported the in vivo antioxidant and anti-aging effects of a soy-based fermented beverage (0.2 mL per day for 8 weeks) using a murine model with aging induced by D-galactose. Their results showed an increase in glutathione and total antioxidant capacity; additionally, the enzyme activities of superoxide dismutase, glutathione peroxidase, and catalase were increased in the serum, brain, and liver of animals administered the fermented beverage. In a similar study, Deeseenthum, Luang-In, John, Chottanom, and Chunchom [131] reported that rice-based fermented (150 mg dissolved in PBS per day for 10 days) beverages exhibited a in vivo antioxidant effect on a rat colitis model through the decrease in lipid peroxidation and increase in superoxide dismutase in serum. On the other hand, Aparicio-García, Martínez-Villaluenga, Frias, Crespo Perez, Fernández, Alba, Rodríguez, and Peñas [129] reported an antihyperlipidemic effect on celiac patients via the administration (200 mL per day for 6 months) of an oat-based fermented beverage. The authors observed decreased total cholesterol and triglycerides levels in the serum of treated patients. Similarly, Algonaiman, Alharbi, and Barakat [130], using a murine model, reported that, compared to the control group, oat-based fermented beverage extracts (7 mL per day for 6 weeks) were significantly ($p < 0.05$) more effective at improving serum lipid profiles in the animals. Moreover, the administration of the fermented beverage efficiently increased the amount of antioxidant enzymes and decreased lipid peroxidation levels. In another study, Wang, Wang, Han, Liang, Zhang, Bai, Yue, and Gao [134] found that the administration of fermented apple juice (10 mL per day for 4 weeks) to type 2 diabetic mice decreased fasting blood glucose and insulin levels and regulated blood lipid metabolism.

Finally, the administration of a sweet cucumber-based fermented beverage had a noticeable antihypertensive effect when using a spontaneously hypertensive rat model [133].

In summary, most of the plant-based fermented beverage studies reported in the literature were performed using in vitro assays, perhaps because in vitro studies are a more economical way to determine preliminary characterization and biological properties compared with in vivo studies. However, there is a lack of sufficient in vivo animal and/or clinical trials to determine the effectiveness of these fermented beverages with different potential beneficial effects towards humans.

5. Conclusions and Future Directions

Plant-based beverages have drawn attention because they are a good alternative for people with lactose intolerance, milk allergies, and a prevalence of hypercholesterolemia. Also, consumers' demand for a more sustainable, plant-based diet has led to their rise in popularity. The fermentation processes, using different microorganisms, can improve their nutritional composition and overall sensory characteristics. However, while these beverages have been proposed as alternatives to dairy milk, they lack equal nutritional value. Although certain plant-based beverages, like those derived from soy, offer comparable nutrients to dairy milk, other strategies such as blending different plant sources, mixing with dairy milk, or supplementing with additional ingredients (e.g., fruits) or additives are recommended to enhance their overall nutritional value. In addition, the fermentation process can decrease the content of antinutritive factors, thereby preserving their nutritional quality. Research into fermented plant-based beverages faces significant challenges related to improving processing techniques, including exploring emerging non-thermal processing technologies that can ensure the safety of the product while maintaining nutritional and sensory qualities. In addition, plant-based fermented beverages have been demonstrated to have diverse compounds with different biological properties toward human health. Further characterization of these bioactive compounds, including metabolomics and proteomics, as well as animal studies and/or human clinical trials must be conducted to support the health claims associated with plant-based fermented beverages. Additional research is needed to optimize the production and commercial feasibility of bioactive compounds from plant-based fermented beverages.

Author Contributions: Conceptualization, A.M.L. and J.E.A.-T.; formal analysis, J.E.A.-T.; resources, T.E.-S. and M.L.Z.-Z.; writing—original draft preparation, B.H.-F. and E.d.J.-J.; writing—review and editing, A.M.L. and J.E.A.-T.; visualization, A.d.J.C.-H., O.S.-C., T.E.-S., R.M.G.-R. and M.L.Z.-Z.; bibliographic search, A.d.J.C.-H., O.S.-C., T.E.-S., R.M.G.-R. and M.L.Z.-Z.; funding acquisition, A.M.L. and J.E.A.-T. All authors have read and agreed to the published version of the manuscript.

Funding: This work is supported by the USDA National Institute of Food and Agriculture, Hatch Act formula fund project 1019794.

Institutional Review Board Statement: Not applicable.

Informed Consent Statement: Not applicable.

Data Availability Statement: Not applicable.

Acknowledgments: The authors gratefully acknowledge the resources and databases provided by Purdue University and Metropolitan Autonomous University. The authors Hidalgo-Fuentes and de Jesús-José are undergraduate students of the Food Science and Technology program at Metropolitan Autonomous University, Campus Lerma. This work was derived from both of their thesis/final projects used to obtain their bachelor's degrees.

Conflicts of Interest: The authors declare no conflicts of interest.

References

1. Sethi, S.; Tyagi, S.K.; Anurag, R.K. Plant-based milk alternatives an emerging segment of functional beverages: A review. *J. Food Sci. Technol.* **2016**, *53*, 3408–3423. [CrossRef] [PubMed]
2. Penha, C.B.; Santos, V.D.P.; Speranza, P.; Kurozawa, L.E. Plant-based beverages: Ecofriendly technologies in the production process. *Innov. Food Sci. Emerg. Technol.* **2021**, *72*, 102760. [CrossRef]
3. Mongi, R.J.; Gomezulu, A.D. Descriptive sensory analysis, consumer acceptability, and conjoint analysis of beef sausages prepared from a pigeon pea protein binder. *Heliyon* **2022**, *8*, e10703. [CrossRef] [PubMed]
4. Polyak, E.; Breitenbach, Z.; Frank, E.; Mate, O.; Figler, M.; Zsalig, D.; Simon, K.; Szijarto, M.; Szabo, Z. Food and sustainability: Is it a matter of choice? *Sustainability* **2023**, *15*, 7191. [CrossRef]
5. Willits-Smith, A.; Aranda, R.; Heller, M.C.; Rose, D. Addressing the carbon footprint, healthfulness, and costs of self-selected diets in the USA: A population-based cross-sectional study. *Lancet Planet. Health* **2020**, *4*, e98–e106. [CrossRef] [PubMed]
6. Xu, X.; Sharma, P.; Shu, S.; Lin, T.-S.; Ciais, P.; Tubiello, F.N.; Smith, P.; Campbell, N.; Jain, A.K. Global greenhouse gas emissions from animal-based foods are twice those of plant-based foods. *Nat. Food* **2021**, *2*, 724–732. [CrossRef]
7. Poore, J.; Nemecek, T. Reducing food's environmental impacts through producers and consumers. *Science* **2018**, *360*, 987–992. [CrossRef]
8. STATISTA. Which (Plant) Milk Is the Most Sustainable? 2023. Available online: https://www.statista.com/chart/22659/cows-milk-plant-milk-sustainability/ (accessed on 27 February 2024).
9. STATISTA. Milk Substitutes—Worldwide. 2023. Available online: https://es.statista.com/outlook/cmo/food/dairy-products-eggs/milk-substitutes/worldwide (accessed on 1 June 2023).
10. USDA. Estimated Fluid Milk Products Sales Report. United States Department of Agriculture. EFMS-0423. 2023. Available online: https://www.ams.usda.gov/sites/default/files/media/EstimatedFluidProductsMilkSales.pdf (accessed on 20 June 2023).
11. Deep, Y.N.; Sangita, B.; Arvind, J.K.; Ranjeet, S. Plant based dairy analogues: An emerging food. *Agric. Res. Technol.* **2017**, *10*, 555781. [CrossRef]
12. Silva, A.R.A.; Silva, M.M.N.; Ribeiro, B.D. Health issues and technological aspects of plant-based alternative milk. *Food Res. Int.* **2020**, *131*, 108972. [CrossRef]
13. Mäkinen, O.E.; Wanhalinna, V.; Zannini, E.; Arendt, E.K. Foods for special dietary needs: Non-dairy plant-based milk substitutes and fermented dairy-type products. *Crit. Rev. Food Sci. Nutr.* **2016**, *56*, 339–349. [CrossRef]
14. Nawaz, M.A.; Singh, T.K.; Jegasothy, H.; Buckow, R. UHT treatment on the stability of faba bean protein emulsion. *Proceedings* **2021**, *70*, 29. [CrossRef]
15. McClements, D.J. Development of next-generation nutritionally fortified plant-based milk substitutes: Structural design principles. *Foods* **2020**, *9*, 421. [CrossRef]
16. Nawaz, M.A.; Tan, M.; Øiseth, S.; Buckow, R. An emerging segment of functional legume-based beverages: A review. *Food Rev. Int.* **2022**, *38*, 1064–1102. [CrossRef]
17. Grau-Fuentes, E.; Rodrigo, D.; Garzón, R.; Rosell, C.M. Understanding the marketed plant-based beverages: From ingredients technological function to their nutritional value. *J. Funct. Foods* **2023**, *106*, 105609. [CrossRef]
18. Braun, K.; Hanewald, A.; Vilgis, T.A. Milk emulsions: Structure and stability. *Foods* **2019**, *8*, 483. [CrossRef] [PubMed]
19. Patra, T.; Rinnan, Å.; Olsen, K. The physical stability of plant-based drinks and the analysis methods thereof. *Food Hydrocoll.* **2021**, *118*, 106770. [CrossRef]
20. FDA. Title 21—Food and Drugs Chapter I-Food and Drug Administration Department of Health and HUMAN Services Subchapter B—Food for Human Consumption. Part 131—Milk and Cream. Subpart B—Requirements for Specific Standardized Milk and Cream. 2023. Available online: https://www.accessdata.fda.gov/scripts/cdrh/cfdocs/cfcfr/CFRSearch.cfm?fr=131.110&SearchTerm=milk (accessed on 15 June 2023).
21. FDA. *Labeling of Plant-Based Milk Alternatives and Voluntary Nutrient Statements; Draft Guidance for Industry (FDA-2023-D-0451)*; Food and Drug Administration: Silver Spring, MD, USA, 2023.
22. *NOM-155-SCFI-2012*; Leche-Denominaciones, Especificaciones Fisicoquímicas, Información Comercial y Métodos de Prueba. Diario Oficial de la Federación: Mexico City, México, 2012.
23. Ji, G.-Z.; Li, X.-M.; Dong, Y.; Shi, Y.-D. Composition, formation mechanism, and removal method of off-odor in soymilk products. *J. Food Sci.* **2022**, *87*, 5175–5190. [CrossRef] [PubMed]
24. Gani, A.; Wani, S.M.; Masoodi, F.A.; Salim, R. Characterization of rice starches extracted from Indian cultivars. *Food Sci. Technol. Int. Cienc. Y Tecnol. De Los Aliment. Int.* **2013**, *19*, 143–152. [CrossRef] [PubMed]
25. Kumar, M.; Tomar, M.; Punia, S.; Dhakane-Lad, J.; Dhumal, S.; Changan, S.; Senapathy, M.; Berwal, M.K.; Sampathrajan, V.; Sayed, A.A.S.; et al. Plant-based proteins and their multifaceted industrial applications. *LWT Food Sci. Technol.* **2022**, *154*, 112620. [CrossRef]
26. Tangyu, M.; Muller, J.; Bolten, C.J.; Wittmann, C. Fermentation of plant-based milk alternatives for improved flavour and nutritional value. *Appl. Microbiol. Biotechnol.* **2019**, *103*, 9263–9275. [CrossRef]
27. Pontonio, E.; Rizzello, C.G. Milk alternatives and non-dairy fermented products: Trends and challenges. *Foods* **2021**, *10*, 222. [CrossRef] [PubMed]

28. Marco, M.L.; Heeney, D.; Binda, S.; Cifelli, C.J.; Cotter, P.D.; Foligné, B.; Gänzle, M.; Kort, R.; Pasin, G.; Pihlanto, A.; et al. Health benefits of fermented foods: Microbiota and beyond. *Curr. Opin. Biotechnol.* **2017**, *44*, 94–102. [CrossRef] [PubMed]
29. Basinskiene, L.; Cizeikiene, D. Chapter 3—Cereal-Based Nonalcoholic Beverages. In *Trends in Non-Alcoholic Beverages*; Galanakis, C.M., Ed.; Academic Press: Cambridge, MA, USA, 2020; pp. 63–99.
30. Jonas da Rocha Esperança, V.; Corrêa de Souza Coelho, C.; Tonon, R.; Torrezan, R.; Freitas-Silva, O. A review on plant-based tree nuts beverages: Technological, sensory, nutritional, health and microbiological aspects. *Int. J. Food Prop.* **2022**, *25*, 2396–2408. [CrossRef]
31. Siddiqui, S.A.; Alvi, T.; Sameen, A.; Khan, S.; Blinov, A.V.; Nagdalian, A.A.; Mehdizadeh, M.; Adli, D.N.; Onwezen, M. Consumer acceptance of alternative proteins: A systematic review of current alternative protein sources and interventions adapted to increase their acceptability. *Sustainability* **2022**, *14*, 15370. [CrossRef]
32. Nawaz, M.A.; Buckow, R.; Katopo, L.; Stockmann, R. Chapter 6—Plant-based beverages. In *Engineering Plant-Based Food Systems*; Prakash, S., Bhandari, B.R., Gaiani, C., Eds.; Academic Press: Cambridge, MA, USA, 2023; pp. 99–129.
33. Montemurro, M.; Pontonio, E.; Coda, R.; Rizzello, C.G. Plant-based alternatives to Yogurt: State-of-the-art and perspectives of new biotechnological challenges. *Foods* **2021**, *10*, 316. [CrossRef]
34. Silva, J.G.S.; Rebellato, A.P.; Caramês, E.T.d.S.; Greiner, R.; Pallone, J.A.L. In vitro digestion effect on mineral bioaccessibility and antioxidant bioactive compounds of plant-based beverages. *Food Res. Int.* **2020**, *130*, 108993. [CrossRef]
35. Cichońska, P.; Ziarno, M. Legumes and legume-based beverages fermented with lactic acid bacteria as a potential carrier of probiotics and prebiotics. *Microorganisms* **2022**, *10*, 91. [CrossRef]
36. Sithole, T.R.; Ma, Y.-X.; Qin, Z.; Liu, H.-M.; Wang, X.-D. Influence of peanut varieties on the sensory quality of peanut butter. *Foods* **2022**, *11*, 3499. [CrossRef]
37. Chalupa-Krebzdak, S.; Long, C.J.; Bohrer, B.M. Nutrient density and nutritional value of milk and plant-based milk alternatives. *Int. Dairy J.* **2018**, *87*, 84–92. [CrossRef]
38. Fructuoso, I.; Romão, B.; Han, H.; Raposo, A.; Ariza-Montes, A.; Araya-Castillo, L.; Zandonadi, R.P. An overview on nutritional aspects of plant-based beverages used as substitutes for mow's milk. *Nutrients* **2021**, *13*, 2650. [CrossRef]
39. Nimbkar, S.; Negi, A.; Thirukumaran, R.; Moses, J.A.; Sinija, V.R. Effect of thermal and nonthermal techniques on the physico-chemical quality of high-fat coconut cream. *J. Food Process Eng.* **2023**, *46*, e14462. [CrossRef]
40. Bagheri, H. Application of infrared heating for roasting nuts. *J. Food Qual.* **2020**, *2020*, 8813047. [CrossRef]
41. Zhang, K.; Dong, R.; Hu, X.; Ren, C.; Li, Y. Oat-based foods: Chemical constituents, glycemic index, and the effect of processing. *Foods* **2021**, *10*, 1304. [CrossRef]
42. Cichońska, P.; Ziębicka, A.; Ziarno, M. Properties of rice-based beverages fermented with lactic acid bacteria and propionibacterium. *Molecules* **2022**, *27*, 2558. [CrossRef] [PubMed]
43. Patra, M.; Bashir, O.; Amin, T.; Wani, A.W.; Shams, R.; Chaudhary, K.S.; Mirza, A.A.; Manzoor, S. A comprehensive review on functional beverages from cereal grains-characterization of nutraceutical potential, processing technologies and product types. *Heliyon* **2023**, *9*, e16804. [CrossRef] [PubMed]
44. Paucar-Menacho, L.M.; Simpalo-López, W.D.; Castillo-Martínez, W.E.; Esquivel-Paredes, L.J.; Martínez-Villaluenga, C. Reformulating bread using sprouted pseudo-cereal grains to enhance its nutritional value and sensorial attributes. *Foods* **2022**, *11*, 1541. [CrossRef] [PubMed]
45. Shukla, S. *Plant-Based India: Nourishing Recipes Rooted in Tradition*; Hachette UK: London, UK, 2022.
46. Adams, A. *The Dairy-Free Kitchen: 100 Recipes for All the Creamy Foods You Love—Without Lactose, Casein, or Dairy*; Fair Winds Press: Beverly, MA, USA, 2014.
47. Mridula, D.; Sharma, M. Development of non-dairy probiotic drink utilizing sprouted cereals, legume and soymilk. *LWT Food Sci. Technol.* **2015**, *62*, 482–487. [CrossRef]
48. Bridges, M. Moo-ove over, cow's milk: The rise of plant-based dairy alternatives. *Pract. Gastroenterol.* **2018**, *42*, 20–27.
49. Röös, E.; Garnett, T.; Watz, V.; Sjörs, C. The role of dairy and plant based dairy alternatives in sustainable diets. In *SLU Future Food Reports 3*; Swedish University of Agricultural Sciences: Uppsala, Sweden, 2018; Available online: https://pub.epsilon.slu.se/16016/1/roos_e_et_al_190304.pdf (accessed on 15 December 2023).
50. Singhal, S.; Baker, R.D.; Baker, S.S. A comparison of the nutritional value of cow's milk and nondairy beverages. *J. Pediatr. Gastroenterol. Nutr.* **2017**, *64*, 799–805. [CrossRef]
51. Vaikma, H.; Kaleda, A.; Rosend, J.; Rosenvald, S. Market mapping of plant-based milk alternatives by using sensory (RATA) and GC analysis. *Future Foods* **2021**, *4*, 100049. [CrossRef]
52. Pérez-González, M.; Gallardo-Chacón, J.J.; Valencia-Flores, D.; Ferragut, V. Optimization of a headspace SPME GC-MS methodology for the analysis of processed almond beverages. *Food Anal. Methods* **2015**, *8*, 612–623. [CrossRef]
53. Bhatla, S.C. Secondary Metabolites. In *Plant Physiology, Development and Metabolism*; Bhatla, S.C., A. Lal, M., Eds.; Springer: Singapore, 2018; pp. 1099–1166.
54. Jeske, S.; Bez, J.; Arendt, E.K.; Zannini, E. Formation, stability, and sensory characteristics of a lentil-based milk substitute as affected by homogenisation and pasteurisation. *Eur. Food Res. Technol.* **2019**, *245*, 1519–1531. [CrossRef]
55. Sharma, R.; Garg, P.; Kumar, P.; Bhatia, S.K.; Kulshrestha, S. Microbial fermentation and its role in quality improvement of fermented foods. *Fermentation* **2020**, *6*, 106. [CrossRef]

56. Maicas, S. The role of yeasts in fermentation processes. *Microorganisms* **2020**, *8*, 1142. [CrossRef] [PubMed]
57. Harper, A.R.; Dobson, R.C.J.; Morris, V.K.; Moggré, G.-J. Fermentation of plant-based dairy alternatives by lactic acid bacteria. *Microb. Biotechnol.* **2022**, *15*, 1404–1421. [CrossRef]
58. Paul, A.A.; Kumar, S.; Kumar, V.; Sharma, R. Milk analog: Plant based alternatives to conventional milk, production, potential and health concerns. *Crit. Rev. Food Sci. Nutr.* **2020**, *60*, 3005–3023. [CrossRef] [PubMed]
59. Xiong, Y.; Zhang, P.; Warner, R.D.; Shen, S.; Fang, Z. Cereal grain-based functional beverages: From cereal grain bioactive phytochemicals to beverage processing technologies, health benefits and product features. *Crit. Rev. Food Sci. Nutr.* **2022**, *62*, 2404–2431. [CrossRef] [PubMed]
60. Liu, H.; Xu, X.; Cui, H.; Xu, J.; Yuan, Z.; Liu, J.; Li, C.; Li, J.; Zhu, D. Plant-based fermented beverages and key emerging processing technologies. *Food Rev. Int.* **2022**, *39*, 5844–5863. [CrossRef]
61. Ewe, J.-A.; Yeo, S.-K. Fermented soymilk as a nutraceutical. In *Beneficial Microorganisms in Food and Nutraceuticals*; Liong, M.-T., Ed.; Springer International Publishing: Cham, Switzerland, 2015; pp. 133–159.
62. Alexandre, E.M.C.; Aguiar, N.F.B.; Voss, G.B.; Pintado, M.E. Properties of fermented beverages from food wastes/by-products. *Beverages* **2023**, *9*, 45. [CrossRef]
63. Gil-Serna, J.; Vázquez, C.; Patiño, B. Mycotoxins in functional beverages: A review. *Beverages* **2020**, *6*, 52. [CrossRef]
64. Siddiqui, S.A.; Erol, Z.; Rugji, J.; Taşçı, F.; Kahraman, H.A.; Toppi, V.; Musa, L.; Di Giacinto, G.; Bahmid, N.A.; Mehdizadeh, M.; et al. An overview of fermentation in the food industry—Looking back from a new perspective. *Bioresour. Bioprocess.* **2023**, *10*, 85. [CrossRef]
65. Solanki, P.; Putatunda, C.; Kumar, A.; Bhatia, R.; Walia, A. Microbial proteases: Ubiquitous enzymes with innumerable uses. *3 Biotech* **2021**, *11*, 428. [CrossRef]
66. Santiago-López, L.; Aguilar-Toalá, J.E.; Hernández-Mendoza, A.; Vallejo-Cordoba, B.; Liceaga, A.M.; González-Córdova, A.F. Invited review: Bioactive compounds produced during cheese ripening and health effects associated with aged cheese consumption. *J. Dairy Sci.* **2018**, *101*, 3742–3757. [CrossRef]
67. Jeyakumar, E.; Lawrence, R. 10—Microbial fermentation for reduction of antinutritional factors. In *Current Developments in Biotechnology and Bioengineering*; Rai, A.K., Singh, S.P., Pandey, A., Larroche, C., Soccol, C.R., Eds.; Elsevier: Amsterdam, The Netherlands, 2022; pp. 239–260.
68. Samtiya, M.; Aluko, R.E.; Dhewa, T. Plant food anti-nutritional factors and their reduction strategies: An overview. *Food Prod. Process. Nutr.* **2020**, *2*, 6. [CrossRef]
69. Licandro, H.; Ho, P.H.; Nguyen, T.K.C.; Petchkongkaew, A.; Nguyen, H.V.; Chu-Ky, S.; Nguyen, T.V.A.; Lorn, D.; Waché, Y. How fermentation by lactic acid bacteria can address safety issues in legumes food products? *Food Control* **2020**, *110*, 106957. [CrossRef]
70. Madeira, J.V.; Macedo, J.A.; Macedo, G.A. Detoxification of castor bean residues and the simultaneous production of tannase and phytase by solid-state fermentation using Paecilomyces variotii. *Bioresour. Technol.* **2011**, *102*, 7343–7348. [CrossRef] [PubMed]
71. Dhiman, S.; Mukherjee, G.; Singh, A.K. Recent trends and advancements in microbial tannase-catalyzed biotransformation of tannins: A review. *Int. Microbiol.* **2018**, *21*, 175–195. [CrossRef] [PubMed]
72. Miller, A.W.; Dearing, D. The metabolic and ecological interactions of oxalate-degrading bacteria in the mammalian gut. *Pathogens* **2013**, *2*, 636–652. [CrossRef] [PubMed]
73. Pontonio, E.; Raho, S.; Dingeo, C.; Centrone, D.; Carofiglio, V.E.; Rizzello, C.G. Nutritional, functional, and technological characterization of a novel gluten- and lactose-free yogurt-style snack produced with selected lactic acid bacteria and leguminosae flours. *Front. Microbiol.* **2020**, *11*, 1664. [CrossRef] [PubMed]
74. Vila-Real, C.; Pimenta-Martins, A.; Mbugua, S.; Hagrétou, S.-L.; Katina, K.; Maina, N.H.; Pinto, E.; Gomes, A.M.P. Novel synbiotic fermented finger millet-based yoghurt-like beverage: Nutritional, physicochemical, and sensory characterization. *J. Funct. Foods* **2022**, *99*, 105324. [CrossRef]
75. Sanni, A.I.; Onilude, A.A.; Adeleke, E.O. Preparation and characteristics of lactic acid fermented cowpea milk. *Z. Für Leb. Und-Forsch. A* **1999**, *208*, 225–229. [CrossRef]
76. Rekha, C.R.; Vijayalakshmi, G. Bioconversion of isoflavone glycosides to aglycones, mineral bioavailability and vitamin B complex in fermented soymilk by probiotic bacteria and yeast. *J. Appl. Microbiol.* **2010**, *109*, 1198–1208. [CrossRef] [PubMed]
77. Lai, L.-R.; Hsieh, S.-C.; Huang, H.-Y.; Chou, C.-C. Effect of lactic fermentation on the total phenolic, saponin and phytic acid contents as well as anti-colon cancer cell proliferation activity of soymilk. *J. Biosci. Bioeng.* **2013**, *115*, 552–556. [CrossRef] [PubMed]
78. Hou, J.-W.; Yu, R.-C.; Chou, C.-C. Changes in some components of soymilk during fermentation with bifidobacteria. *Food Res. Int.* **2000**, *33*, 393–397. [CrossRef]
79. Bernat, N.; Cháfer, M.; Chiralt, A.; Laparra, L.M.; González-Martínez, C. Almond milk fermented with different potentially probiotic bacteria improves iron uptake by intestinal epithelial (Caco-2) cells. *Int. J. Food Stud.* **2015**, *4*, 49–60. [CrossRef]
80. Santos, C.C.A.d.A.; Libeck, B.d.S.; Schwan, R.F. Co-culture fermentation of peanut-soy milk for the development of a novel functional beverage. *Int. J. Food Microbiol.* **2014**, *186*, 32–41. [CrossRef]
81. Orts, A.; Revilla, E.; Rodriguez-Morgado, B.; Castaño, A.; Tejada, M.; Parrado, J.; García-Quintanilla, A. Protease technology for obtaining a soy pulp extract enriched in bioactive compounds: Isoflavones and peptides. *Heliyon* **2019**, *5*, e01958. [CrossRef]
82. Karovičová, J.; Kohajdová, Z.; Minarovičová, L.; Lauková, M.; Greifová, M.; Greif, G.; Greif, J. Utilisation of quinoa for development of fermented beverages. *Slovak J. Food Sci.* **2020**, *14*, 465–472. [CrossRef]

83. Jeske, S.; Zannini, E.; Cronin, M.F.; Arendt, E.K. Impact of protease and amylase treatment on proteins and the product quality of a quinoa-based milk substitute. *Food Funct.* **2018**, *9*, 3500–3508. [CrossRef]
84. Zheng, Y.; Fei, Y.; Yang, Y.; Jin, Z.; Yu, B.; Li, L. A potential flavor culture: *Lactobacillus harbinensis* M1 improves the organoleptic quality of fermented soymilk by high production of 2,3-butanedione and acetoin. *Food Microbiol.* **2020**, *91*, 103540. [CrossRef]
85. de Moraes Filho, M.L.; Busanello, M.; Prudencio, S.H.; Garcia, S. Soymilk with okara flour fermented by *Lactobacillus acidophilus*: Simplex-centroid mixture design applied in the elaboration of probiotic creamy sauce and storage stability. *LWT Food Sci. Technol.* **2018**, *93*, 339–345. [CrossRef]
86. Nissen, L.; Demircan, B.; Taneyo-Saa, D.L.; Gianotti, A. Shift of aromatic profile in probiotic hemp drink formulations: A metabolomic approach. *Microorganisms* **2019**, *7*, 509. [CrossRef]
87. Menezes, A.G.T.; Ramos, C.L.; Dias, D.R.; Schwan, R.F. Combination of probiotic yeast and lactic acid bacteria as starter culture to produce maize-based beverages. *Food Res. Int.* **2018**, *111*, 187–197. [CrossRef]
88. Horáčková, S.; Mühlhansová, A.; Sluková, M.; Schulzová, V.; Plocková, M. Fermentation of soymilk by yoghurt and bifidobacteria strains. *Czech J. Food Sci.* **2015**, *33*, 313–319. [CrossRef]
89. Demarinis, C.; Verni, M.; Pinto, L.; Rizzello, C.G.; Baruzzi, F. Use of selected lactic acid bacteria for the fermentation of legume-based water extracts. *Foods* **2022**, *11*, 3346. [CrossRef]
90. Salmerón, I.; Thomas, K.; Pandiella, S.S. Effect of potentially probiotic lactic acid bacteria on the physicochemical composition and acceptance of fermented cereal beverages. *J. Funct. Foods* **2015**, *15*, 106–115. [CrossRef]
91. Kundu, P.; Dhankhar, J.; Sharma, A. Development of non dairy milk alternative using soymilk and almond milk. *Curr. Res. Nutr. Food Sci.* **2018**, *6*, 203–210. [CrossRef]
92. Antoine, A.A.; Koua, A.; Djeneba, O.H.; Ruth D'Avila, G.T. Primary characterization of a novel soymilk-cashew fermented with an improving of its antioxidant and anti-Inflammatory contents. *Food Nutr. Sci.* **2023**, *14*, 421–435. [CrossRef]
93. El-Sayed, H.; Ramadan, M.F. Production of probiotic-fermented rice milk beverage fortified with cactus pear and physalis pulp. *Zagazig J. Agric. Res.* **2020**, *47*, 165–177. [CrossRef]
94. Świąder, K.; Wegner, K.; Piotrowska, A.; Tan, F.J.; Sadowska, A. Plants as a source of natural high-intensity sweeteners: A review. *J. Appl. Bot. Food Qual.* **2019**, *92*, 160–171.
95. Pinto, T.; Vilela, A.; Cosme, F. Chemical and sensory characteristics of fruit juice and fruit fermented beverages and their consumer acceptance. *Beverages* **2022**, *8*, 33. [CrossRef]
96. Ruiz Rodríguez, L.G.; Zamora Gasga, V.M.; Pescuma, M.; Van Nieuwenhove, C.; Mozzi, F.; Sánchez Burgos, J.A. Fruits and fruit by-products as sources of bioactive compounds. Benefits and trends of lactic acid fermentation in the development of novel fruit-based functional beverages. *Food Res. Int.* **2021**, *140*, 109854. [CrossRef]
97. Battistini, C.; Gullón, B.; Ichimura, E.S.; Gomes, A.M.P.; Ribeiro, E.P.; Kunigk, L.; Moreira, J.U.V.; Jurkiewicz, C. Development and characterization of an innovative synbiotic fermented beverage based on vegetable soybean. *Braz. J. Microbiol.* **2018**, *49*, 303–309. [CrossRef]
98. Vanga, S.K.; Raghavan, V. How well do plant based alternatives fare nutritionally compared to cow's milk? *J. Food Sci. Technol.* **2018**, *55*, 10–20. [CrossRef] [PubMed]
99. Tonolo, F.; Moretto, L.; Folda, A.; Scalcon, V.; Bindoli, A.; Bellamio, M.; Feller, E.; Rigobello, M.P. Antioxidant properties of fermented soy during shelf life. *Plant Foods Hum. Nutr.* **2019**, *74*, 287–292. [CrossRef] [PubMed]
100. Undhad Trupti, J.; Das, S.; Solanki, D.; Kinariwala, D.; Hati, S. Bioactivities and ACE-inhibitory peptides releasing potential of lactic acid bacteria in fermented soy milk. *Food Prod. Process. Nutr.* **2021**, *3*, 10. [CrossRef]
101. de Queirós, L.D.; de Ávila, A.R.A.; Botaro, A.V.; Chirotto, D.B.L.; Macedo, J.A.; Macedo, G.A. Combined isoflavones biotransformation increases the bioactive and antioxidant capacity of soymilk. *Appl. Microbiol. Biotechnol.* **2020**, *104*, 10019–10031. [CrossRef] [PubMed]
102. Gaya, P.; Peirotén, Á.; Medina, M.; Landete, J.M. Isoflavone metabolism by a collection of lactic acid bacteria and bifidobacteria with biotechnological interest. *Int. J. Food Sci. Nutr.* **2016**, *67*, 117–124. [CrossRef] [PubMed]
103. Villares, A.; Rostagno, M.A.; García-Lafuente, A.; Guillamón, E.; Martínez, J.A. Content and 'profile of isoflavones in soy-based foods as a function of the production process. *Food Bioprocess Technol.* **2011**, *4*, 27–38. [CrossRef]
104. Łopusiewicz, Ł.; Drozłowska, E.; Siedlecka, P.; Mężyńska, M.; Bartkowiak, A.; Sienkiewicz, M.; Zielińska-Bliźniewska, H.; Kwiatkowski, P. Development, characterization, and bioactivity of non-dairy kefir-like fermented beverage based on flaxseed oil cake. *Foods* **2019**, *8*, 544. [CrossRef] [PubMed]
105. Łopusiewicz, Ł.; Drozłowska, E.; Trocer, P.; Kwiatkowski, P.; Bartkowiak, A.; Gefrom, A.; Sienkiewicz, M. The effect of fermentation with kefir grains on the physicochemical and antioxidant properties of beverages from blue lupin (*Lupinus angustifolius* L.) seeds. *Molecules* **2020**, *25*, 5791. [CrossRef] [PubMed]
106. Zahrani, A.J.A.; Shori, A.B. Improve the antioxidant activity and viability of *B. longum* and *B. animalis* subsp *lactis* in fermented soy and almond milk. *Food Sci. Technol.* **2023**, *43*, e118122. [CrossRef]
107. Łopusiewicz, Ł.; Waszkowiak, K.; Polanowska, K.; Mikołajczak, B.; Śmietana, N.; Hrebień-Filisińska, A.; Sadowska, J.; Mazurkiewicz-Zapałowicz, K.; Drozłowska, E. The effect of yogurt and kefir starter cultures on bioactivity of fermented industrial by-roduct from *Cannabis sativa* production-Hemp press cake. *Fermentation* **2022**, *8*, 490. [CrossRef]

108. Kittibunchakul, S.; Yuthaworawit, N.; Whanmek, K.; Suttisansanee, U.; Santivarangkna, C. Health beneficial properties of a novel plant-based probiotic drink produced by fermentation of brown rice milk with GABA-producing *Lactobacillus pentosus* isolated from Thai pickled weed. *J. Funct. Foods* **2021**, *86*, 104710. [CrossRef]
109. Uruc, K.; Tekin, A.; Sahingil, D.; Hayaloglu, A.A. An alternative plant-based fermented milk with kefir culture using apricot (*Prunus armeniaca* L.) seed extract: Changes in texture, volatiles and bioactivity during storage. *Innov. Food Sci. Emerg. Technol.* **2022**, *82*, 103189. [CrossRef]
110. Qadi, W.S.M.; Mediani, A.; Benchoula, K.; Wong, E.H.; Misnan, N.M.; Sani, N.A. Characterization of physicochemical, biological, and chemical changes associated with coconut milk fermentation and correlation revealed by 1H NMR-mased metabolomics. *Foods* **2023**, *12*, 1971. [CrossRef]
111. Sanjukta, S.; Rai, A.K.; Muhammed, A.; Jeyaram, K.; Talukdar, N.C. Enhancement of antioxidant properties of two soybean varieties of Sikkim Himalayan region by proteolytic *Bacillus subtilis* fermentation. *J. Funct. Foods* **2015**, *14*, 650–658. [CrossRef]
112. Sharma, P.; Sharma, D.; Kaur, S.; Borah, A. Optimization of flaxseed milk fermentation for the production of functional peptides and estimation of their bioactivities. *Food Sci. Technol. Int. Cienc. Y Tecnol. De Los Aliment. Int.* **2021**, *27*, 585–597. [CrossRef]
113. Abadl, M.M.T.; Mohsin, A.Z.; Sulaiman, R.; Abas, F.; Muhialdin, B.J.; Meor Hussin, A.S. Biological activities and physiochemical properties of low-fat and high-fat coconut-based kefir. *Int. J. Gastron. Food Sci.* **2022**, *30*, 100624. [CrossRef]
114. Akan, E.; Karakaya, S.; Eda Eker Özkacar, M.; Kinik, Ö. Effect of food matrix and fermentation on angiotensin-converting enzyme inhibitory activity and β-glucan release after in vitro digestion in oat-based products. *Food Res. Int.* **2023**, *165*, 112508. [CrossRef]
115. Marazza, J.A.; Nazareno, M.A.; de Giori, G.S.; Garro, M.S. Enhancement of the antioxidant capacity of soymilk by fermentation with *Lactobacillus rhamnosus*. *J. Funct. Foods* **2012**, *4*, 594–601. [CrossRef]
116. Dai, C.; Ma, H.; He, R.; Huang, L.; Zhu, S.; Ding, Q.; Luo, L. Improvement of nutritional value and bioactivity of soybean meal by solid-state fermentation with *Bacillus subtilis*. *LWT Food Sci. Technol.* **2017**, *86*, 1–7. [CrossRef]
117. Yeo, S.-K.; Liong, M.-T. Angiotensin I-converting enzyme inhibitory activity and bioconversion of isoflavones by probiotics in soymilk supplemented with prebiotics. *Int. J. Food Sci. Nutr.* **2010**, *61*, 161–181. [CrossRef]
118. Liu, Y.; Zhang, Y.; Ro, K.-S.; Li, H.; Wang, L.; Xie, J.; Wei, D. Gastrointestinal survival and potential bioactivities of *Lactobacillus curieae* CCTCC M2011381 in the fermentation of plant food. *Process Biochem.* **2020**, *88*, 222–229. [CrossRef]
119. He, Z.; Zhang, H.; Wang, T.; Wang, R.; Luo, X. Effects of five different lactic acid bacteria on bioactive components and volatile compounds of oat. *Foods* **2022**, *11*, 3230. [CrossRef]
120. Łopusiewicz, Ł.; Kwiatkowski, P.; Drozłowska, E. Production and characterization of yogurt-like fermented beverage based on camelina (*Camelina sativa* L.) seed press cake. *Appl. Sci.* **2022**, *12*, 1085. [CrossRef]
121. Costa Fernandes, A.B.; Marcolino, V.A.; Silva, C.; Barão, C.E.; Pimentel, T.C. Potentially synbiotic fermented beverages processed with water-soluble extract of Baru almond. *Food Biosci.* **2021**, *42*, 101200. [CrossRef]
122. Hew, J.H.; Shafie, S.R.; Sulaiman, N. Proximate composition and anti-microbial activity of kefir produced from cow's and almond milk. *J. Trop. Life Sci.* **2023**, *13*, 287–296. [CrossRef]
123. Merenkova, S.; Zinina, O.; Potoroko, I. Fermented plant beverages stabilized with microemulsion: Confirmation of probiotic properties and antioxidant activity. *Fermentation* **2022**, *8*, 723. [CrossRef]
124. Kaprasob, R.; Kerdchoechuen, O.; Laohakunjit, N.; Sarkar, D.; Shetty, K. Fermentation-based biotransformation of bioactive phenolics and volatile compounds from cashew apple juice by select lactic acid bacteria. *Process Biochem.* **2017**, *59*, 141–149. [CrossRef]
125. Chen, J.; Wu, Y.; Yang, C.; Xu, X.; Meng, Y. Antioxidant and hypolipidemic effects of soymilk fermented via *Lactococcus acidophilus* MF204. *Food Funct.* **2017**, *8*, 4414–4420. [CrossRef]
126. Miraghajani, M.; Zaghian, N.; Mirlohi, M.; Feizi, A.; Ghiasvand, R. The Impact of probiotic soy milk consumption on oxidative stress among type 2 diabetic kidney disease patients: A randomized controlled clinical trial. *J. Ren. Nutr.* **2017**, *27*, 317–324. [CrossRef] [PubMed]
127. Tiss, M.; Souiy, Z.; Abdeljelil, N.b.; Njima, M.; Achour, L.; Hamden, K. Fermented soy milk prepared using kefir grains prevents and ameliorates obesity, type 2 diabetes, hyperlipidemia and liver-kidney toxicities in HFFD-rats. *J. Funct. Foods* **2020**, *67*, 103869. [CrossRef]
128. Hu, T.; Chen, R.; Qian, Y.; Ye, K.; Long, X.; Park, K.-Y.; Zhao, X. Antioxidant effect of *Lactobacillus fermentum* HFY02-fermented soy milk on D-galactose-induced aging mouse model. *Food Sci. Hum. Wellness* **2022**, *11*, 1362–1372. [CrossRef]
129. Aparicio-García, N.; Martínez-Villaluenga, C.; Frias, J.; Crespo Perez, L.; Fernández, C.F.; Alba, C.; Rodríguez, J.M.; Peñas, E. A novel sprouted oat fermented beverage: Evaluation of safety and health benefits for celiac individuals. *Nutrients* **2021**, *13*, 2522. [CrossRef]
130. Algonaiman, R.; Alharbi, H.F.; Barakat, H. Antidiabetic and hypolipidemic efficiency of *Lactobacillus plantarum* fermented oat (*Avena sativa*) extract in streptozotocin-induced diabetes in rats. *Fermentation* **2022**, *8*, 267. [CrossRef]
131. Deeseenthum, S.; Luang-In, V.; John, S.M.; Chottanom, P.; Chunchom, S. Effects of kefir fermentation on antioxidation activities (in vitro) and antioxidative stress (in vivo) of three thai rice milk varieties prepared by ultrasonication technique. *Pharmacogn. J.* **2018**, *10*, 1061–1066. [CrossRef]
132. Ardiansyah; David, W.; Handoko, D.D.; Kusbiantoro, B.; Budijanto, S.; Shirakawa, H. Fermented rice bran extract improves blood pressure and glucose in stroke-prone spontaneously hypertensive rats. *Nutr. Food Sci.* **2019**, *49*, 844–853. [CrossRef]

133. Chang, V.H.-S.; Chiu, T.-H.; Fu, S.-C. In vitro anti-inflammatory properties of fermented pepino (*Solanum muricatum*) milk by γ-aminobutyric acid-producing *Lactobacillus brevis* and an in vivo animal model for evaluating its effects on hypertension. *J. Sci. Food Agric.* **2016**, *96*, 192–198. [CrossRef]
134. Wang, X.; Wang, Y.; Han, M.; Liang, J.; Zhang, M.; Bai, X.; Yue, T.; Gao, Z. Evaluating the changes in phytochemical composition, hypoglycemic effect, and influence on mice intestinal microbiota of fermented apple juice. *Food Res. Int.* **2022**, *155*, 110998. [CrossRef]

MDPI AG
Grosspeteranlage 5
4052 Basel
Switzerland
Tel.: +41 61 683 77 34

Foods Editorial Office
E-mail: foods@mdpi.com
www.mdpi.com/journal/foods

www.ingramcontent.com/pod-product-compliance
Lightning Source LLC
Chambersburg PA
CBHW040810250926
51850CB00007BA/1129
* 9 7 8 3 7 2 5 8 7 5 0 4 7 *